I0065499

Mineral Surface Reactions at the Nanoscale

Mineral Surface Reactions at the Nanoscale

Special Issue Editor

Christine V. Putnis

MDPI • Basel • Beijing • Wuhan • Barcelona • Belgrade

MDPI

Special Issue Editor
Christine V. Putnis
University of Münster
Germany
Curtin University
Australia

Editorial Office
MDPI
St. Alban-Anlage 66
4052 Basel, Switzerland

This is a reprint of articles from the Special Issue published online in the open access journal *Minerals* (ISSN 2075-163X) from 2018 to 2019 (available at: https://www.mdpi.com/journal/minerals/special_issues/Surface_Reactions)

For citation purposes, cite each article independently as indicated on the article page online and as indicated below:

LastName, A.A.; LastName, B.B.; LastName, C.C. Article Title. *Journal Name* **Year**, *Article Number*, Page Range.

ISBN 978-3-03897-896-1 (Pbk)
ISBN 978-3-03897-897-8 (PDF)

Cover image courtesy of Christine V. Putnis.
Modified from an AFM image of a brucite surface dissolving in the presence of a solution containing cadmium—see Hövelmann et al., page 31.

© 2019 by the authors. Articles in this book are Open Access and distributed under the Creative Commons Attribution (CC BY) license, which allows users to download, copy and build upon published articles, as long as the author and publisher are properly credited, which ensures maximum dissemination and a wider impact of our publications.
The book as a whole is distributed by MDPI under the terms and conditions of the Creative Commons license CC BY-NC-ND.

Contents

About the Special Issue Editor

Christine V. Putnis is a research scientist in the Institut für Mineralogie at the University of Münster, Germany, where she has worked for the past 23 years. Previously, she worked as a research associate in the Department of Earth Sciences, University of Cambridge, UK. Christine's research interests involve the investigation of reactions at mineral–fluid interfaces at the nanoscale, predominately using in situ atomic force microscopy (AFM). Reactions at mineral surfaces commonly involve the coupling between dissolution and precipitation, and observations that confirm this mechanism of mineral behavior have enabled a further understanding of how processes such as weathering, toxic element sequestration, and mineral replacement reactions can be achieved. Christine has published widely on this topic. Christine also holds an adjunct professor position in the Department of Chemistry, Curtin University, Perth, Australia, where she has continued nanoscale research for several months each year for the past 5 years.

minerals

Editorial

Editorial for Special Issue "Mineral Surface Reactions at the Nanoscale"

Christine V. Putnis [1,2]

[1] Institut für Mineralogie, University of Münster, 48149 Münster, Germany; putnisc@uni-muenster.de
[2] The Institute for Geoscience Research, Department of Chemistry, Curtin University, Perth 6845, Australia

Received: 14 March 2019; Accepted: 15 March 2019; Published: 17 March 2019

check for updates

Abstract: Reactions at mineral surfaces are central to all geochemical processes. As minerals comprise the rocks of the Earth, the processes occurring at the mineral–aqueous fluid interface control the evolution of the rocks and, hence, the structure of the crust of the Earth during such processes at metamorphism, metasomatism, and weathering. In recent years, focus has been concentrated on mineral surface reactions made possible through the development of advanced analytical techniques, such as atomic force microscopy (AFM), advanced electron microscopies (SEM and TEM), phase shift interferometry, confocal Raman spectroscopy, advanced synchrotron-based applications, complemented by molecular simulations, to confirm or predict the results of experimental studies. In particular, the development of analytical methods that allow direct observations of mineral–fluid reactions at the nanoscale have revealed new and significant aspects of the kinetics and mechanisms of reactions taking place in fundamental mineral–fluid systems. These experimental and computational studies have enabled new and exciting possibilities to elucidate the mechanisms that govern mineral–fluid reactions, as well as the kinetics of these processes, and, hence, to enhance our ability to predict potential mineral behavior. In this Special Issue "Mineral Surface Reactions at the Nanoscale", we present 12 contributions that highlight the role and importance of mineral surfaces in varying fields of research.

When a mineral comes into contact with an aqueous solution, the mineral is out of equilibrium and so will start to dissolve. As a mineral surface has a complex energy landscape of steps and kinks, mineral dissolution is now known to be a process dominated by a spectrum of dissolution rates over a continuously changing surface. In their article, "Temporal Evolution of Calcite Surface Dissolution Kinetics", Bibi et al. [1] quantify the kinetics of dissolution on a calcite surface using vertical scanning interferometry that allows 3D imaging of a changing mineral surface during the reaction, resulting in a temporal evolution of a calcite surface and presenting insights into the role of various surface components during dissolution.

If an aqueous solution is supersaturated with respect to a mineral phase, that phase may precipitate, that is, a mineral grows from a supersaturated solution. Crystal growth has been studied continuously for well over a century, but it is only in recent times that the classical mechanisms of crystal growth have been questioned and new concepts developed of non-classical nucleation, that challenge existing concepts of mineral growth [2]. In the article, "How can Additives Control the Early Stages of Mineralization", Denis Gebauer [3] presents an overview, as well as new observations, on the interaction of additive molecules and potential mineral precursor species at the mineral–fluid interface.

Historically, mineral dissolution and growth thermodynamics and kinetics have been investigated separately. However, there is increasing evidence that in the Earth, dissolution is often accompanied by the growth of a new phase [4]. The feedback mechanisms involved in this coupled dissolution–precipitation scenario present interesting and important information for the understanding of Earth processes, as well as enabling conditions to be determined for the control of crystal growth through replacement reactions, to enable potential environmental remediation strategies, as well as the design of preferential crystal growth. Several articles focus on the coupling of dissolution and precipitation at the mineral–fluid interface. Toxic elements may be sequestered into a more stable phase through this coupled process. Hövelmann et al. [5] describe AFM experiments where potentially polluting heavy metal ions, released from industrial processes, are precipitated in a new phase on a brucite $Mg(OH)_2$ surface. Direct in situ observations allow the elucidation of the reaction mechanism. Such experimental research can be complemented by molecular dynamics simulations, as described in the article by Garcia et al. [6], "Water Structure, Dynamics and Ion Adsorption at the Aqueous {010} Brushite Surface". A computational approach to the understanding of mineral–surface reactions can also predict and confirm experimental observations.

Di Lorenzo et al. [7] present a mineral carbonation process as a potential mechanism for CO_2 sequestration in their article "The Carbonation of Wollastonite: A Model Reaction to Test Natural and Biomimetic Catalysts for Enhanced CO_2 Sequestration". With ever-increasing amounts of CO_2 in the atmosphere, as a result of anthropogenic activities, it is essential to explore possible sequestration methods, and the formation of stable carbonate minerals is one potential solution. Other contributions also explore coupled reactions at the mineral–fluid interface in terms of potential element sequestration or immobilization of pollutants, such as phosphate by Wang et al. [8] or rare earth elements by Fei et al. [9].

The identification of mineral surface processes is further explored by Silva-Quiñones et al. [10] using polarization microscopy and X-ray photoelectron spectroscopy (XPS). The effective and efficient extraction of metals from ores requires knowledge of the changes occurring at mineral surfaces during ore processing. It is only by understanding mineral surface reactions that industrial processes can be more energy efficient as well as effective for successful ore treatment. In this case, advanced analytical techniques are employed. King and Geisler [11] in "Tracing Mineral Reactions using Confocal Raman Spectroscopy" present a review of the increasing use of Raman spectroscopy together with isotope tracing to identify phases at the mineral–fluid interface during coupled reactions.

Detailed micro-structural evidence is presented by Greiner et al. [12] to introduce biomineral reactivity during the fluid-mediated replacement of biological aragonite (in bivalve, coral, and cuttlebone skeletons) to apatite (the mineral of bone material). This is a pseudomorphic replacement as a result of a tight interface-coupled dissolution–precipitation process. The authors emphasize the formation of porosity during the replacement process, allowing for the infiltration of fluid within the parent aragonite. The development of porosity has been recognized as an essential feature of an interface-coupled dissolution–precipitation process [13].

Albite is a common rock-forming mineral. During metasomatic processes, such as weathering, albite is replaced by other Na-bearing minerals. In natural and experimental samples, the replacement of albite by Na-rich secondary phases, such as sodalite and nepheline, has been investigated by Drüppel and Wirth [14]. Their observations indicate that the replacement is formed by an interface-coupled dissolution–reprecipitation mechanism. Their results also add to the understanding of element mobility, including trace element mobilization, and; therefore, indicate important implications for the interpretation of mineral reactivity in the presence of a fluid phase.

The final contribution by Zhao and Pring [15] presents an overview of the replacement of gold-(silver) telluride minerals that form a significant part of important gold-containing deposits across the world. Under hydrothermal conditions, gold-silver alloys replace the telluride minerals. Experiments highlight the replacement process and detailed textural analysis of the product phases allows for an interpretation of the reaction mechanism.

All the contributions in this special issue of *Minerals* are connected by mineral surface reactions at the nanoscale and as such indicate the importance of detailed and careful analysis of mineral surfaces before, during, and after reactions in aqueous fluids. The knowledge gained from interpretations of mineral surface reactivity spans a wide range of important Earth processes as well as potential

Minerals **2019**, *9*, 185

environmental remediation. The mechanism highlighted in these contributions, whereby one mineral phase is replaced by another more stable phase, is interface-coupled dissolution–precipitation.

Author Contributions: C.V.P. wrote the paper.

Funding: C.V.P. acknowledges funding (grant numbers: 290040, 317235, 316889) received through the EU 7th Framework ITNs Minsc, CO$_2$ React, and Flowtrans at the University of Münster, Germany.

Conflicts of Interest: The author declares no conflicts of interest.

References

1. Bibi, I.; Arvidson, R.S.; Fischer, C.; Luttge, A. Temporal evolution of calcite surface dissolution kinetics. *Minerals* **2018**, *8*, 256. [CrossRef]
2. Gebauer, D.; Kellermeier, M.; Gale, J.D.; Bergström, L.; Cölfen, H. Pre-nucleation clusters as solute precursors in crystallization. *Chem. Soc. Rev.* **2014**, *43*, 2348. [CrossRef] [PubMed]
3. Gebauer, D. How can additives control the early stages of mineralization. *Minerals* **2018**, *8*, 179. [CrossRef]
4. Ruiz-Agudo, E.; Putnis, C.V.; Putnis, A. Coupled dissolution and precipitation at mineral-fluid interfaces. *Chem. Geol.* **2014**, *383*, 132–146. [CrossRef]
5. Hövelmann, J.; Putnis, C.V.; Benning, L.G. Metal sequestration through coupled dissolution-precipitation at the brucite-water interface. *Minerals* **2018**, *8*, 346. [CrossRef]
6. Garcia, N.A.; Raiteri, P.; Vlieg, E.; Gale, J.G. Water structure, dynamics and ion adsorption at the aqueous {010} brushite surface. *Minerals* **2018**, *8*, 334. [CrossRef]
7. Di Lorenzo, F.; Ruiz-Agudo, C.; Ibañez-Velasco, A.; Gil-Sian Millán, R.; Navarro, J.A.R.; Ruiz-Agudo, E.; Rodriguez-Navarro, C. The carbonation of wollastonite: A model reaction to test natural and biomimetic catalysts for enhanced CO$_2$ sequestration. *Minerals* **2018**, *8*, 209. [CrossRef]
8. Wang, L.; Putnis, C.V.; Hövelmann, J.; Putnis, A. Interfacial precipitation of phosphate on hematite and goethite. *Minerals* **2018**, *8*, 207. [CrossRef]
9. Fei, Y.; Hua, J.; Liu, C.; Li, F.; Zhu, Z.; Xiao, T.; Chen, M.; Gao, T.; Wei, Z.; Hao, L. Aqueous Fe(II)-induced phase transformation of ferrihydrite coupled adsorption/immobilization of rare earth elements. *Minerals* **2018**, *8*, 357. [CrossRef]
10. Silva-Quiñones, D.; He, C.; Jacome-Collazos, M.; Benndorf, C.; Teplyakov, A.V.; Rodriguez-Reyes, J.C.F. Identification of surface processes in individual minerals of a complex ore through the analysis of polished sections using polarization microscopy and X-ray photoelectron spectroscopy (XPS). *Minerals* **2018**, *8*, 427. [CrossRef]
11. King, H.E.; Geisler, T. Tracing mineral reactions using confocal Raman spectroscopy. *Minerals* **2018**, *8*, 158. [CrossRef]
12. Greiner, M.; Férnandez-Diaz, L.; Griesshaber, E.; Zenkert, M.N.; Yin, X.; Ziegler, A.; Veintemillas-Verdaguer, S.; Schmahl, W.W. Biomineral reactivity: The kinetics of the replacement reaction of biological aragonite to apatite. *Minerals* **2018**, *8*, 315. [CrossRef]
13. Putnis, A.; Putnis, C.V. The mechanism of reequilibration of solids in the presence of a fluid phase. *J. Solid Sate Chem.* **2007**, *180*, 1783–1786. [CrossRef]
14. Drüppel, K.; Wirth, R. Metasomatic replacement of albite in nature and experiments. *Minerals* **2018**, *8*, 214. [CrossRef]
15. Zhao, J.; Pring, A. Mineral transformations in gold-(silver) tellurides in the presence of fluids: Nature and experiment. *Minerals* **2019**, *9*, 167. [CrossRef]

© 2019 by the author. Licensee MDPI, Basel, Switzerland. This article is an open access article distributed under the terms and conditions of the Creative Commons Attribution (CC BY) license (http://creativecommons.org/licenses/by/4.0/).

Article

Temporal Evolution of Calcite Surface Dissolution Kinetics

Irshad Bibi [1,2,*], Rolf S. Arvidson [1,3], Cornelius Fischer [4] and Andreas Lüttge [1,3]

[1] MARUM and Fachbereich Geowissenschaften, Universität Bremen, D-28334 Bremen, Germany; rsa4046@uni-bremen.de (R.S.A.); andrluet@uni-bremen.de (A.L.)
[2] Institute of Soil and Environmental Sciences, University of Agriculture Faisalabad, Faisalabad 38040, Pakistan
[3] Department of Earth, Environmental and Planetary Sciences, Rice University, Houston, TX 77005, USA
[4] Helmholtz-Zentrum Dresden-Rossendorf, Institute of Resource Ecology, Reactive Transport Department, D-04318 Leipzig, Germany; c.fischer@hzdr.de
* Correspondence: irshad.niazi81@gmail.com or irshad.niazi@uaf.edu.pk; Tel.: +92-333-6597761

Received: 13 April 2018; Accepted: 12 June 2018; Published: 16 June 2018

Abstract: This brief paper presents a rare dataset: a set of quantitative, topographic measurements of a dissolving calcite crystal over a relatively large and fixed field of view (~400 μm^2) and long total reaction time (> 6 h). Using a vertical scanning interferometer and patented fluid flow cell, surface height maps of a dissolving calcite crystal were produced by periodically and repetitively removing reactant fluid, rapidly acquiring a height dataset, and returning the sample to a wetted, reacting state. These reaction-measurement cycles were accomplished without changing the crystal surface position relative to the instrument's optic axis, with an approximate frequency of one data acquisition per six minutes' reaction (~10/h). In the standard fashion, computed differences in surface height over time yield a detailed velocity map of the retreating surface as a function of time. This dataset thus constitutes a near-continuous record of reaction, and can be used to both understand the relationship between changes in the overall dissolution rate of the surface and the morphology of the surface itself, particularly the relationship of (*a*) large, persistent features (e.g., etch pits related to screw dislocations; (*b*) small, short-lived features (e.g., so-called pancake pits probably related to point defects); (*c*) complex features that reflect organization on a large scale over a long period of time (i.e., coalescent "super" steps), to surface normal retreat and step wave formation. Although roughly similar in frequency of observation to an in situ atomic force microscopy (AFM) fluid cell, this vertical scanning interferometry (VSI) method reveals details of the interaction of surface features over a significantly larger scale, yielding insight into the role of various components in terms of their contribution to the cumulative dissolution rate as a function of space and time.

Keywords: calcite; surface; kinetics; rate spectra; retreat velocity; dissolution

1. Introduction

The thermodynamic reactions on minerals surface determines their stability and fate in a given environment. As these reactions drive a series of natural (e.g., nutrient release and availability, soil pH buffering, carbon cycling, water acidification) and industrial processes (e.g., geological CO$_2$ sequestration, cement hydration, weathering, and corrosion of industrial structures), understanding their kinetics is a key focus of environmental science. Despite an ongoing debate over uncertainties in predicting long-term mineral dissolution rates in a wide spectrum of environments, there is general agreement that such predictive capabilities depend on understanding fundamental reaction mechanisms at the mineral-fluid interface, as well as their variation in time and space [1].

Prior to the advent of atomic force microscopy (AFM), vertical scanning interferometry (VSI), and related surface microscopic techniques, distinguishing potential sources of variation in dissolution rate using conventional bulk-powder experiments (data are summarized in refs. [2–4] was generally not feasible. The currently widespread use of these instruments has yielded a detailed understanding of the interaction of minerals with their ambient solution (e.g., ref. [5]). Both microscopes afford *direct observation* of mineral surfaces, and in situ AFM [6–12] and VSI [1,13–17] have greatly expanded the understanding of reaction kinetics for a diverse range of carbonates, silicates, and other important phases. Calcite has been a favorite AFM and VSI target, due to its clear importance in environmental systems, its simple composition, and perfect cleavage. For example, early in situ AFM work determined the crystallographic control of the velocity of atomic scale steps in pure solutions [18], documenting the anisotropic specificity of kink and step orientations. These site-specific properties reflect in part the oblique intersection of the hexagonal unit cell with the $(10\bar{1}4)$ cleavage surface. This obliquity generates unique kink sites and step edges (i.e., the $\langle \bar{4}41 \rangle, \langle 48\bar{1} \rangle$ step directions are crystallographically equivalent, but opposite-facing step pairs are not), and specific geometries in calcium-oxygen coordination give rise to corresponding orientation-specific properties. In the study of calcite dissolution kinetics, a major AFM research focus has been the relationship of step velocity, spacing, geometry, and related observables to solution properties (pH, dissolved lattice ions, impurity components, supporting electrolytes, ionic strength, etc.; see citations above). This detailed work, spanning the early 1990's to the present day, has yielded a rich set of observations and measurements, now augmented with significant thermodynamics modeling and kinetics simulation [19–22]. As a result, calcite is one of the best-studied crystals of environmental significance in terms of the relationship between its dissolution rate and the physics and chemical structure of its surface. However, despite the overall simplicity of calcite's structure, these datasets have also revealed significant complexity in step movement and morphology. This complexity implies that the general morphology of the surface is a sensitive reflector, not only of its interactions with the surrounding solution, but with itself as well (see e.g., ref. [23]). Variations in morphology are tied to heterogeneities in rate, phenomena we have termed *intrinsic* variations (tied to crystallographic and microstructural factors), to distinguish them from those arising from environmental inputs of purely extrinsic origin [24]. As rate controls, the latter environmental factors are no less important, and significant effort has been devoted to understanding their influence [25]. Here, however, we wish to focus entirely on intrinsic variation in rate expressed over the calcite surface.

As mentioned above, detailed AFM studies of dissolving calcite have made critical advances in quantifying the dynamics of surface features at or close to the atomic scale [26,27], and have focused largely on step movement. The AFM studies conducted in situ have easily detailed the anisotropy in step velocities, quantifying the variation of crystallographically inequivalent "obtuse" $\left([\bar{4}41]_{+}, [48\bar{1}]_{+} \right)$ and "acute" $\left([\bar{4}41]_{-}, [48\bar{1}]_{-} \right)$ steps as a function of saturation state, impurity burden, pH, ionic strength, and other controls (work reviewed in ref. [11]). From these data, the overall rate of dissolution has often been quantified by essentially integrating step velocity over time, e.g., $v_{\perp} = v_s(h_s/\Lambda)$, where the surface normal retreat velocity ($v_{\perp}$) is computed from the step velocity (v_s), step height (h_s), and the standard terrace width (Λ) separating a step from its neighbor; the rate in units of moles per unit area per unit time can then be given as $r = v_{\perp}/\overline{V}$, where $\overline{V}$ is the molar volume. Alternatively, the mean rate over an AFM field of view can be computed as $r = \overline{V}^{-1} \, \overline{v}_s \overline{\left(\frac{\partial z}{\partial x} \right)}$, where $\overline{\left(\frac{\partial z}{\partial x} \right)}$ is the mean surface gradient, and $\overline{v}_s$ is the mean step velocity. More elaborate means of achieving surface normal retreat rates may also involve quantification of the volume and number of etch pits (also reviewed in ref. [11]), but these methods may ignore the background layer-by-layer removal of material (the "global" rate [15]), as well as the inherent coupling between etch pits associated with deep screw dislocations and the surface normal deflation rate, conceptually formalized in the step wave model [28]. We have made this point before, i.e., that while AFM's in situ capabilities provide exquisite and invaluable detail of the dynamics of surface features in minerals over small areas of interest, cantilever constraints do limit the depth—and thus the area—that can be effectively measured. This limit makes the study of variation in dissolution rates problematic, as it is easy to imagine how local step velocities and spacing may be influenced by events or features outside the AFM's immediate field of view. VSI, with a significantly larger field of view, supplies rapid and robust acquisition of surface height data at high vertical resolution, and height changes are absolute over time when acquisition of data is made in conjunction with a masked or coated reference surface. In terms of measurement of dissolution rates, VSI's limitations are its lateral resolution—a function of the objective's numerical aperture (see ref. [4], for discussion), and the complexity of direct, truly in situ measurements, which must be compensated for the fluid's contribution to the optic signal (see also in situ digital holographic microscopy in ref. [17]). These limits are often the motivation for using AFM and VSI instruments in concert as complementary tools.

Here, as a means of approaching the basic problem of spatial and temporal rate variation raised in earlier papers [1,15,24], we wish to explore what can be learned from continuous observation of a single crystal in a simple solution for an extended period. Using an open fluid cell and large calcite single crystal, we developed a so-called near in-situ VSI cell, in which a fluid of fixed composition flows rapidly over the surface, a surface that is fixed relative to the instrument's optic path. During reaction, all relevant environmental conditions (flow rate, temperature, pH) are also held constant. Periodically, fluid is rapidly removed from the surface for a period sufficient for data acquisition, after which flow is restored. The sample is again allowed to react, and the sequence repeated. In this manner, a high number of near-continuous measurements (80) were acquired over a large (414 × 313 μm^2, ~0.13 mm^2), fixed region of the surface for 6 h reaction time. In many AFM and VSI experiments, a common protocol is to cleave the sample prior to solution contact, thus introducing a "pristine" initial condition. In contrast, we reacted the surface for ~2 days under flow conditions prior to first measurements. In our view, the resulting observational record constitutes a unique dataset, and allows us to quantify how a statistically significant area of the surface changes over time as dissolution proceeds. We can thus begin to address a basic question: does a steady state surface exist? Can the *concept* of steady state dissolution, underpinning much of our understanding and experimental approach, be reasonably defended? Does a steady state surface have a statistical signature? These questions constitute our basic motivations.

2. Methods

The reactant solution used here was identical in composition to that used in previous work [15], prepared by dissolving sufficient reagent grade Na_2CO_3 within a large volume of deionized water (30 L, 18.2 MΩ-cm) to give a total alkalinity of 4.4 meq/kg-H_2O. After preparation, this solution was sparged with water-saturated laboratory air to attain equilibrium with respect to atmospheric pCO_2 and a stable pH (8.82 at 22 °C, measured by Orion semimicro combination glass electrode). Constant sparging was sufficient to agitate the solution over the course of the experiment and produced a pH stable over 25 h (within ±0.02 pH units). This solution contained essentially no calcium and was thus highly under-saturated with respect to calcite. Input solution temperature was maintained at 22 °C. The sample crystal was prepared by gentle cleaving of a large synthetic calcite crystal with a razor to produce a (10$\bar{1}$4) cleavage fragment, which was then affixed to the cell surface between the flow path of the input and scavenging ports (Figure 1).

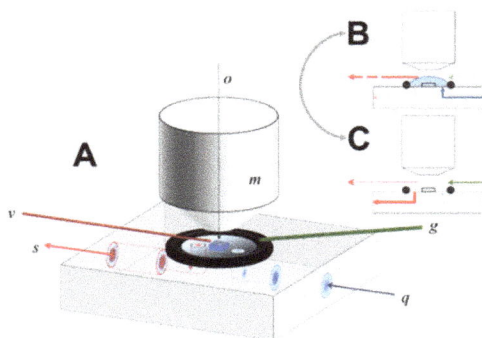

Figure 1. Fluid cell operation. (**A**) Principal cell components include input flow port (*q*), scavenging pump port (*s*), capillary vacuum aspirator (*v*), and N$_2$ dry gas line (*g*). Sample crystal is positioned under the optic path (*o*) of the Mirau interferometric objective (*m*). During reaction (**B**), fluid is introduced through the flow port and fills the volume bounded by the O-ring, immersing the sample crystal. The scavenging port is closed, and fluid is removed by the vacuum aspirator, whose lateral position controls the height of the fluid envelope. Input flow rate is sufficient for rapid exchange within the cell's open volume. Prior to data collection (**C**), the input port is closed and scavenging port is opened, allowing fluid to drain rapidly from the cell. The sample surface is simultaneously dried via N$_2$. Once data collection is completed, the cell is returned to flow configuration (**B**). Total cycle time is 20–30 s [29].

Similar to much of our previous work (e.g., refs. [15,30]), we used a MicroXAM (ADE Phase Shift, now KLA-Tencor) commercial vertical scanning interferometer (KLA-Tencor Co., Milpitas, CA, USA) equipped with Nikon Mirau objectives (10×, 20×, 50×, and 100× magnification), white light source (550 nm center wavelength), and high-resolution CCD (750 × 480 pixels). This instrument was mounted on an air-suspension table in a vibration-isolated laboratory. Details regarding the general theory of operation, acquisition and treatment of data using this instrument are available in Luttge and Arvidson [31].

The fluid cell is extremely simple in design, fabricated from polyether ether ketone (PEEK) and consisting of internal ports for fluid input to and removal from an open reaction fluid volume [29]. Reactant fluid was delivered to the cell by a peristaltic pump at a volumetric flow rate of ~6 mL per minute, thus giving well-mixed fluid residence times of less than 8 s, sufficiently brief to eliminate transport dependency and maintain far-from-equilibrium conditions based on comparison with previous experiments [15]. As shown in Figure 1, fluid introduced to the cell's open surface through the input channel (*q*) is confined laterally by a Viton O-ring, resulting in complete immersion of the sample crystal in the process.

During reaction mode (Figure 1B), the fluid parcel is maintained and continuously withdrawn by a capillary vacuum aspirator suspended a few mms over the sample surface. The total fluid volume is thus to some extent variable (although typically <0.5 mL), a function of the balance between the input flow pump rate versus the position and pressure drop of the capillary; no fluid is removed by the scavenging port during reaction. The capillary position is micro-adjusted so that the crystal sample is mechanically undisturbed, preventing contact of the upper surface of the fluid with the Mirau objective, thus maintaining registration of the sample surface with respect to the optic path of the interferometer during the entire experiment. Drift of the sample position with respect to the optic path was not rigorously quantified, but is in the range of a single pixel, verified by ad hoc comparison of the pixel position of fiducial landmarks over time.

To shift to data acquisition mode (Figure 1B,C), the input flow was diverted, the scavenging port opened to a vacuum line, and fluid rapidly removed from the entire cell. The instrument's focal condition and fringe position is then zeroed, and data acquisition completed using a 10 μm scan length.

After data acquisition was complete, the cell was returned to flow operation (Figure 1B). In practice this cycle (wet-dry-wet) could be accomplished in ~20 s.

After the mounting of the sample in the cell but prior to collection of the data shown in this paper, the sample was reacted continuously at pH 8.82 for more than 48 h, thus allowing substantial dissolution of the initially prepared surface. The field of view was then fixed and focused using a 20× objective, and topographic data collected at roughly 5-min intervals for 6 h using the above control cycle, yielding 80 total datasets (field of view 414 × 313 μm^2). These measurements and subsequent calculations constitute the total dataset presented in this paper. Because measurements are obtained over the same field of view, the resulting sequence can be used to evaluate the overall morphological changes in surface topography. In addition, we subtracted various height maps to a produce a corresponding difference map, showing the spatial distribution of dissolution flux. These changes were also evaluated by comparison of surface gradient maps, i.e., the surface derivative or change in slope within a single dataset. A gradient or derivative map effectively flattens the surface allowing us to highlight surface features that otherwise differed in absolute height by hundreds of nanometers up to microns. For example, with this approach we can thus reveal the distribution of shallow, pancake pits forming at point defects simultaneously with deep etch pits forming at the hollow cores of screw dislocations. Comparison of gradient maps also allowed us to detect and confirm areas of the surface over which changes in absolute height due to dissolution were minimal, and thus could effectively serve as reference areas. This was necessary as no mask was used to explicitly create a non-reactive reference area in this experiment; in practice, these flux measurements thus represent a minimum rate, as some surface retreat may have occurred between measurements. Given our purpose, this issue is of limited importance: these errors will be small because of the correspondingly small vertical z-contribution of global retreat (lateral changes in surface topography also afford a check on the spatial distribution of reactivity) and will not interfere with our focus on the spatial distribution of rates and their statistical distribution.

Potential Topography Evolution of Smooth Surface Plateau Sections

Plateau sections are considered to be unreactive in such cases that no evolution of nanotopography (>1 nm, precision of the method) has been observed using interferometry techniques.

This assumption is incorrect if multiple calcite layers (i) show a high density and homogeneous distribution of point defects, and (ii) their dissolution results in a very smooth surface (roughness <1 nm) after several tens of minutes of reaction time. In such cases (not observed in our data sets), we need to correct for the position of the calculated rate spectra with respect to the absolute rate values (x-axis). Such a correction always results in an increase of the overall rate, cf. Figure 2.

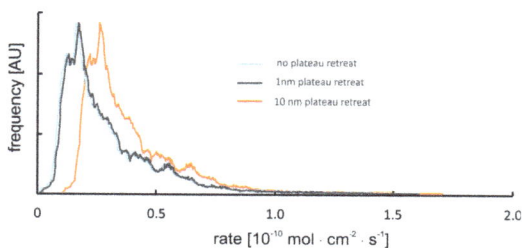

Figure 2. Visualization of rate data correction due to surface normal retreat of smooth surface plateaus. The precision of the instrument would allow for a reproducibility that is defined by the difference between the blue-colored and the gray-colored rate curves.

The sensitivity of the instrument and the resulting uncertainty of calculated rate spectra are illustrated by the positions of the blue vs. gray rate curves. In the case of a surface normal retreat, that is one order of magnitude bigger than the sensitivity of the instrument, the rate curve would shift to the position of the orange-colored line.

The occurrence of such a situation and a resulting error as shown is, however, unlikely. Any bigger retreat of surface plateaus, that are considered to be unreactive, would result in a few surface portions of the rate maps that show surface growth. We did, however, not observe such behavior thus assuming the uncertainty of reported rate curves is rather close to the 1 nm roughness situation instead of the 10 nm retreat rate curves, as shown in Figure 2.

3. Results

3.1. Dissolution Rate Maps and Rate Spectra

The initial height map is shown in Figure 3 (i.e., prior to the start of the sequence but after 48 h reaction), and shows several large, well-developed etch pits (the outcrop of deep screw dislocations), low relief terraces with only minor pits, and a large step-like feature ("M" in Figure 3) we shall discuss later, giving a total height range of 1.7 μm. Figure 4 shows the sequence of rate (difference) maps computed from VSI surface topography measurements at eight consecutive time steps: $t = 0.75, 1.50, 2.25, 3.00, 3.75, 4.5, 5.25,$ and 6.00 h. The first time step, Figure 4a, thus represents the difference in height between data collected at $t = 0$ and $t = 0.75$ h, with the rate (denoted in the color map) computed by dividing these difference map by the elapsed time and molar volume; similarly, the second time step is the difference map between $t = 0.75$ and $t = 1.50$ h, and so on. These maps illustrate the large spatial variability in dissolution rate, as well as its variation over time. Differences exceeding an order of magnitude are immediately apparent, with a total range of ~0.1×10^{-10} mol·cm^{-2}·s^{-1} and 1.6×10^{-10} mol·cm^{-2}·s^{-1}. Lower rates ($\leq 0.3 \times 10^{-10}$ mol·cm^{-2}·s^{-1}) are distributed over large portions of the surface (terraces areas 'T' in Figure 3) that otherwise lack deep etch pits. Intermediate rates (between $>0.3 \times 10^{-10}$ and $<0.5 \times 10^{-10}$ mol·cm^{-2}·s^{-1}) are typical of areas of surface normal retreat (upper left corner) and the steep, interior walls of large etch pits. The highest rates, ranging between $\geq 0.5 \times 10^{-10}$ mol·cm^{-2}·s^{-1} and $\leq 0.8 \times 10^{-10}$ mol·cm^{-2}·s^{-1}, are found at the upper shoulder of large pits (i.e., at the change in slope that marks the macroscopic pit boundary with the surrounding terrace), as well as the stepped feature ("M" in Figure 3). This stepped feature is the most unusual aspect of the surface observed in the entire dataset, present topographically as a large scarp (here we use the term *super-step*), that also exhibits the highest absolute rate and the largest distribution thereof over the surface.

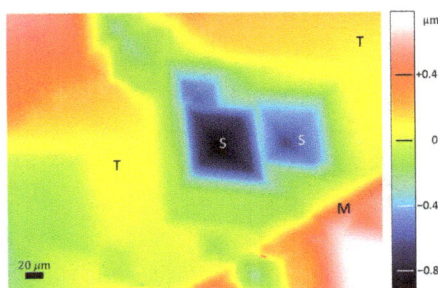

Figure 3. False color height map of surface at onset of reaction sequence, showing well-developed etch pits ("S"), terraces ("T"), and super-step feature ("M") discussed in text. Field of view 413.6 × 313.3 μm², 20× Mirau objective, 20 μm horizontal scale bar, total height range 1.7 μm. Note that the intercept (z_0) of the interferometer's height scale is arbitrarily set, and negative versus positive values are simply heights relative to a datum, and do not imply growth or otherwise positive changes in surface elevation. Comparisons of height data obtained at various reaction times (e.g., subtraction or difference maps) compute a pairwise correction in these intercepts to yield a consistent analysis (see discussion in text).

A significant acceleration in dissolution rate occurs over the second interval (Figure 4b). Terrace areas that previously exhibited minimum rates (blue colors in Figure 4a) are the locus for nucleation of many new small etch pits over this interval (Figure 4b). This nucleation event gives rise to an increase in the maximum rate (1.2×10^{-10} mol·cm^{-2}·s^{-1} versus 0.8×10^{-10} mol·cm^{-2}·s^{-1} in Figure 4a). The lateral expansion of large etch pits and the deepening of the screw dislocations also show increases in rate ($\leq 1.2 \times 10^{-10}$ mol·cm^{-2}·s^{-1}; Figure 4b). The dissolution rate continues to increase over the third reaction interval as illustrated by an increase in the highest rate to 1.4×10^{-10} mol·cm^{-2}·s^{-1} (Figure 4c). Much of this increase is associated with coordinated movement of the super-step itself (note the high reaction rates, yellows and reds, along this linear feature). In addition, note that the fluxes from the etch pits themselves are not spatially uniform: most of the flux is associated either with the core region and the uppermost steps, whereas the interior walls of the pit itself show fluxes similar to the surrounding terraces (t = 2.25 h, Figure 4c).

By the end of the fourth reaction interval (t = 3.00 h, Figure 4d), high rates are again located at upper most margins of large etch pits (obtuse step edges), whereas the acute counterparts have reduced fluxes compared to the previous interval (Figure 4c). The flux associated with the super-step movement, although still a major component, also shows reduced activity over this interval. The highest observed rate dropped to ~1×10^{-10} mol·cm^{-2}·s^{-1}.

From t = 3.75 to the end of the sequence at t = 6.00 h, the distribution of surface flux shows small variations in the above pattern. The terraces are primarily dominated by the nucleation and coalescence of small etch pits, contributing fluxes ranging from ~0.1 to 0.3×10^{-10} mol·cm^{-2}·s^{-1}. Large etch pits centered at screw dislocations continue to deepen, with pulses of activity confined to either the central core region or their uppermost shoulders at the pit margin [3]; in comparison, the vicinal faces constituting the sides of the pits show relatively low fluxes (e.g., Figure 4e). The uppermost margins of large etch pits again undergo episodes of enhanced anisotropy in flux distributions (e.g., t = 6 h, Figure 4h), with the highest rates associated with obtuse step edges (those facing north and east), versus acute step edges (south, west). Movement of the super-step makes a persistently high contribution to the overall flux, to the extent that it cannibalizes ongoing dissolution processes in its path. The flux is sufficiently high to effectively decapitate a nascent screw dislocation that first appears at t = 1.50 h (Figure 4b, arrow), and is annihilated by the last time step (corresponding arrow in Figure 4h).

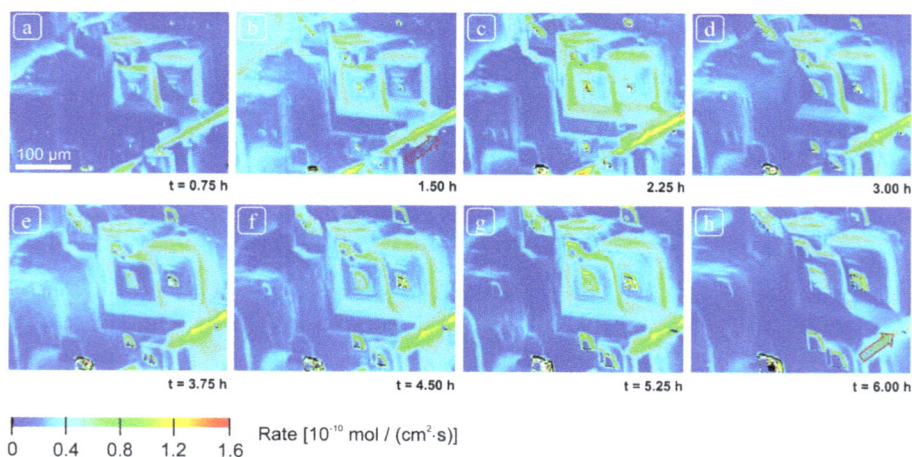

Figure 4. Rate maps of calcite reacted surface after reaction intervals ranging between 0.75 h and 6 h (**a–h**). Obtuse steps associated with deep etch pits propagate to the north and east, acute steps to the south and west, respectively, from their associated dislocation centers. See text for description.

Histogram analysis of flux maps in the reaction rate spectra can provide information about frequency of rate contributors over overall surface reaction rate. Figure 5 shows the corresponding sequence of rate spectra obtained by using the difference map data shown in Figure 4. These rate spectra contain two types of information. First, the shape of the function (single peak versus discrete peaks) is relevant to surface energy range and distribution; second, the peak heights provide information about the frequency of distinct energetic sites [32]. The overall calculated rate ranges between $\leq 0.1 \times 10^{-10}$ mol·cm^{-2}·s^{-1} and $\geq 1.4 \times 10^{-10}$ mol·cm^{-2}·s^{-1} (Figure 5). The highest peak (frequency maximum, a single discrete peak) corresponds to the largest number of surface sites and represents the lowest rates (i.e., ~0.15×10^{-10} mol·cm^{-2}·s^{-1}, Figure 5). The highest rates (>0.4×10^{-10} mol·cm^{-2}·s^{-1}, green colors towards the tail of the spectrum) correspond to the retreat from highly reactive sites; for example, large, deep etch pits at screw dislocations in the central part of the calcite crystal (Figure 5a) and from the super-step feature.

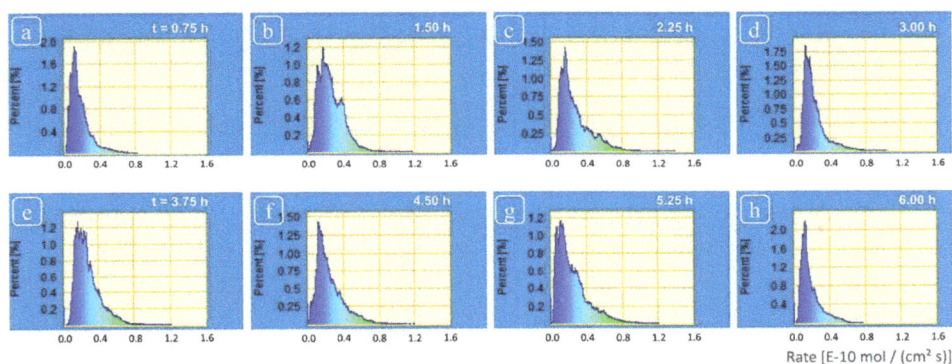

Figure 5. Rate spectra sequence showing spatial and temporal heterogeneity of dissolution rates of calcite surface after reaction intervals ranging between 0.75 h and 6 h (**a–h**). Although the horizontal scale (rate) is fixed, note that the *y*-scale is variable.

After the completion of second reaction interval (1.5 h), a secondary peak of higher rates appears and reaches its maximum frequency (Figure 5b). This enhanced contribution of higher rates also leads to an increase in the highest observed rate, from 0.8×10^{-10} mol·cm^{-2}·s^{-1} (Figure 5a) to 1.2×10^{-10} mol·cm^{-2}·s^{-1} (Figure 5b). Enhanced activity at highly reactivity sites (etch pits at screw dislocations and the super-step) continues in the third reaction interval, leading to further increase in the maximum observed rate to 1.4×10^{-10} mol·cm^{-2}·s^{-1} (Figure 5c). In the fourth reaction interval, the height of frequency maximum increases again, however secondary rate contributions are reduced significantly, leading to a reduction in the maximum dissolution rate (Figure 5d). However, higher rates increased again during the fifth interval as indicated by reappearance of secondary peaks along tail end of the rate spectrum (Figure 5e). Significantly, the rate spectrum after the eighth or last reaction interval (Figure 5h) showed the highest rate of the order of ~0.8×10^{-10} mol·cm^{-2}·s^{-1} that was similar to those observed after the first reaction interval (Figure 5a).

3.2. Evolution of Surface Topography

Where Figures 4 and 5 detail rates distribution over the surface over a time interval, Figure 6 shows how absolute topography (Figure 6T) and surface gradient (Figure 6G) change as a series of snapshots, corresponding to the start of each time step in Figures 4 and 5, beginning with $t = 0$ h (Figure 6T(a)), and ending at $t = 5.25$ h (Figure 6T(h)). In Figure 6T, the total height range associated with the color bar is also given ($dz = \ldots$), and the lateral and vertical expansion of etch pits with reaction time is clearly observable in this sequence. With continued reaction, new etch pits can clearly be seen to nucleate to the northwest of the largest pits, giving rise to complex coalescence, and the super-step can be seen to migrate towards the southeast. The nucleation of these new pits is even more salient in gradient maps (Figure 6G), demonstrating how the spatial derivative easily identifies the nucleation and growth of new etch pits, regardless of the nature of the generating defect. These data confirm the simultaneous growth and nucleation of etch pits at screw dislocations in tandem with shallow, monolayer pits on terraces whose rapid coalescence also generates surface normal retreat.

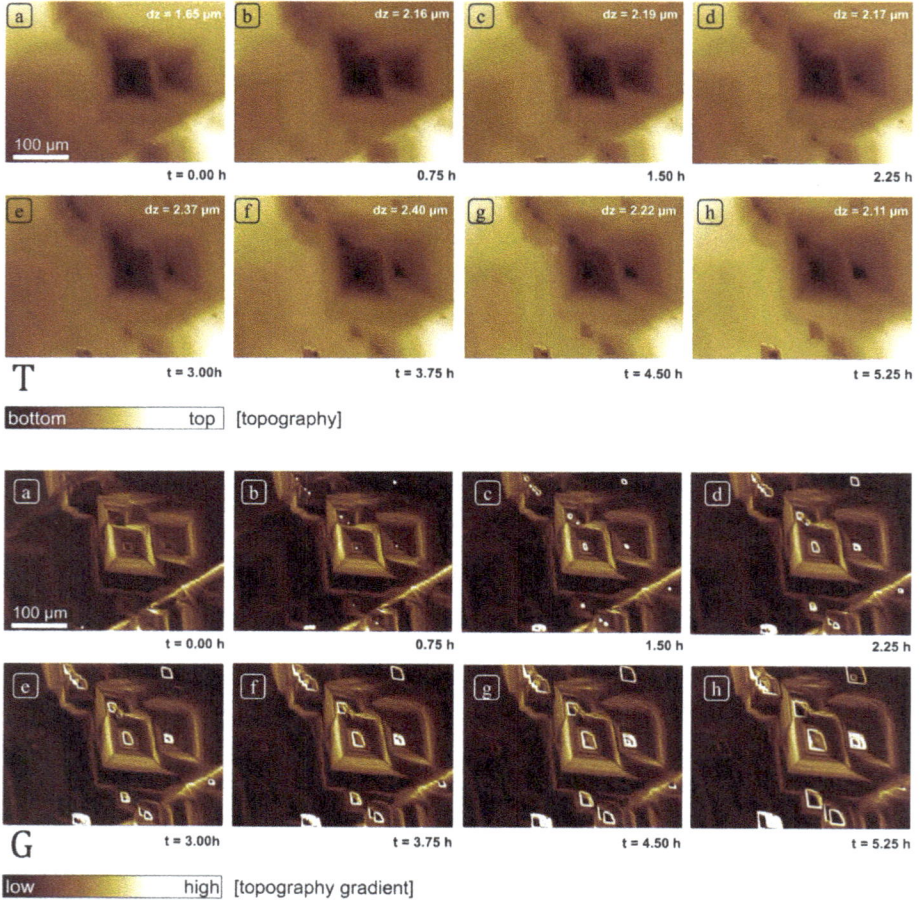

Figure 6. Reaction sequence data. Height (**T**, **top**) and gradient (**G**, **bottom**) maps of calcite surface prior to the reaction (**a**) and after reaction duration of 0.75 h to 5.25 h (**b–h**).

4. Discussion

These data illustrate several important aspects of crystal dissolution kinetics. First, as has been demonstrated previously in many AFM and VSI studies, the instantaneous dissolution rate is clearly variable over the crystal surface, attaining maxima in deep etch pits whose morphology is consistent with screw dislocations, and its minima over terrace sites traversed by steps associated with screw dislocations (step waves) as well as shallow pits likely nucleated at point defects (see [33]). However, in addition to showing how these processes are simultaneously distributed over a significant area of the crystal surface, these data also record rate variation for a significant period in fine grained temporal detail, an aspect not addressed in many of our previous papers on calcite dissolution (e.g., see ref. [1]). Our data are consistent with step velocities and other AFM data pertaining to calcite dissolution are consistent to the extent that dissolution occurs at monolayer steps [11].

However, these data also show, quite serendipitously, that the range of rates and associated morphologic features one observes on the surface, may indeed reflect how long one spends observing it. We suggest that our super-step feature is likely a result of convergence of intersecting $[48\bar{1}]_-$ and $[\bar{4}41]_+$ steps, roughly parallel to $[22\bar{1}]$ and moving normal to $[010]$ directions. Such a feature is clearly not a step itself, but the complex, dynamic product of a large number of coalescent etch pits, thus requiring both significant *time* to develop, and demanding that dissolution operate over a significant *area* as well. We cannot assess the distribution of such features. Their transit over the surface removes significant amounts of material very efficiently, recording dissolution fluxes competitive with or exceeding screw dislocations (although clearly related to them genetically), but their lifetime may also be quite limited. Nevertheless, our point is that such phenomena are clearly highly scale- and time-dependent.

As reported previously in many surface investigations on calcite under similar experimental conditions, calcite crystal dissolved by the formation of rhombohedral etch pits [5,7,11,15]. The size of these pits was highly variable. Other than these well-developed features, the formation of several small and shallow etch pits was also observed. The reactivity of all these surface features was observed to vary as a function of time, however these changes do not point to any consistent, linear change in reactivity over time.

The heterogeneity of the surface reaction rates at the spatial scale also points to the operation of concurrent reaction mechanisms, as opposed to a single mechanism assumed in "mean rate" method. This conclusion rules out the use of a mean rate for predicting long-term dissolution behavior of mineral materials. Rate spectra [1] preserving contribution from all reaction mechanisms as range of rates, rather than a mean value.

4.1. On Spatial Rate Variability

The rate spectra obtained revealed a wide variability in rate distribution ranging from the highest rates (from highly reactive sites due to screw dislocations located at deep etch pits and macro-step surface) versus the slowest rates (from regions of slow reactivity, terrace sites) (Figures 5 and 6). A wide range in spatial rates was observed from these experiments: between 0.03×10^{-10} mol·cm^{-2}·s^{-1} and 1.4×10^{-10} mol·cm^{-2}·s^{-1} (Figure 4). This range covers almost all the different values of calcite dissolution rates obtained under similar solution conditions in previous studies [15]. The spatial rate variability of crystalline material has also been reported in some previous investigations. For example, Arvidson et al. [15] reported large variation in distribution of etch pits and hence a potentially large variation in reactivity of a calcite crystal in a VSI monitored investigation. These authors further demonstrated that attainment of steady state in a dissolution experiment was rather a fortuitous occurrence, as the surface rates were dependent on the nucleation; widening and deepening of etch pits, which is a continuous process. The attainment of a steady state/constant rate, with insignificant variation would require a constant dislocation density which may not be easily fulfilled.

From the differences in absolute rates of two calcite samples (a large single crystal and a microcrystalline calcite) under exactly similar experimental conditions, Fischer et al. [1] ascribed the rate variability to differences in surface reactivity of the two samples. Moreover, there were also significant differences in distribution of these surface rates from two calcite samples. For instance, the highest rates were contributed by screw dislocations located at center of rarely found deep etch pits and the lowest rates were contributed by large number of shallow etch pits on single crystal calcite. These authors further commented that the averaging of such enormously variable rates to render a single mean value was misleading and inappropriate as the mean rate would inadequately represent the range of rates or the diversity of mechanisms [1].

4.2. On Temporal Rate Variability

There are two major reasons why the temporal surface rate variability was not addressed significantly in earlier microscopic rate investigations: (*i*) majority of the studies on morphologic evolution of dissolving crystals and the quantification of associated dissolution rates were carried out via AFM, which has a severe limitation of small scan area, and (*ii*) surface rate measurement was not accomplished at frequent intervals to establish any changes in the surface reactivity and derived dissolution rates with reaction progress [4,11].

In this study, we resolved the issue of temporal rate variability of calcite through frequently measured rates (after every 5 min) and surface topography changes using VSI. The dataset presented here is unique as it comprises of a set of frequently collected observations over a large surface area (414 × 313 μm^2) of calcite crystal, which was large enough to observe the evolution of a large variety of features (monolayer etch pits, multilayer etch pits, screw dislocations, macro-step) through long experimental duration (6 h) on a pre-dissolved single crystal.

There are only few studies where the problem of temporal variability in surface reactivity was discussed to an appreciable extent due to the limitations associated with the use of AFM for these surface studies (as discussed in previous sections). For example, from K-feldspar surface rate measurements, Pollet-Villard et al. [34] found that despite increased etch pit distribution, surface retreat was a linear function of time. Furthermore, these authors reached the conclusion that anisotropic mineral reactivity lead to a continuous modification of the dissolution rate as a function of reaction time. Pollet-Villard et al. [34] also concluded that because of a significant variation in (intrinsic) surface reactivity, it was inappropriate to use the "mean rate" approach to quantify the reactivity of crystalline materials. In contrast, some researchers have reported an intrinsic decrease of long-term dissolution rate of the considered surface due to the development of pit walls with lower surface energy [35,36].

Although we observed changes in distribution of reactivity, no systematic, secular change in reactivity could be established (Figures 5 and 6). These reactivity differences are thus related to structurally controlled factors at the mineral-water interface, characteristic of a complex and heterogeneous crystal surface.

4.3. Interpretation of Reaction Mechanism via Rate Spectral Analysis

The statistical analysis of the flux and rate maps in the frequency domain generates spectra that are linked to surface reaction mechanism (Figure 5), although this linkage is not well understood. Nevertheless, the spectral representation of rate maps ("rate spectra", [1]) does preserve the contribution of all surface features to the overall rate [25], and has seen limited but increasing adoption in the literature [12,17]. It can be clearly seen in Figure 5 that the spectra generated from calcite dissolution rate maps (see also Figure 4) are an asymmetric distribution, one that is consistent with a heterogeneous, Boltzmann distribution of reactive surface sites. These results showed the dynamic nature of calcite surface which after interaction of several hours with fluid maintained the inherent variability and did not show attainment of surface rate stability. The asymmetry exhibited by these rate spectra was related to the presence of surface sites with huge variations in reactivity, i.e., poorly reactive sites (devoid of screw dislocations); and highly reactive sites (deep etch pits, super-step surface) on the other hand. Based on these results, we could not demonstrate any spatial dependence, however, it would be more accurate to relate these changes to the intrinsic heterogeneity of the crystal surface (which is discussed in detail in some recent investigations on surface dissolution kinetics by Luttge et al. [24].

Several studies have recently indicated the fundamental variability of mineral surface rates, mainly due to the occurrence of heterogeneous reactive surface sites [1,12,37,38]. A detailed discussion on heterogeneity of mineral surfaces and its relationship with surface rate variability could be found in a recent review [25]. A major conclusion that has been reached in current and several recent investigations on surface rates is that, instead of considering a mean rate, a stochastic treatment of dissolution kinetics should be preferred where a rate distribution constituting all the possible rate ranges is reported [24].

5. Conclusions

In this study, we present a unique set of VSI observations from the dissolution of a single calcite sample. This near in situ dissolution experiment provided insights on evolution of the surface morphology of a pre-dissolved calcite surface under alkaline solution conditions. By combining the carefully collected VSI data and its quantitative analysis using newly introduced rate spectra approach, we demonstrate that development of a large diversity of coevolving reactive surface features, e.g., ranging from nucleation, growth and annihilation of shallow pits (nm), long term development of complex, deep screw dislocations (d > 1 μm) and poorly understood super-steps (~1 μm) is responsible for the highly variable reaction rates of calcite in this long-term experiment.

Calcite continued to retreat and showed potential variability in reactivity of different surface sites on spatial and temporal scales. Sites that contributed the highest rates were those associated with deep, nested etch pits and the super-step feature. Slower rates were contributed by shallow or newly evolving monolayer etch pits. On this highly dynamic surface the integrated flux of these interacting processes contributes to a spectrum of rates. These asymmetric rate spectra do not reflect a single, dominant, reaction mechanism and are also variable in time and space.

Our VSI data demonstrated no specific trend of surface rate with reaction time. Variability in reactivity of the surface sites could be more realistically related to inherent heterogeneity of the crystal surface. The results also indicate towards the complexity of surface reactions which could not be oversimplified by taking simple means of the reaction rate at any given time step. Thus a stochastic treatment of dissolution kinetics should be preferred over deriving simple means from rate data to preserve the individual contribution of these highly variable surface sites to the overall rate.

Author Contributions: I.B. prepared and finalized paper draft; R.S.A. conceived and designed the original experimental setup, and performed all laboratory experiments; C.F. did data reduction and developed figures; A.L. read and edited the paper; I.B. finalized the paper along with all co-authors.

Acknowledgments: The senior author thanks the Alexander von Humboldt Foundation for a Postdoctoral Research Fellowship (Ref 3.5—PAK—1164117—GFHERMES-P) at the University of Bremen, Germany. All authors thank Christine Putnis for the opportunity to submit this paper. The experimental work was originally performed at the CNST Laboratory at Rice University, and we acknowledge the university's generous support. Lastly, RSA and AL acknowledge support of the US Federal Highway Authority (US FHWA grant DTFH61-12-H-00003).

Conflicts of Interest: The authors declare no conflict of interest.

References

1. Fischer, C.; Arvidson, R.S.; Luttge, A. How predictable are dissolution rates of crystalline material? *Geochim. Cosmochim. Acta* **2012**, *98*, 177–185. [CrossRef]

2. Morse, J.W.; Arvidson, R.S. The dissolution kinetics of major sedimentary carbonate minerals. *Earth Sci. Rev.* **2002**, *58*, 51–84. [CrossRef]

3. Fischer, C.; Luttge, A. Pulsating dissolution of crystalline matter. *Proc. Nat. Academy Sci.* **2018**. [CrossRef] [PubMed]

4. Arvidson, R.S.; Morse, J.W. *Formation and Diagenesis of Carbonate Sediments, In Treatise on Geochemistry*, 2nd ed.; Elsevier Ltd.: Amsterdam, the Netherlands, 2014; Volume 9, pp. 61–101.

5. Ruiz-Agudo, E.; Putnis, C.V.; Jimenez-Lopez, C.; Rodriguez-Navarro, C. An AFM study of calcite dissolution in concentrated saline solutions: The role of magnesium ions. *Geochim. Cosmochim. Acta* **2009**, *73*, 3201–3217. [CrossRef]

6. Hillner, P.E.; Gratz, A.J.; Manne, S.; Hansma, P.K. Atomic-scale imaging of calcite growth and dissolution in real time. *Geology* **1992**, *20*, 359–362. [CrossRef]

7. Dove, P.M.; Platt, F.M. Compatible real-time rates of mineral dissolution by Atomic Force Microscopy (AFM). *Chem. Geol.* **1996**, *127*, 331–338. [CrossRef]

8. Liang, Y.; Baer, D.R.; McCoy, J.M.; Amonette, J.E.; Lafemina, J.P. Dissolution kinetics at the calcite-water interface. *Geochim. Cosmochim. Acta* **1996**, *60*, 4883–4887. [CrossRef]
9. Liang, Y.; Baer, D.R.; McCoy, J.M.; Lafemina, J.P. Interplay between step velocity and morphology during the dissolution of CaCO$_3$ surface. *J. Vac. Sci. Technol. A* **1996**, *14*, 1368–1375. [CrossRef]
10. Jordan, G.; Higgins, S.R.; Eggleston, C.M.; Knauss, K.G.; Schmahl, W.W. Dissolution kinetics of magnesite in acidic aqueous solution, a hydrothermal atomic force microscopy (HAFM) study: Step orientation and kink dynamics. *Geochim. Cosmochim. Acta* **2001**, *65*, 4257–4266. [CrossRef]
11. Ruiz-Agudo, E.; Putnis, C.V. Direct observations of mineral-fluid reactions using atomic force microscopy: The specific example of calcite. *Mineral. Mag.* **2012**, *76*, 227–253. [CrossRef]
12. Emmanuel, S. Mechanisms influencing micron and nanometer-scale reaction rate patterns during dolostone dissolution. *Chem. Geol.* **2014**, *363*, 262–269. [CrossRef]
13. Luttge, A.; Bolton, E.W.; Lasaga, A.C. An interferometry study of the dissolution kinetics of anorthite: The role of reactive surface area. *American Journal of Science* **1999**, *299*, 652–678. [CrossRef]
14. Luttge, A.; Winkler, U.; Lasaga, A.C. Interferometric study of the dolomite dissolution: a new conceptual model for mineral dissolution. *Geochim. et Cosmochim. Acta* **2003**, *67*, 1099–1116. [CrossRef]
15. Arvidson, R.S.; Ertan, I.E.; Amonette, J.E.; Luttge, A. Variation in calcite dissolution rates: A fundamental problem. *Geochim. Cosmochim. Acta* **2003**, *67*, 1623–1634. [CrossRef]
16. Daval, D.; Hellmann, R.; Saldi, G.D.; Wirth, R.; Knauss, K.G. Linking nm-scale measurements of the anisotropy of silicate surface reactivity to macroscopic dissolution rate laws: New insights based on diopside. *Geochim. Cosmochim. Acta* **2013**, *107*, 121–134. [CrossRef]
17. Brand, A.S.; Feng, P.; Bullard, J.W. Calcite dissolution rate spectra measured by in situ digital holographic microscopy. *Geochim. Cosmochim. Acta* **2017**, *213*, 317–329. [CrossRef] [PubMed]
18. Liang, Y.; Baer, D.R. Anisotropic dissolution at the CaCO$_3$(10$\bar{1}$4)-water interface. *Surf. Sci.* **1997**, *373*, 275–287. [CrossRef]
19. De Leeuw, N.H.; Parker, S.C.; Harding, J.H. Molecular dynamics simulation of crystal dissolution from calcite steps. *Phys. Rev. B Condens. Matter Mater. Phys.* **1999**, *60*, 13792–13799. [CrossRef]
20. De Leeuw, N.H.; Parker, S.C. Modeling absorption and segregation of magnesium and cadmium ions to calcite surfaces: Introducing MgCO$_3$ and CdCO$_3$ potential models. *J. Chem. Phys.* **2000**, *112*, 4326–4333. [CrossRef]
21. Kerisit, S.; Parker, S.C.; Harding, J.H. Atomistic simulation of the dissociative adsorption of water on calcite surfaces. *J. Phys. Chem. B* **2003**, *107*, 7676–7682. [CrossRef]
22. Kurganskaya, I.; Luttge, A. Kinetic Monte Carlo approach to study carbonate dissolution. *J. Phys. Chem. C* **2016**, *120*, 6482–6492. [CrossRef]
23. Jordan, G.; Rammensee, W. Dissolution rates of calcite (1014) obtained by scanning force microscopy: Microtpography-based dissolution kinetics on surfaces with anisotropic step velocities. *Geochim. Cosmochim. Acta* **1998**, *62*, 941–947. [CrossRef]
24. Luttge, A.; Arvidson, R.S.; Fischer, C. A stochastic treatment of crystal dissolution kinetics. *Elements* **2013**, *9*, 183–188. [CrossRef]
25. Fischer, C.; Kurganskaya, I.; Schafer, T.; Luttge, A. Variability of crystal surface reactivity: What do we know? *Appl. Geochem.* **2014**, *43*, 132–157. [CrossRef]
26. Söngen, H.; Nalbach, M.; Adam, H. Kühnle, A. Three-dimensional atomic force microscopy mapping at the solid-liquid interface with fast and flexible data acquisition. *Rev. Sci. Instrum.* **2016**, *87*, 063704. [CrossRef] [PubMed]
27. Miyata, K.; Tracey, J.; Miyazawa, K.; Haapasilta, V.; Spijker, P.; Kawagoe, Y.; Foster, A.S.; Tsukamoto, K.; Fukuma, T. Dissolution processes at step edges of calcite in water investigated by high-speed frequency modulation atomic force microscopy and simulation. *Nano Lett.* **2017**, *17*, 4083–4089. [CrossRef] [PubMed]
28. Lasaga, A.C.; Luttge, A. Variation of calcite dissolution rate based on a dissolution stepwave model. *Science* **2001**, *291*, 2400–2404. [CrossRef] [PubMed]
29. Arvidson, R.S.; Luttge, A. System and Method of Fluid Exposure and Data Acquisition. US Patent No. 8164756, 24 April 2012; United States Patent and Trademark Office.
30. Vinson, M.D.; Luttge, A. Multiple length-scale kinetics: An integrated study of calcite dissolution rates and strontium inhibition. *Am. J. Sci.* **2005**, *305*, 119–146. [CrossRef]

31. Luttge, A.; Arvidson, R.S. Reactions at surfaces: A new approach integrating interferometry and kinetic simulations. *J. Am. Ceram. Soc.* **2010**, *93*, 3519–3530. [CrossRef]

32. Fischer, C.; Luttge, A. Beyond the conventional understanding of water-rock reactivity. *Earth Planet. Sci. Lett.* **2017**, *457*, 100–105. [CrossRef]

33. Luttge, A.; Arvidson, R.S.; Fischer, C. Kinetic concepts for quantitative prediction of fluid-solid interactions. *Chem. Geol.* **2018**, in review.

34. Pollet-Villard, M.; Daval, D.; Ackerer, P.; Saldi, G.D.; Wild, B.; Knauss, K.G.; Fritz, B. Does crystallographic anisotropy prevent the conventional treatment of aqueous mineral reactivity? A case study based on K-feldspar dissolution kinetics. *Geochim. Cosmochim. Acta* **2016**, *190*, 294–308. [CrossRef]

35. Godinho, J.R.A.; Piazolo, S.; Balic-Zunic, T. Importance of surface structure on dissolution of fluorite: Implications for surface dynamics and dissolution rates. *Geochim. Cosmochim. Acta* **2014**, *126*, 398–410. [CrossRef]

36. Smith, M.E.; Knauss, K.G.; Higgins, S.R. Effects of crystal orientation on the dissolution of calcite by chemical and microscopic analysis. *Chem. Geol.* **2013**, *360–361*, 10–21. [CrossRef]

37. Harries, D.; Pollok, K.; Langenhorst, F. Oxidative dissolution of 4C- and NC-pyrrhotite: Intrinsic reactivity differences, pH dependence, and the effect of anisotropy. *Geochim. Cosmochim. Acta* **2013**, *102*, 23–44. [CrossRef]

38. Morse, J.W.; Arvidson, R.S.; Luttge, A. Calcium carbonate formation and dissolution. *Chem. Rev.* **2007**, *107*, 342–381. [CrossRef] [PubMed]

© 2018 by the authors. Licensee MDPI, Basel, Switzerland. This article is an open access article distributed under the terms and conditions of the Creative Commons Attribution (CC BY) license (http://creativecommons.org/licenses/by/4.0/).

Commentary

How Can Additives Control the Early Stages of Mineralisation?

Denis Gebauer

Department of Chemistry, Physical Chemistry, University of Konstanz, 78457 Konstanz, Germany;
denis.gebauer@uni-konstanz.de; Tel.: +49-(0)7531-88-2169

Received: 10 April 2018; Accepted: 23 April 2018; Published: 26 April 2018

Abstract: The interactions between additives and mineral precursors and intermediates are at the heart of additive-controlled crystallisation, which is of high importance for various fields. In this commentary, we reflect on potential modes of additive control according to classical nucleation theory on one hand, and from the viewpoint of the so-called pre-nucleation cluster pathway on the other. This includes a brief review of the corresponding literature. While the roles of additives are discussed generally, i.e., without specific chemical or structural details, corresponding properties are outlined where possible. Altogether, our discussion illustrates that "non-classical" nucleation pathways promise an improved understanding of additive-controlled scenarios, which could be utilised in targeted applications in various fields, ranging from scale inhibition to materials chemistry.

Keywords: additives; biomineralisation; classical nucleation theory; interfaces; liquid precursors; minerals; mesocrystals; non-classical nucleation; pre-nucleation clusters; polymorphs

1. Introduction

Additive-controlled mineralization, in the broadest sense, is central to various fields [1–3], ranging from biomineralisation [4–8] to scale inhibition [9–11]. It also touches upon materials chemistry, where the realisation of target-oriented synthetic routes to hybrid materials with advanced properties is every researcher's dream [12]. In additive-controlled processes, especially the early stages of particle formation are important, as they can determine particle size, morphology, or polymorphism. In other words, the basic question is, how can additives control nucleation?

It is obvious that possible controls over the early stages of mineralisation processes are based on interactions of the additives with mineral precursors, intermediates and nascent particles. These may or may not include interfacial interactions. In order to be able to contemplate potential modes of additive control (or, interference) during nucleation processes, it is thus important to consider the structural and thermodynamic speciation of (pre-) nuclei, particle precursors and intermediates, which can interact with the additives and thereby change the nucleation pathway when compared to additive-free scenarios. There are at least two fundamentally distinct ways to look at this. First (Figure 1, top), according to classical nucleation theory (CNT), the monomeric chemical constituents of the nascent particles (i.e., in case of minerals, mostly ions) undergo stochastic collisions in supersaturated solutions, which lead to the formation of thermodynamically unstable pre-critical nuclei. Upon reaching a critical size, the probability of which depends on the level of supersaturation, the transition to post-critical nuclei that can grow without limit, given a sufficient supply of ions, is facilitated. The central notion of CNT is indeed that this nucleus of critical size, which serves as the relevant transition state formed within a metastable state of equilibrium (similar to the activated complex invoked in chemical kinetics [13]), governs the nucleation rate. Importantly, within CNT, pre-critical, critical, and post-critical nuclei are considered to be characterised by a solid-liquid interfacial surface, which is assumed to correspond to that of the macroscopic interface (capillary assumption). In the case where different forms or polymorphs exist, their accessibility depends on the level of supersaturation. If more than one polymorph is accessible, the fastest nucleation rate will determine which form will occur initially. Second (Figure 1, bottom), according to the so-called pre-nucleation cluster (PNC) pathway [14], ions form thermodynamically stable PNCs, which are characterised by a Flory-like [15], broad, decaying cluster size distribution and are highly dynamic. PNCs thus do not have a interfacial surface and must be considered as solutes [16]. Upon reaching a certain ion activity product (IAP), the clusters can change their structural form, leading to a significant decrease of the dynamics and rendering them phase-separated nano-droplets. Driven by the reduction of interfacial surface area, the as-formed nano-droplets undergo aggregation and/or coalescence, yielding larger liquid intermediate phases. These dehydrate and solidify toward amorphous intermediates, which eventually transform into crystals. Here, the major difference when compared to CNT is that the fundamental precursors are thermodynamically stable and form independent of the level of supersaturation. Note again, these PNCs are solutes and thus formally do not have a phase interface. The relevant transition state is then not primarily based on the size of the species forming, but instead on a significant decrease of cluster dynamics, which becomes possible upon exceeding a certain specific IAP, namely the liquid-liquid binodal limit. In other words, liquid–liquid interfacial surfaces between the nascent mineral and the solution are formed initially, which then become solid-liquid interfaces upon dehydration and solidification according to the PNC pathway.

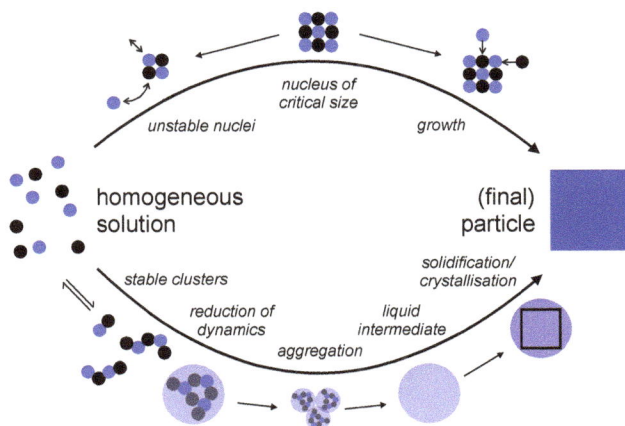

Figure 1. Schematic illustration of the mechanism of nucleation according to classical nucleation theory (CNT, top) and the pre-nucleation cluster (PNC) pathway (bottom). For different polymorphs or forms, the accessibility depends on the level of supersaturation. The sizes of the different species are system specific and cannot be fully generalized; the critical size (top, middle) for realistic supersaturation levels is typically within tens of ions, i.e., smaller than approximately 3–4 nm in diameter [17]. PNCs are similar in size but thermodynamically stable, and thus significantly more abundant than classical (pre-) critical nuclei. The smallest sizes of phase separated nano-droplets, which directly emerge from the PNC precursors, are thus also in the lower nanometer regime. Upon aggregation and coalescence, dense liquid droplets with sizes up to several hundred nanometers can be formed [18]. Consequently, depending on the kinetics of aggregation and dehydration, which can also be influenced by the presence of additives (see below), the size of solid amorphous intermediates can range from ca. 20 nm to hundreds of micrometer in size [19]. For further explanation, see the text.

Here we contemplate the possibilities of potential additive controls over nucleation from the viewpoint of CNT on one hand, and according to the PNC pathway on the other. This is done for the generic case of "an additive", i.e., without specific chemical or structural properties, but where common properties become apparent for designing ion-additive or additive-mineral interactions for realising corresponding control patterns, these are specified as far as possible. We stress that our considerations are theoretical, that is, beyond speculation on principle, and hope to provide researchers with alternative explanations of additive effects for original experimental observations. In general, additives that can influence the early stages of mineralisation range from simple ions and small molecules to macromolecules. Altogether, the discussion below will show that the modes of additive-control over nucleation are rather limited within CNT, but highly diverse from the point of view of the PNC pathway, allowing to explain experimental observations. Last, but not least, we provide such examples of experimental observations that can be rationalised based on the PNC pathway in a straightforward manner but seem incompatible with CNT. In our opinion, this underpins the value of "non-classical" nucleation theories for an improved understanding of additive-controlled nucleation leading to the potential for a broad impact on various scientific fields including biomineralisation, scale inhibition, and materials chemistry.

2. Possible Modes of Additive Control within the Notions of Classical Nucleation Theory

According to CNT, the nucleation rate J can be written as [20];

$$J = A \cdot \exp(-E_A/RT) \cdot \exp\left(-\Delta G_c^0/RT\right) \tag{1}$$

with a pre-exponential factor, A; a general activation energy, E_A; the universal gas constant, R; absolute temperature, T; and the standard free energy associated with the formation of the critical nucleus from free ions, ΔG_c^0. The pre-exponential factor takes the geometrical shape of the critical nucleus into account, which is often assumed to be spherical as other shapes introduce a significant penalty for nucleation [21]. This pre-factor A also depends on the molecular volume of the nascent nuclei's monomeric chemical constituents, i.e., for the case of mineralisation, the ions. However, A is independent of supersaturation. The general activation energy, E_A, arises from desolvation of the ions or rearrangements within the forming nuclei that may be required for the generation of the critical nucleus, but these contributions are typically neglected, as they are regarded to be minor (and are also difficult to quantify). ΔG_c^0 gives rise to a thermodynamic barrier to nucleation, which is the central quantity derived within CNT;

$$\Delta G_c^0 = \beta \cdot v^2 \cdot \sigma^3 / \phi^2 \tag{2}$$

where β is a geometrical factor (that has its lowest value for a sphere, for which $\beta_s = 16\pi/3$), v is the molecular volume of the single ions, σ is the interfacial tension between the macroscopic mineral and the solution, and ϕ is the so-called affinity. In case of minerals, the affinity can be generally formulated as;

$$\phi = RT \cdot \ln(IAP/K_{sp}) \tag{3}$$

with the solubility product of the nascent phase K_{sp} and a correspondingly consistent IAP of the relevant ions. Please note that as opposed to the sign convention of ΔG, phase separation is impossible and possible for $\phi < 0$ and $\phi > 0$, respectively. The quotient of IAP and K_{sp} is the supersaturation ratio S;

$$S = IAP/K_{sp} \tag{4}$$

Highlighting that the standard free energy of critical nuclei (Equation (2)) depends on supersaturation (Equations (3) and (4)), thereby governing the nucleation rate (Equation (1)). The dependence of the nucleation barrier on supersaturation is the hallmark of CNT; as opposed to generic chemical kinetics that can be understood within the theory of the activated complex [13].

This brief recapitulation of CNT enables us to identify possible effects of additives interacting with (pre-) critical nuclei, thereby influencing nucleation. In doing so, we will also neglect any potential effects of the additives on the kinetic barrier (i.e., E_A; Equation (1)) as well as the pre-exponential factor, A, and focus on the thermodynamic barrier (i.e., ΔG_c^0; Equations (1) and (2)). In any case, it has to be realised that the concentration of (pre-) critical nuclei is minuscule [21], owing to their excess free energy, so additive concentrations have to be in general relatively high in order to enable interactions with these rare and short-lived transition states to nucleation.

2.1. Additive Incorporation into the Nascent Phase

Additives could be incorporated into forming nuclei, affecting their bulk thermodynamic and structural properties. Since bulk and surface effects can be discussed separately within CNT, we here also assume that additive incorporation into the bulk would not affect the nuclei's interfacial properties. Whether or not this assumption is justified is certainly debatable; however, as soon as interfacial properties of nuclei are affected by additives, they very likely dominate the effects on the nucleation barrier because the surface free energy (surface tension) is cubed in the numerator when calculating ΔG_c^0 (Equation (2)). In this case, the effects of additive incorporation into the bulk of nuclei are most likely negligible and the combined effects of additive adsorption and incorporation can be discussed based on the interfacial effects alone, which are addressed in the next section.

Additive incorporation into nuclei upon stochastic solute clustering would only occur to any significant extent for a negative standard free energy of interaction between the monomeric ions and the additive. Due to lowering a given IAP* characterising a certain supersaturated state to IAP′ upon ion binding by the additive, this would thus also reduce the supersaturation ratio (Equation (4)) to S′, and with it, the affinity (Equation (3)) to ϕ' when compared to the additive-free scenario. On the other hand, additive incorporation into nuclei would, if anything, increase the molecular volume v of the ions due to the additive adsorption to v′, and may introduce deviations from a spherical shape, increasing β to β′. Spontaneous additive incorporation into nuclei thus implies S′ < S*, v′ > v, and β′ > β and increases the thermodynamic barrier to nucleation of the additive-containing phase, $\Delta G^0_c{}'$, with respect to that of the additive free phase, ΔG^0_c, according to;

$$\Delta G_c^0(S') = \beta \cdot v^2 \cdot \sigma^3 / \phi'^2 < \Delta G_c^{0\prime}(S') = \beta' \cdot v'^2 \cdot \sigma^3 / \phi'^2 \qquad (5)$$

However, additives that can be incorporated into nuclei would be (weak) nucleation inhibitors mostly based on the lowered supersaturation ratio S′; Equation (5) shows that the formation of additive-containing nuclei is indeed improbable, as the barrier for the formation of additive-free critical nuclei ΔG_c^0 remains lower also at the lowered supersaturation ratio S′. In turn, this consideration implies that from the viewpoint of CNT, the inclusion of additives into crystal lattices [22] rather happens during the subsequent growth stages. Here, the incorporation of additives affects the thermodynamic stability of the nascent phase, and thus K_{sp}. Since additive incorporation impairs the interactions between the ions in the crystal lattice, this would almost certainly always be a thermodynamic destabilisation, increasing the solubility threshold from K_{sp} to K_{sp}'. However, the additive-free phase would still be accessible during homogeneous nucleation, in principle, and the driving force for nucleation of the pure macroscopic phase would remain unchanged (Equation (3)). Altogether, our considerations show that the pure mineral phase would be initially nucleated, homogeneously, and the additive be incorporated during subsequent growth, then inhibiting this process owing to a lowered effective supersaturation due to the increased solubility, K_{sp}' [23].

2.2. Interfacial Adsorption of Additives on Nascent Nuclei

As already outlined in the previous section, any potential effects of additives on the interfacial properties of the nascent nuclei would dominate the nucleation behaviour because the interfacial free energy is cubed for calculating the magnitude of ΔG_c^0 (Equation (2)), probably outweighing the contribution of any other effects. The interfacial tension upon additive adsorption, σ', can be estimated according to the Young equation, which is valid for planar interfaces;

$$\sigma' = \sigma - k(\sigma_{as} - \sigma_{ac}) \qquad (6)$$

where σ is the interfacial tension between the nucleus and the solution (which is assumed to be equal to that between the macroscopic phases as already mentioned above), σ_{as} and σ_{ac} are the interfacial tensions between the additive and the solution and the additive and the nucleus, respectively, and k is a factor that depends on the aspect ratio of the nucleus [17]. While Equation (6) provides merely an estimate of effects of interfacial adsorption of additives based on oversimplifying assumptions (flat interfaces), it should be noted that corresponding interfacial properties are difficult to determine experimentally, or only with significant errors [24]. Note that these uncertainties would then be cubed (Equation (2)) and enter the exponent for calculating corresponding nucleation rates (Equation (1)), causing even larger quantitative uncertainties in measurable nucleation parameters. Thus, we prefer to focus on a qualitative discussion of corresponding effects here. In any case, for $\sigma_{as} < \sigma_{ac}$, the interfacial tension in the presence of the additive would be increased, $\sigma' > \sigma$, however, there is then no spontaneous adsorption of the additive and nucleation proceeds homogeneously. For $\sigma_{as} = \sigma_{ac}$, there is the limiting case where the interfacial tension between the nucleus and the solution with and without the additive is equal, $\sigma' = \sigma$, and the additive would have no effect due to insignificant adsorption. For $\sigma_{as} > \sigma_{ac}$, the additive would spontaneously adsorb to the nucleus and promote nucleation, due to the lowered interfacial tension with the adsorbed additive, $\sigma' < \sigma$. This highlights that again, significant additive effects can only be expected for spontaneous additive adsorption on the nuclei's interfacial surface, which always lowers the interfacial free energy, promoting nucleation by analogy to heterogeneous nucleation in presence of surfaces with favourable interfacial properties.

In the case where different polymorphs or forms of a mineral are accessible, these are characterised by distinct solubilities. For example, consider three polymorphs A, B, C with solubilities $K_{sp}(A) < K_{sp}(B) < K_{sp}(C)$, i.e., A, B, and C are stable, metastable, and unstable polymorphs, respectively. It can be argued that the less stable structures B and C have a lower cohesive energy than A, and hence also a lower surface free energy, suggesting that $\sigma(A) > \sigma(B) > \sigma(C)$ [25,26]. In this case, the nucleation rate of the unstable form would be the highest (Equations (1) and (2)). Thus, for a supersaturation ratio at which all polymorphs are accessible, this consideration is a CNT rationalisation of the well-known Ostwald–Volmer rule. In the case where additives bind specifically to one of the polymorphic forms, the interfacial tension of the stable forms could be lowered below that of the un- and meta-stable polymorphs, leading to their direct nucleation instead of ripening according to Ostwald's rule of stages. In this way, CNT can explain the observation that the nucleation of metastable polymorphs can be poisoned by the presence of additives or impurities; however, it seems impossible to explain scenarios where stable polymorphs cannot be obtained as opposed to metastable forms in presence of certain additives or impurities because specific additive binding can only promote nucleation, and an efficient (post-nucleation) inhibition of ripening towards the stable form would be required. Within a CNT perspective, such observations would then rather imply the complete inhibition of growth by the additive or impurity, not of nucleation.

2.3. Additive Mesophases as Environments for Nucleation

Additives may form mesophases, such as micelles, in the case where they have amphiphilic properties, depending on the additive concentration, as well as that of the ions [27]. When the formation of additive mesophases is induced by the binding of ions that constitute mineral phases, the overall ion concentration within the formed mesophase could be locally increased [28]. Formally, this may lead to a metastable intermediate, which is consistent with two-step nucleation [29,30], but here, rather a "heterogeneous two-step nucleation" scenario.

Whether or not nucleation in this ion-enriched environment proceeds via critical nuclei and according to the notions of CNT or not, however, is debatable. It has to be pointed out that the binding of ions in an additive matrix does not increase the concentration of *free* ions in the mesophase, and stable bound states are not available for stochastic collisions of free ions toward the formation of critical nuclei according to CNT [31]. This is because in the case where mineral ions can induce the formation of additive mesophases, then the underlying ion-additive interactions are spontaneous and associated with a negative standard free energy. Even if the ion-additive interactions within the mesophase are successively replaced by the counter-ion of the relevant nascent mineral [28], the ion complexes formed are necessarily in a state of even lower free energy and are thus in a free energy trap for the formation of critical nuclei [31]. In this sense, the situation is rather analogous to additive incorporation into the mineral phase, where the main effect is the lowering of supersaturation and a weak inhibition of nucleation, as discussed in Section 2.1. Also, in such mesophases, confinement effects can become important [32–34], and the increased viscosity may play an additional role toward inhibition effects. In our opinion, the observed facilitation of nucleation in such phases [28] can be better explained within the notions of "non-classical" nucleation theories as described below (Section 3.2).

3. Additive Control Patterns in "Non-Classical" Nucleation

As opposed to (pre-) critical nuclei, stable PNCs are more abundant species and interactions with additives appear in general more likely. Also, in the multi-step PNC pathway (Figure 1, bottom), interactions between the mineral precursors and intermediates can control the early stages of particle formation through various points of attack, which can allow a better rationalisation of the multiple roles of additives in crystallisation processes [35] as discussed in the following. Generally, it should be noted that according to the PNC pathway, amorphous intermediates form first but may be highly transient. In this sense, also the PNC pathway rationalises the Ostwald–Volmer rule (cf. Section 2.2).

3.1. Adsorption of Free Ions and Pre-Nucleation Clusters

Additives can bind single ions, which occurs to a significant extent only for a corresponding negative standard free energy of interaction (also see Section 2.1). For a very strong interaction and/or high additive concentrations, this can lead to the formation of additive mesophases or coacervates (Figure 2, top) if the additive structure and chemistry allows. These do not provide a pathway to crystals, as the relevant ions are bound in stable states and there is, if any, only moderate supersaturation within the coacervate (only the free ions contribute to the supersaturation ratio, Equation (4)). The chemical functional groups of the additive that allow for interactions with free ions also facilitate PNC binding. Since there is competition between additive binding and counter-ion association, PNC binding becomes more probable for ion-additive interactions that are weaker than cation-anion interactions and/or lower additive concentrations. The pathway toward particles from these states (Figure 2, middle) will be discussed in more detail in the following section.

Additives can influence ion association even without direct binding by modulating the water structure and introducing kosmotropic or chaotropic effects [36,37]. Due to the direct impact on the entropic balance of ion association through the release of hydration waters [38], these effects can increase or weaken ion-ion and/or ion-additive interactions [39], providing additional means to tune the strength of corresponding intermolecular interactions. In combination, the binding of PNCs by additives can thermodynamically stabilise or destabilise the ion associates. In the case where there are polyamorphic forms, as for calcium carbonate [19], the link between pre- and post-nucleation speciation may thus influence the type of polyamorphic solid formed by the presence of additives. In cases where no amorphous polymorphism exists, it is possible that additives induce distinct short-range structural features or motifs in amorphous intermediates (also see Section 3.3).

The binding of PNCs by additives can also have kinetic effects. For instance, it has been proposed that the binding of PNCs can lead to favourable three-dimensional arrangements of ions facilitating the change in the dynamics of the solute clusters underlying liquid-liquid separation [40]. In this case, the presence of the additive would drive phase separation, which may also be associated with the reduction of the IAP corresponding to the locus of the liquid-liquid binodal limit compared to the additive-free case (see blue arrow in Figure 3). The inverse scenario is also possible, where the strong binding of PNCs by additives would lead to configurations that do not allow for the change in speciation toward nano-droplets; however, this case cannot be distinguished from coacervate formation. In the case of weak binding of PNCs by the additive and/or low additive concentrations, inhibition of phase separation via binding of PNCs in configurations that impair nano-droplet formation appear difficult, as phase separation can still occur via non-adsorbed precursors.

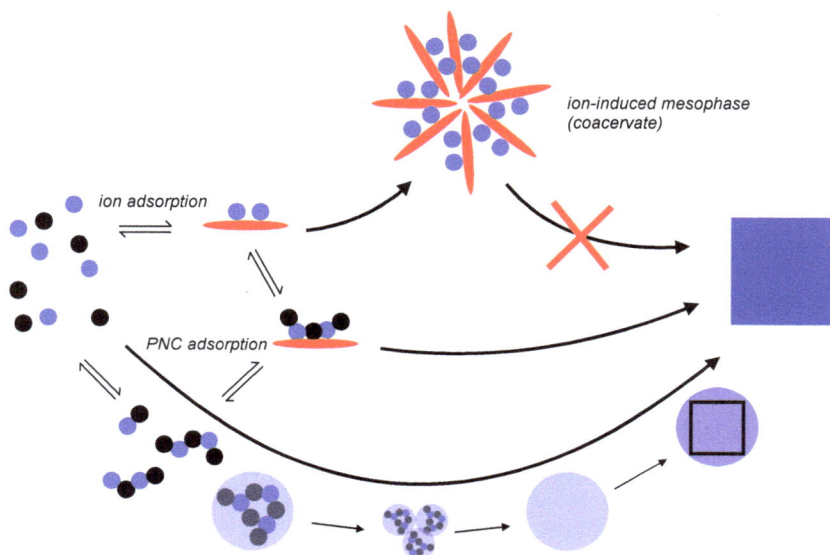

Figure 2. Schematic illustration of the mechanism of nucleation according to the PNC pathway (bottom, cf. Figure 1) with potential effects of ion (top) and PNC (middle) adsorption by an additive (red ellipsoid). For strong ion-additive interactions and/or high additive concentrations, the formation of coacervates is expected (if the additive chemistry and structure allows), which inhibit mineral formation entirely. Corresponding size regimes span the whole colloidal domain depending on the type of mesophase formed [27]. For weaker ion-additive interactions and/or low additive concentrations, PNC binding by the additive (middle) becomes probable, and the process can proceed toward particle formation. Possibilities for additive control during the latter process are discussed in more detail in Section 3.2, also see Figure 3. For explanation see the text.

3.2. Additives and Liquid Intermediates

In case of calcium carbonate, the occurrence of liquid intermediates is well-known and their use in bio-inspired materials chemistry has been thoroughly demonstrated [41]. However, the so-called polymer-induced liquid precursors (PILPs) are polymer-stabilised rather than polymer-induced states, and the notions of the PNC pathway can explain their occurrence in a straightforward manner (Figure 3). Since the PNC pathway seems to be more general for particle formation from aqueous solutions than previously thought [14,42], the occurrence of liquid intermediates may be exploited for a number of compounds. For moderate or weak interactions between the additive and PNCs, and/or low additive concentrations [43], the stabilisation of liquid intermediates becomes possible (Figure 3). Consistently, when weaker cation-binding properties of additives are exploited, additive concentrations need to be increased for sufficient PILP stabilisation [44]. However, it appears to be fundamentally required that the additive additionally stabilises the liquid intermediates against dehydration [45], or the nano-droplets against aggregation and/or coalescence by means of colloidal stabilisation [46]. At this point, interfacially active properties of the additives can also become important. These may be influenced by the presence of additional species and control or tune the wettability of the additive-stabilised liquid intermediates on distinct surfaces [47]. This phenomenon can be exploited in target-oriented materials synthesis [48].

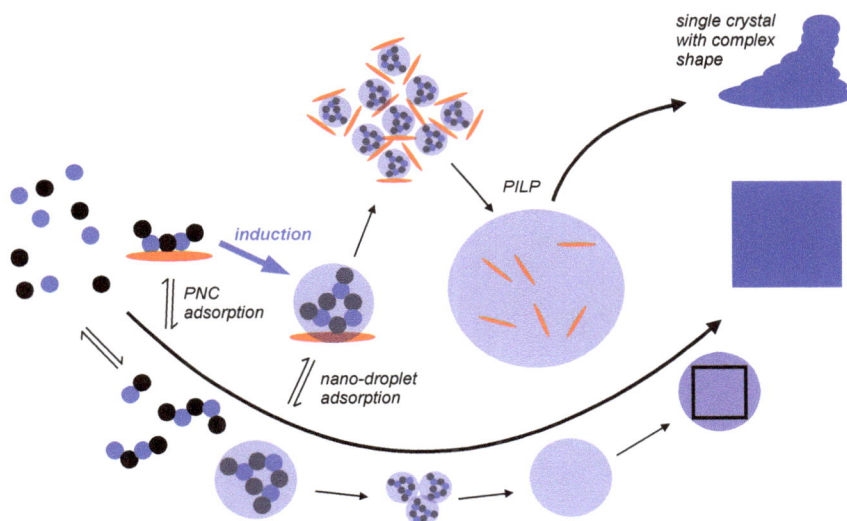

Figure 3. Schematic illustration of the mechanism of nucleation according to the PNC pathway (bottom, cf. Figure 1) with the effects of PNC and nano-droplet adsorption by an additive (red ellipsoid). For strong ion-additive interactions and low additive concentrations (or medium-to-low-strength interaction but high additive concentrations), the additive becomes incorporated in the liquid intermediates, and may kinetically stabilise these intermediate states that can grow into macroscopic "polymer-induced" liquid precursors (PILPs; note that these species are rather polymer-stabilised than polymer-induced states [45,49]), reaching sizes of hundreds of micrometres that can be observed by means of light microscopy [41]. This pathway is expected for additives that also inhibit dehydration and/or coalescence of nano-droplets. Eventually, single crystals with complex shapes can be obtained in this PILP-mediated process [41]. As discussed in Section 3.1, the binding of PNCs by additives in favourable configurations can also induce liquid-liquid separation as indicated by the bold blue arrow. For further explanation see the text.

3.3. Additives and Nascent Solid Amorphous Intermediates

When an additive can bind PNCs and nano-droplets but cannot stabilise the latter against dehydration and/or aggregation/coalescence to any significant extent, the formation of mesostructured amorphous solids is possible (Figure 4). This may be based on either nano-droplet aggregation, or the aggregation of already formed amorphous, solid nanoparticles. While it may be difficult to discriminate between the two possibilities experimentally, it can be argued that a nano-droplet-based pathway would yield amorphous solids structured on shorter length scales (ca. 2–5 nm) than the aggregation of amorphous nanoparticles that emerge later. However, depending on the kinetics of dehydration, aggregation and crystallisation, a complex convoluted pathway involving both liquid and solid amorphous precursors may also be viable. In any case, an insufficient capability of the additive to act against dehydration (or, even a promoted dehydration of the liquid intermediates due to the additive) seems to be required. It is conceivable that tuning the kinetics of interface-coupled re-dissolution/precipitation pathways [50] of this intermediate, or direct crystallisation of the mesostructured amorphous solid via solid-state transformation [51], can then yield mesocrystals [2,52,53] upon crystallisation (Figure 4, top).

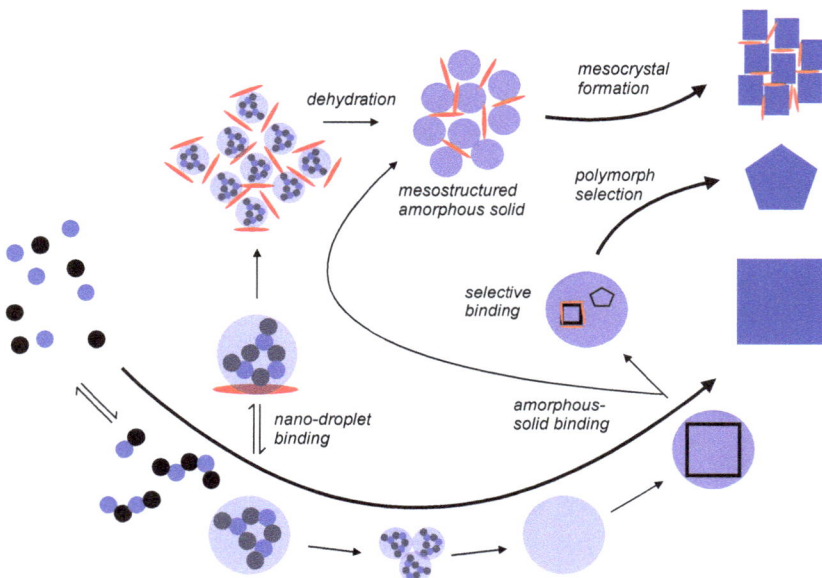

Figure 4. Schematic illustration of the mechanism of nucleation according to the PNC pathway (bottom, also cf. Figure 1) with potential effects of nano-droplet and amorphous-solid adsorption by additives (red ellipsoid) that do not kinetically stabilise liquid intermediates, at least to any significant extent. For explanation see the text.

Additives could also exhibit specific binding capabilities to certain short-range structural motifs in amorphous intermediates [54]. On the one hand, this may induce the formation of polyamorphic forms that normally do not form under the given conditions, or even induce the formation of distinct short-range structural motifs in amorphous intermediates where polyamorphism does not exist for the pure systems. This could be based on the interfacial binding of additives or structural incorporation. Note, however, that it is difficult to discuss the phenomenon of amorphous polymorphism for phases where additives are structurally incorporated, as the compositions vary and the phenomenon strictly applies only to the pure compounds with unchanged compositions. In any case, the specific binding of distinct short-range structural domains by additives may inhibit their crystallisation, so that other environments crystallise, thereby facilitating polymorph selection (Figure 4, middle) [35,55]. It is also conceivable that specific interactions of certain amorphous short-range structural motifs with additives would trigger the crystallisation of only these environments, enabling a distinct mechanism of polymorph selection from amorphous intermediates.

4. Conclusions

The above discussion shows that classical nucleation theory (CNT) can rationalise only a limited number of additive controls over nucleation, i.e., a rather weak inhibition of nucleation based on the binding of free ions, or the promotion of nucleation due to the lowering of interfacial free energies upon additive binding. The latter may also facilitate polymorph selection for some scenarios. One might argue that distinct additive controls could be rationalised within CNT by strong but unspecific effects on the pre-exponential factor, A, and/or the generic activation energy, E_A, (Equation (1)) but this would be a rather hand-waving and actually unqualified argument given that these factors are typically neglected within CNT. In contrast, there are several different mechanisms for additive control during "non-classical" nucleation according to the pre-nucleation cluster (PNC) pathway. These can explain strong inhibition, as well as promotion of nucleation, not only for stable crystalline forms but also for amorphous intermediates. The kinetic stabilisation of liquid and amorphous intermediates can provide pathways to crystals with complex shapes via PILPs, as well as mesocrystals, depending on the capabilities of the additives for, e.g., binding water and thereby tuning the kinetics of the PNC pathway after the emergence of phase separated nanodroplets. Also, the thermodynamic stabilisation and destabilisation of PNCs and amorphous intermediates is possible, which can give rise to the development of distinct short-range structural motifs in the presence of an additive. In turn, specific interactions between additives and distinct short-range structural environments in amorphous intermediates can also provide control over polymorph selection.

Last, but not least, we would like to point out two experimental observations that are difficult to explain based on CNT, but straightforward to understand from the viewpoint of the PNC pathway. Ion potential measurements in combination with THz spectroscopy showed a very strong inhibition of calcium carbonate nucleation by minuscule amounts of polycarboxylates in the low µg/mL regime [45], which was already literature known [35]. While the experiments, for the first time, revealed the locus of the liquid-liquid binodal limit, at which solute PNCs can change their speciation and become phase-separated nanodroplets, the barrier for classical nucleation under corresponding conditions is formidable even in absence of the additive, rendering nucleation via critical nuclei highly improbable [45]. Also, the very strong inhibitory capacity of the low polymer concentration can in our opinion not be explained by CNT. First, owing to the minor polymer concentration, there is no detectable binding of free calcium ions by the polymer and thus, no significant effect on the level of supersaturation that could explain the observed inhibition. Second, as shown in Section 2.2, an increase in interfacial free energy upon additive adsorption—that would be required for inhibition of nucleation from a classical perspective—would render additive adsorption on (pre-) critical nuclei thermodynamically unfavourable (Equation (6)), and nucleation would proceed homogeneously, i.e., without significant additive effects. It is impossible to rationalize the inhibition of nucleation by effects on the interfacial free energy of nuclei upon additive adsorption within CNT. Only spontaneous additive adsorption on (pre-) critical nuclei can have effects on nucleation from the viewpoint of CNT, which always promotes nucleation due to a lowered interfacial tension. The incorporation of the polycarboxylates into liquid intermediates and the inhibition of dehydration due to their superadsorbent properties, on the other hand, explains the highly effective inhibition from the point of view of the PNC pathway as outlined above. In contrast, the same additive molecules can strongly promote the nucleation of iron(III) (oxyhydr)oxides as shown by potentiometric titrations in combination with turbidity measurements at different pH values [40]; this effect, however, is hardly due to the lowering of the interfacial free energy, as it only occurs under conditions where olation PNCs can form in the system [40]. Rather, the three-dimensional configuration of olation PNCs adsorbed on the polymer appears to facilitate the formation of oxo-bridged iron(III) centres in PNCs at much lower concentrations when compared to the additive-free case, which is the fundamental change in reaction mechanism underlying the event of phase separation towards a reduction of cluster dynamics due to the formation of stronger bonds [56]. The orthogonal role of the same additives in distinct systems highlights that additive control patterns are highly dependent on the distinct chemistries, consistent with the PNC pathway, rather than generic physicochemical parameters like supersaturation that are central to CNT. The combination of additives may further bring about synergistic effects [57], which eventually allow the realisation of sophisticated, tuneable control patterns over particle formation. Future research will show if a generalised theory for additive control over particle formation based on the PNC pathway is possible and whether predictable outcomes can be exploited in a target-oriented manner.

Acknowledgments: D.G. is a Research Fellow of the Zukunftskolleg of the University of Konstanz. We thank Julian Gale for stimulating discussions and feedback on the manuscript.

Conflicts of Interest: The author declares no conflict of interest.

References

1. Ruiz-Agudo, E.; Putnis, C.V.; Rodriguez-Navarro, C. Reactions between minerals and aqueous solutions. In *Mineral Reaction Kinetics: Microstructures, Textures, Chemical and Isotopic Signatures*; Mineralogical Society of Great Britain & Ireland: Chantilly, GA, USA, 2017; pp. 419–467.
2. Cölfen, H.; Antonietti, M. *Mesocrystals and Nonclassical Crystallization*; John Wiley & Sons, Ltd.: Chichester, UK, 2008.
3. Meldrum, F.C.; Cölfen, H. Controlling Mineral Morphologies and Structures in Biological and Synthetic Systems. *Chem. Rev.* **2008**, *108*, 4332–4432. [CrossRef] [PubMed]
4. Evans, J.S. 'Apples' and 'oranges': Comparing the structural aspects of biomineral- and ice-interaction proteins. *Curr. Opin. Colloid Interface Sci.* **2003**, *8*, 48–54. [CrossRef]
5. Evans, J.S. "Liquid-like" biomineralization protein assemblies: A key to the regulation of non-classical nucleation. *CrystEngComm* **2013**, *15*, 8388–8394. [CrossRef]
6. Evans, J. Polymorphs, Proteins, and Nucleation Theory: A Critical Analysis. *Minerals* **2017**, *7*, 62. [CrossRef]
7. Evans, J.S. Principles of Molecular Biology and Biomacromolecular Chemistry. *Rev. Min. Geochem.* **2003**, *54*, 31–56. [CrossRef]
8. Ibsen, C.J.S.; Birkedal, H. Pyrophosphate-Inhibition of Apatite Formation Studied by In Situ X-Ray Diffraction. *Minerals* **2018**, *8*, 65. [CrossRef]
9. Boulahlib-Bendaoud, Y.; Ghizellaoui, S.; Tlili, M. Inhibition of $CaCO_3$ scale formation in ground waters using mineral phosphates. *Desal. Water Treat.* **2012**, *38*, 382–388. [CrossRef]
10. Ruiz-Agudo, E.; Burgos-Cara, A.; Ruiz-Agudo, C.; Ibañez-Velasco, A.; Cölfen, H.; Rodriguez-Navarro, C. A non-classical view on calcium oxalate precipitation and the role of citrate. *Nat. Commun.* **2017**, *8*, 1–10. [CrossRef] [PubMed]
11. Rodriguez-Navarro, C.; Benning, L.G. Control of Crystal Nucleation and Growth by Additives. *Elements* **2013**, *9*, 203–209. [CrossRef]
12. Mann, S. *Biomimetic Materials Chemistry*; VCH: New York, NY, USA, 1996.
13. Eyring, H. The activated complex in chemical reactions. *Chem. Rev.* **1935**, *17*, 65–77. [CrossRef]
14. Gebauer, D.; Kellermeier, M.; Gale, J.D.; Bergström, L.; Cölfen, H. Pre-nucleation clusters as solute precursors in crystallisation. *Chem. Soc. Rev.* **2014**, *43*, 2348–2371. [CrossRef] [PubMed]
15. Flory, P.J. Molecular size distribution in linear condensation polymers. *J. Am. Chem. Soc.* **1936**, *58*, 1877–1885. [CrossRef]
16. Demichelis, R.; Raiteri, P.; Gale, J.D.; Quigley, D.; Gebauer, D. Stable prenucleation mineral clusters are liquid-like ionic polymers. *Nat. Commun.* **2011**, *2*, 590. [CrossRef] [PubMed]
17. Hu, Q.; Nielsen, M.H.; Freeman, C.L.; Hamm, L.M.; Tao, J.; Lee, J.R.I.; Han, T.Y.J.; Becker, U.; Harding, J.H.; Dove, P.M.; et al. The thermodynamics of calcite nucleation at organic interfaces: Classical vs. non-classical pathways. *Faraday Discuss.* **2012**, *159*, 509–523. [CrossRef]

18. Smeets, P.J.M.; Finney, A.R.; Habraken, W.J.E.M.; Nudelman, F.; Friedrich, H.; Laven, J.; De Yoreo, J.J.; Rodger, P.M.; Sommerdijk, N.A.J.M. A classical view on nonclassical nucleation. *Proc. Natl. Acad. Sci. USA* **2017**, *114*, E7882–E7890. [CrossRef] [PubMed]

19. Cartwright, J.H.E.; Checa, A.G.; Gale, J.D.; Gebauer, D.; Sainz-Díaz, C.I. Calcium carbonate polyamorphism and its role in biomineralization: How many amorphous calcium carbonates are there? *Angew. Chem. Int. Ed.* **2012**, *51*, 11960–11970. [CrossRef] [PubMed]

20. De Yoreo, J.J.; Vekilov, P.G. Principles of crystal nucleation and growth. *Rev. Min. Geochem.* **2003**, *54*, 57–93. [CrossRef]

21. Nielsen, A.E. *Kinetics of Precipitation*; Pergamon Press: New York, NY, USA, 1964.

22. Li, H.; Xin, H.L.; Muller, D.A.; Estroff, L.A. Visualizing the 3D Internal Structure of Calcite Single Crystals Grown in Agarose Hydrogels. *Science* **2009**, *326*, 1244–1247. [CrossRef] [PubMed]

23. Davis, K.J.; Dove, P.M.; De Yoreo, J.J. The Role of Mg^{2+} as an Impurity in Calcite Growth. *Science* **2000**, *290*, 1134–1137. [CrossRef] [PubMed]

24. Dörfler, H.D. *Grenzflächen- und Kolloidchemie [Interface and Colloid Chemistry]*; VCH: Weinheim, Germany, 1994.

25. Sun, W.; Jayaraman, S.; Chen, W.; Persson, K.A.; Ceder, G. Nucleation of metastable aragonite $CaCO_3$ in seawater. *Proc. Natl. Acad. Sci. USA* **2015**, *112*, 3199–3204. [CrossRef] [PubMed]

26. Navrotsky, A. Nanoscale Effects on Thermodynamics and Phase Equilibria in Oxide Systems. *ChemPhysChem* **2011**, *12*, 2207–2215. [CrossRef] [PubMed]

27. Israelachvili, J.N. *Intermolecular and Surface Forces*, 3rd ed.; Academic Press: Burlington, MA, USA, 2011.

28. Smeets, P.J.M.; Cho, K.R.; Kempen, R.G.E.; Sommerdijk, N.A.J.M.; De Yoreo, J.J. Calcium carbonate nucleation driven by ion binding in a biomimetic matrix revealed by in situ electron microscopy. *Nat. Mater.* **2015**, *14*, 394–399. [CrossRef] [PubMed]

29. Vekilov, P.G. Nucleation. *Cryst. Growth Des.* **2010**, *10*, 5007–5019. [CrossRef] [PubMed]

30. Vekilov, P.G. The two-step mechanism of nucleation of crystals in solution. *Nanoscale* **2010**, *2*, 2346–2357. [CrossRef] [PubMed]

31. De Yoreo, J.J. Crystal nucleation: More than one pathway. *Nat. Mater.* **2013**, *12*, 284–285. [CrossRef] [PubMed]

32. Stephens, C.J.; Ladden, S.F.; Meldrum, F.C.; Christenson, H.K. Amorphous Calcium Carbonate is Stabilized in Confinement. *Adv. Funct. Mater.* **2010**, *20*, 2108–2115. [CrossRef]

33. Kröger, R. Andreas Verch Liquid Cell Transmission Electron Microscopy and the Impact of Confinement on the Precipitation from Supersaturated Solutions. *Minerals* **2018**, *8*, 21. [CrossRef]

34. Mao, L.B.; Xue, L.; Gebauer, D.; Liu, L.; Yu, X.F.; Liu, Y.Y.; Cölfen, H.; Yu, S.H. Anisotropic nanowire growth via a self-confined amorphous template process: A reconsideration on the role of amorphous calcium carbonate. *Nano Res.* **2016**, *9*, 1334–1345. [CrossRef]

35. Gebauer, D.; Cölfen, H.; Verch, A.; Antonietti, M. The multiple roles of additives in $CaCO_3$ crystallization: A quantitative case study. *Adv. Mater.* **2009**, *21*, 435–439. [CrossRef]

36. Rao, A.; Gebauer, D.; Cölfen, H. Modulating Nucleation by Kosmotropes and Chaotropes: Testing the Waters. *Crystals* **2017**, *7*, 302. [CrossRef]

37. Burgos-Cara, A.; Putnis, C.; Rodriguez-Navarro, C.; Ruiz-Agudo, E. Hydration Effects on the Stability of Calcium Carbonate Pre-Nucleation Species. *Minerals* **2017**, *7*, 126. [CrossRef]

38. Kellermeier, M.; Raiteri, P.; Berg, J.K.; Kempter, A.; Gale, J.D.; Gebauer, D. Entropy Drives Calcium Carbonate Ion Association. *ChemPhysChem* **2016**, *17*, 3535–3541. [CrossRef] [PubMed]

39. Sinn, C.G.; Dimova, R.; Antonietti, M. Isothermal Titration Calorimetry of the Polyelectrolyte/Water Interaction and Binding of Ca^{2+}: Effects Determining the Quality of Polymeric Scale Inhibitors. *Macromolecules* **2004**, *37*, 3444–3450. [CrossRef]

40. Scheck, J.; Drechsler, M.; Ma, X.; Stöckl, M.T.; Konsek, J.; Schwaderer, J.B.; Stadler, S.M.; De Yoreo, J.J.; Gebauer, D. Polyaspartic acid facilitates oxolation within iron(III) oxide pre-nucleation clusters and drives the formation of organic-inorganic composites. *J. Chem. Phys.* **2016**, *145*, 211917–211919. [CrossRef] [PubMed]

41. Gower, L.B. Biomimetic model systems for investigating the amorphous precursor pathway and its role in biomineralization. *Chem. Rev.* **2008**, *108*, 4551–4627. [CrossRef] [PubMed]

42. Gebauer, D. Prenucleation Clusters. In *Encyclopedia of Nanotechnology*; Springer: Dordrecht, The Netherlands, 2015; pp. 1–7.

43. Jiang, Y.; Gower, L.; Volkmer, D.; Cölfen, H. The existence region and composition of a polymer-induced liquid precursor phase for DL-glutamic acid crystals. *Phys. Chem. Chem. Phys.* **2012**, *14*, 914–919. [CrossRef] [PubMed]

44. Cantaert, B.; Kim, Y.Y.; Ludwig, H.; Nudelman, F.; Sommerdijk, N.A.J.M.; Meldrum, F.C. Think positive: Phase separation enables a positively charged additive to induce dramatic changes in calcium carbonate morphology. *Adv. Funct. Mater.* **2012**, *22*, 907–915. [CrossRef]

45. Sebastiani, F.; Stefan, L.P.W.; Born, B.; Luong, T.Q.; Cölfen, H.; Gebauer, D.; Havenith, M. Water Dynamics from THz Spectroscopy Reveal the Locus of a Liquid-Liquid Binodal Limit in Aqueous $CaCO_3$ Solutions. *Angew. Chem. Int. Ed.* **2016**, *56*, 490–495. [CrossRef] [PubMed]

46. Kellermeier, M.; Gebauer, D.; Melero-García, E.; Drechsler, M.; Talmon, Y.; Kienle, L.; Cölfen, H.; García-Ruiz, J.M.; Kunz, W. Colloidal stabilization of calcium carbonate prenucleation clusters with silica. *Adv. Funct. Mater.* **2012**, *22*, 4301–4311. [CrossRef]

47. Berg, J.K.; Jordan, T.; Binder, Y.; Börner, H.G.; Gebauer, D. Mg^{2+} tunes the wettability of liquid precursors of $CaCO_3$: Toward controlling mineralization sites in hybrid materials. *J. Am. Chem. Soc.* **2013**, *135*, 12512–12515. [CrossRef] [PubMed]

48. Khouzani, M.F.; Schütz, C.; Durak, G.M.; Fornell, J.; Sort, J.; Salazar-Alvarez, G.; Bergström, L.; Gebauer, D. A $CaCO_3$/nanocellulose-based bioinspired nacre-like material. *J. Mater. Chem. A* **2017**, *5*, 16128–16133. [CrossRef]

49. Bewernitz, M.A.; Gebauer, D.; Long, J.R.; Cölfen, H.; Gower, L.B. A meta-stable liquid precursor phase of calcium carbonate and its interactions with polyaspartate. *Faraday Discuss.* **2012**, *159*, 291–312. [CrossRef]

50. Ruiz-Agudo, E.; Putnis, C.V.; Putnis, A. Coupled dissolution and precipitation at mineral–fluid interfaces. *Chem. Geol.* **2014**, *383*, 132–146. [CrossRef]

51. Ihli, J.; Wong, W.C.; Noel, E.H.; Kim, Y.Y.; Kulak, A.N.; Christenson, H.K.; Duer, M.J.; Meldrum, F.C. Dehydration and crystallization of amorphous calcium carbonate in solution and in air. *Nat. Commun.* **2014**, *5*, 3169. [CrossRef] [PubMed]

52. Sturm, E.V.; Cölfen, H. Mesocrystals: Past, Presence, Future. *Crystals* **2017**, *7*, 207.

53. Sturm, E.V.; Cölfen, H. Mesocrystals: Structural and morphogenetic aspects. *Chem. Soc. Rev.* **2016**, *45*, 5821–5833. [CrossRef] [PubMed]

54. Addadi, L.; Raz, S.; Weiner, S. Taking advantage of disorder: Amorphous calcium carbonate and its roles in biomineralization. *Adv. Mater.* **2003**, *15*, 959–970. [CrossRef]

55. Gebauer, D.; Verch, A.; Börner, H.G.; Cölfen, H. Influence of selected artificial peptides on calcium carbonate precipitation—A quantitative study. *Cryst. Growth Des.* **2009**, *9*, 2398–2403. [CrossRef]

56. Scheck, J.; Wu, B.; Drechsler, M.; Rosenberg, R.; Van Driessche, A.E.S.; Stawski, T.M.; Gebauer, D. The Molecular Mechanism of Iron(III) Oxide Nucleation. *J. Phys. Chem. Lett.* **2016**, *7*, 3123–3130. [CrossRef] [PubMed]

57. Wolf, S.L.P.; Jähme, K.; Gebauer, D. Synergy of Mg^{2+} and poly(aspartic acid) in additive-controlled calcium carbonate precipitation. *CrystEngComm* **2015**, *17*, 6857–6862. [CrossRef]

© 2018 by the author. Licensee MDPI, Basel, Switzerland. This article is an open access article distributed under the terms and conditions of the Creative Commons Attribution (CC BY) license (http://creativecommons.org/licenses/by/4.0/).

![minerals logo] *minerals*

MDPI

Article

Metal Sequestration through Coupled Dissolution–Precipitation at the Brucite–Water Interface

Jörn Hövelmann [1,*], Christine V. Putnis [2,3] and Liane G. Benning [1,4,5]

[1] German Research Centre for Geosciences (GFZ), Interface Geochemistry, 14473 Potsdam, Germany; benning@gfz-potsdam.de
[2] Institut für Mineralogie, University of Münster, 48149 Münster, Germany; putnisc@uni-muenster.de
[3] Department of Chemistry, The Institute for Geoscience Research (TIGeR), Curtin University, 6845 Perth, Australia
[4] School of Earth and Environment, University of Leeds, Leeds LS2 9JT, UK
[5] Department of Earth Sciences, Freie Universität Berlin, 12249 Berlin, Germany
* Correspondence: jhoevelm@gfz-potsdam.de; Tel.: +49-331-288-28703

Received: 18 July 2018; Accepted: 7 August 2018; Published: 10 August 2018

check for updates

Abstract: The increasing release of potentially toxic metals from industrial processes can lead to highly elevated concentrations of these metals in soil, and ground- and surface-waters. Today, metal pollution is one of the most serious environmental problems and thus, the development of effective remediation strategies is of paramount importance. In this context, it is critical to understand how dissolved metals interact with mineral surfaces in soil–water environments. Here, we assessed the processes that govern the interactions between six common metals (Zn, Cd, Co, Ni, Cu, and Pb) with natural brucite ($Mg(OH)_2$) surfaces. Using atomic force microscopy and a flow-through cell, we followed the coupled process of brucite dissolution and subsequent nucleation and growth of various metal bearing precipitates at a nanometer scale. Scanning electron microscopy and Raman spectroscopy allowed for the identification of the precipitates as metal hydroxide phases. Our observations and thermodynamic calculations indicate that this coupled dissolution–precipitation process is governed by a fluid boundary layer at the brucite–water interface. Importantly, this layer differs in composition and pH from the bulk solution. These results contribute to an improved mechanistic understanding of sorption reactions at mineral surfaces that control the mobility and fate of toxic metals in the environment.

Keywords: dissolution–precipitation; toxic metals; brucite; mineral–water interface

1. Introduction

The presence and behavior of metals in the environment is of increasing concern to society, as their increasing release from industrial processes poses a major threat to ecosystems and human health [1–3]. Consequently, research is needed to present possible remediation strategies [4]. In very low quantities, some metals, such as, iron (Fe), cobalt (Co), zinc (Zn), copper (Cu), manganese (Mn), and molybdenum (Mo), are vital to living organisms, as they are required for various physiological and biochemical processes [5]. However, above certain threshold concentrations they become toxic to all living systems. For some elements, such as copper, there is a very narrow range between dietary (<0.9 mg/day) and toxic effects (>10 mg/day) [6]. Other metals, such as cadmium (Cd), nickel (Ni), tin (Sn), lead (Pb), mercury (Hg), or chromium (Cr), are considered as non-essential elements. Although nickel is known to play an important role for some plants and microorganisms, its requirement as an essential element for higher organisms is disputed and larger doses or chronic exposure to nickel can lead to serious health problems such as lung fibrosis and skin dermatitis [7]. Nickel compounds are also classified by the International Agency for Research on Cancer (IARC) as human carcinogens. Of particular concern to human health are also lead and cadmium, which have been shown to have high toxicities or carcinogenic effects even at low levels of exposure [8]. Lead poisoning, for example, can lead to damages of the brain, kidneys, liver, and the central nervous system [9].

Metals occur as oxides, sulfides, carbonates, or silicates in rocks of the Earth's crust, and they may dissolve into our groundwater through natural processes such as weathering and erosion. However, direct or indirect discharges from industrial activities and wastes can lead to highly elevated metal concentrations in soil and water. Waste containing high metal loads is produced by various industries including mining operations, metal plating facilities, fertilizer industries, and petroleum refining industries, as well as battery, paper, and pigment industries [4,10,11]. All of these processes generate large quantities of hazardous wastewaters, residues, and sludge, whose metal concentrations can reach extremely high levels well in excess of 1000 mg/L [11], and therefore require extensive waste treatment. Particular concerns may also arise from abandoned mining and smelting sites, where, after the cessation of industrial activity, the release of acidic and metalliferous waters continues for many years [12,13]. Metal concentrations in such settings typically range from tens to several hundreds of mg/L [14].

In the context of remediation, it is, however, crucial to understand the speciation, toxicity, mobility, and bioavailability of toxic metals. This is because, to a large degree the behaviour of these metals is controlled by reactions taking place at mineral surfaces in the soil environment, often involving coupled dissolution and precipitation, adsorption, ion exchange, and oxidation/reduction processes. Also, ion-binding to organic matter or metal oxide/hydroxide surfaces has an important impact on the mobility and transport of metals in soils and water [1–3,15]. The rates of these processes depend on multiple factors, such as pH, Eh, temperature, and the presence of complexing agents including organic molecules [1–3,10]. Unlike organic pollutants, metals are not biodegradable and tend to accumulate in living organisms. Therefore, the efficient removal and immobilization of toxic metals from industrial effluents and contaminated groundwater and soil is of enormous importance for protecting the environment and human health. The technologies that are being applied rely on various processes, such as chemical precipitation, ion exchange, adsorption, flotation, or electrochemical deposition [4,10,11]. Chemical precipitation is one of the most widely used methods because of its simplicity and low cost. It typically involves the addition of a precipitant that brings the pH into the basic range to cause the precipitation of metal hydroxides. The most commonly used precipitants include NaOH, lime (CaO), calcite ($CaCO_3$), and portlandite ($Ca(OH)_2$) [16,17]. However, these are not always able to reduce the metal load to an acceptable level [18]. For example, calcite dissolution buffers the pH to values between 6 and 7, which allows for the efficient hydroxide precipitation of trivalent metals such as Fe^{3+} or Cr^{3+}, but is not high enough for divalent metals (including Zn^{2+}, Cd^{2+}, Co^{2+}, Pb^{2+}, Cu^{2+}, and Ni^{2+}). On the other hand, lime, portlandite, and NaOH, when used in excess, can easily raise the pH to levels above 12, which is too high, as the solubility of many metal hydroxides increases again at very high pH values.

Magnesium hydroxide ($Mg(OH)_2$), prepared from the hydration of magnesium oxide (MgO) or in the form of the natural mineral brucite, is considered to be an attractive alternative reagent because its dissolution buffers the solution pH between 8.5 and 10, where most divalent metals are the least soluble [18]. The relatively low solubility of $Mg(OH)_2$ ensures a slow release of OH^-, thus providing a long-term source of alkalinity. It has already been demonstrated in various laboratory and field experiments that $Mg(OH)_2$ can effectively remove a range of metals, such as Cr, Cd, Pb, Ni, Zn, Co, and Cu, from highly contaminated water [14,18–26]. In the case of bulk $Mg(OH)_2$ powder, the major mechanism of metal removal is assumed to be precipitation. On the other hand, adsorption is generally considered to be the principal immobilization mechanism for some recently developed MgO- and $Mg(OH)_2$-based nanomaterials, which are potentially useful in water remediation applications [27–34]. However, direct nanoscale insights into the dissolution–precipitation reactions at the interface between

$Mg(OH)_2$ and metal-containing solutions are so far lacking, because previous studies have mainly focused on the bulk efficiency of metal removal by $Mg(OH)_2$.

In this research, we have studied the interaction between natural brucite and six common metals (Ni, Cu, Co, Zn, Cd, and Pb) that are often found at elevated concentrations in contaminated soil and groundwater. Our main objectives were (1) to characterize and quantify the spatial and temporal coupling between brucite dissolution and the subsequent nucleation and growth of metal precipitates and (2) to assess the role of such coupled dissolution–precipitation reactions in controlling the immobilization of metals. Brucite is of high relevance in this context, not only because of its direct applicability in water remediation, but also because its simple, layered structure is a fundamental building unit in a wide range of minerals. Examples include layered double hydroxides (LDH) as well as trioctahedral phyllosilicates, such as chlorite and saponite clays. These minerals are common constituents of natural soils and have been extensively studied in the past because their large surface areas and cation exchange capacities make them potentially useful adsorbents or ion exchangers for the removal of environmental contaminants [35,36]. Hence, our results may also provide more fundamental insights into how metals are immobilized at mineral surfaces in the environment.

2. Materials and Methods

2.1. Atomic Force Microscopy

All of the experiments were performed using natural brucite crystals (Tallgruvan, Norberg, Sweden). The initial, essentially monomineralic brucite rock sample contained minor amounts of dolomite, magnetite, and pyroaurite, which were avoided during the atomic force microscope (AFM) specimen preparation. Only optically clear brucite crystals were used. Immediately before each experiment, a brucite crystal was cleaved parallel to the (001) cleavage plane to expose a fresh surface. The final dimensions of the brucite specimens were ca. 3 mm × 3 mm × 0.2 mm. The aqueous metal-bearing solutions were prepared by dissolving reagent grade salts of $NiSO_4 \cdot 6H_2O$, $ZnSO_4 \cdot 7H_2O$, $CoSO_4 \cdot 7H_2O$, $CdSO_4 \cdot 8/3H_2O$, $CuCl_2 \cdot 2H_2O$, or $PbCl_2$ into double-deionized water (resistivity >18 MΩ·cm). The adjustments of pH were made by adding 0.01 mol/L HCl. The concentrations were varied between 0.1 and 50 mmol/L to cover a broad range of concentrations typically found in industrial effluents as well as contaminated soil and groundwater [11,14].

The brucite (001) surfaces were imaged at room temperature (22 $\pm$ 1 °C) using a Bruker Multimode atomic force microscope (AFM) operating in contact mode. In situ experiments were performed within an O-ring sealed flow-through cell from Digital Instruments (Bruker, Billerica, MA, USA). The solutions were injected at regular time intervals between each scan (lasting ~1.5 min), giving an effective flow rate of 22 $\mu L \cdot s^{-1}$. The chosen flow rate was to ensure a surface-controlled reaction rather than diffusion control [37]. The time for injections, followed by the scans, was kept constant. The experiments were also performed using continuous flow by gravity feed as well as continuous flow using a pump. The results from all of the methods were similar, except that the image quality is reduced by fluid movement. Therefore, to obtain the best images, the flow was stopped during the actual scan. Si_3N_4 probes (NP-S10, Bruker, Billerica, MA, USA) were used for all of the AFM experiments at a scan rate of 3.81 Hz and samples/lines of 256, and occasionally 512 for better imaging. The set point was kept at 2.000 V. The scan areas were typically 3 μm × 3 μm and 5 μm × 5 μm with smaller areas (1 μm × 1 μm) chosen for closer observations. The scan angle was determined from trace and retrace matching to give accurate scanning. As tip-sample interactions can result in enhanced reaction kinetics at the surface [38], care was always taken to zoom out to a larger scan area after a number of scans in one area, in order to check for the effect of the tip. Also, new areas previously not scanned were always checked for differences or similarities. The images were analyzed using the NanoScope Analysis software (version 1.50). The step retreat velocities or etch pit spreading rates were calculated measuring the length increase of the etch pit step edges (s) per unit time in sequential images scanned in the same direction. For each experimental condition, at least five different etch pits were analyzed in 1–5 pairs of sequential images. Each etch pit spreading rate value thus represents an average of 5–25 individual measurements. In some experiments, the scanning was stopped from time to time and the solution in the fluid cell was kept static for several minutes to check the results of longer time reactions without any tip interference. In these cases, diffusion within the fluid cell took place and the reactions were not solely surface reaction controlled.

Ex situ experiments were performed following in situ AFM experiments. The reacted brucite samples were removed from the AFM fluid cell and were placed in a beaker filled with 7 mL of the different solutions at room temperature, in order to observe further reactions. After 12–48 h, the samples were recovered from the solution and quickly dried using adsorbent paper, and then immediately re-imaged in air in the AFM. We also performed ex situ experiments using the same protocol, but with "fresh" brucite crystals that had not been previously reacted in the AFM flow-through cell. In this case, the crystals were imaged only in air by AFM, before being placed into the reaction beaker, so as to ensure that the pristine surfaces were clean.

2.2. Scanning Electron Microscopy

The samples from the ex situ experiments were also imaged using an Ultra 55 Plus (Carl Zeiss Microscopy GmbH, Oberkochen, Germany) scanning electron microscope (SEM) equipped with an energy dispersive X-ray (EDX) detector for the elemental analysis of the reacted brucite surfaces and newly formed precipitates. Before imaging, all of the samples were coated with a 20 nm-thick layer of carbon and the imaging and analyses were performed at 20 keV.

2.3. Raman Spectroscopy

A confocal Raman spectrometer (LabRAM HR Evolution, Horiba Jobin Yvon, Bensheim, Germany) operating with the 532 nm line of a frequency-doubled Nd: YAG solid state laser was used for the analysis of surface precipitates on brucite, after reaction with metal-bearing solutions. Spectra in the range of 100 and 1200 cm^{-1} were obtained with a 100× objective lens, a 50 μm pinhole, and a grating of 1800 grooves/mm using an acquisition time 2 s × 10 s. The reference spectra of different metal phases were obtained from The RRUFF™ Project database [39].

2.4. Geochemical Modelling

The hydro-geochemical software PHREEQC [40] (version 3.2.0-9820) was used to calculate the chemical speciation and saturation state, with respect to relevant solid phases of the initial and final solutions used in the experiments. All of the calculations were done using the minteq.v4 database. The database was modified to include heterogenite (CoOOH), using a logK value of −7.973 for the reaction $CoOOH + 3H^+ = Co^{3+} + 2H_2O$ (calculated from the standard Gibbs free energies of formation given in Chivot et al. [41]).

3. Results and Discussion

3.1. Brucite Dissolution in the Presence of Dissolved Metals

The freshly exposed brucite (001) surfaces were initially flat, apart from some pre-existing step edges (Figure 1A). Once exposed to metal bearing solutions (pH 3–6), brucite dissolution occurred with the retreat of step edges and the formation of etch pits with the typical equilateral triangular shape that results from the three-fold rotational axis normal to the (001) brucite cleavage surface [42,43] (Figure 1B). Most of the etch pits were initially shallow with depths of ~0.5 nm, corresponding to the thickness of one unit-cell layer (0.47 nm) of the brucite crystal lattice. The lateral spreading of these etch pits eventually caused them to merge, leaving behind small islands that disappeared upon further dissolution. The complete removal of one unit-cell layer was followed by the formation and spreading of new etch pits resulting in a layer-by-layer dissolution. The etch pit density was highly variable, both locally and temporally (ranging from <10 to >100 pits in a scanned area of 5 μm × 5 μm), suggesting a large heterogeneity in the distribution of the crystal defects. Some etch pits developed into deeper, concentric pits, whose slopes consisted of a high density of monolayer steps. Such pits most likely originated from structural defects that penetrate several layers, whereas the shallow monolayer etch pits nucleated at either defect-free sites or point defects [44]. Rows of deep etch pits were also frequently observed. These are likely to be associated with linear defects intersecting at the surface. Over time, the step density and roughness of the dissolving brucite surface increased because of the deepening, spreading, and subsequent coalescence of deep etch pits. Consequently, the monolayer etch pits nucleating on the narrow interstep terraces vanished quickly because of the merging with the adjacent steps.

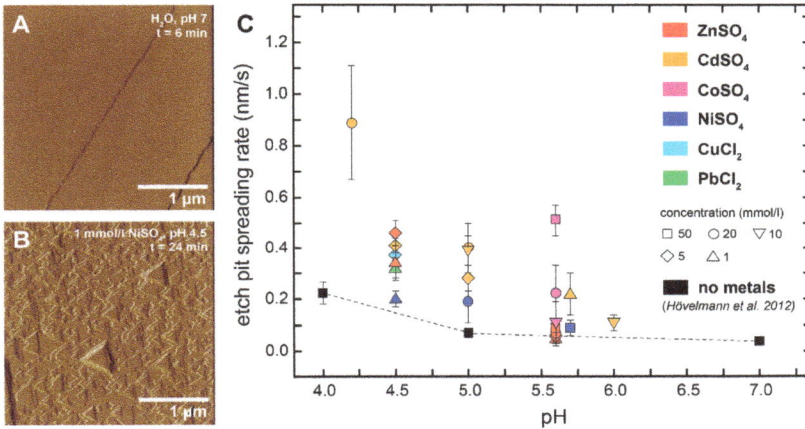

Figure 1. Brucite dissolution in the presence of dissolved metals. (**A**) In situ atomic force microscope (AFM) deflection image of a brucite (001) surface taken after 6 min in contact with pure water. The image shows that brucite surfaces were initially flat, apart from some pre-existing step edges; (**B**) In situ AFM deflection image of a brucite surface taken after 24 min in contact with 1 mmol/L NiSO$_4$, pH 4.5. The dissolution of brucite in acidic metal solutions resulted in the formation of triangular etch pits; (**C**) Comparison of etch pit spreading rates in the absence (black symbols) and presence (colored symbols) of dissolved metals. Error bars indicate the standard deviation of the measured values (Table 1).

Table 1. Compilation of measured etch pit spreading rates in the presence and absence of dissolved metals.

Metal	Concentration (mmol/L)	pH	Spreading Rate (nm/s)	Standard Deviation	n
Zn	1	4.5	0.34	0.07	19
	5	4.5	0.46	0.05	15
	1	5.6	0.05	0.03	24
	5	5.6	0.06	0.03	12
	20	5.6	0.07	0.03	9
	50	5.6	0.09	0.03	6
Cd	20	4.2	0.89	0.22	5
	5	4.5	0.41	0.03	5
	5	5	0.28	0.05	16
	10	5	0.39	0.06	10
	20	5	0.4	0.1	10
	1	5.7	0.22	0.08	5
	10	6	0.11	0.03	10
Co	10	5.6	0.11	0.08	8
	20	5.6	0.22	0.11	17
	50	5.6	0.51	0.06	10

Table 1. *Cont.*

Metal	Concentration (mmol/L)	pH	Spreading Rate (nm/s)	Standard Deviation	*n*
Ni	1	4.5	0.2	0.03	10
	20	5	0.19	0.08	26
	50	5.7	0.09	0.03	18
Cu	5	4.5	0.37	0.1	10
Pb	1	4.5	0.32	0.04	10
no metals *		4	0.22	0.04	8
		5	0.07	0.01	10
		7	0.04	0.01	11

* data from Hövelmann et al. [43].

The average spreading rate (v_s) of the shallow etch pits (expressed as $v_s = 1/(2\sqrt{3})$ ds/dt, where s is the side length of a triangular etch pit) was measured from successive AFM scans for several experiments. The etch pit spreading rates and their standard deviations were calculated from multiple measurements in the pH range between 4 and 6 (Table 1). Figure 1C presents the measured etch pit spreading rates at different metal concentrations between 1 and 50 mmol/L, in comparison to the metal-free rates previously reported by Hövelmann et al. [43]. Our data reveal that all six metals that were tested have a positive effect in increasing the spreading rates. This effect tended to be higher at lower pH values. For example, at pH 5.6, Zn had only a small effect compared to the Zn free system, whereas at pH 4.5, the spreading rates were almost twice as high compared with those in the metal-free solutions. Similarly, Cd at pH 6 showed only a minor effect, but gave four to five times higher rates at pH 4.2. At a given pH, the spreading rates also tended to increase with increasing metal concentration. For example, for Co at pH 5.6, we observed an increase from 0.11 ± 0.08 nm/s in 10 mmol/L to 0.51 ± 0.06 nm/s in 50 mmol/L solutions. Overall, Cd appeared to have the largest effect of all of the six metals. A two to three times dissolution rate increase was even observed at a Cd concentration of 1 mmol/L and a relatively high pH of 5.7. In contrast, Ni showed only a minor effect at this pH, even at very high concentrations of 50 mmol/L.

It has been previously shown that dissolved metals such as Cu^{2+}, Co^{2+}, Ni^{2+}, Cd^{2+}, and Al^{3+} can inhibit both the dissolution and growth of many oxides, carbonates, and silicates [45]. This inhibitory effect is attributed to the formation of inner-sphere binuclear/multinuclear surface complexes that bridge two (or more) dissolution active surface metal centers of the crystal lattice. It is thus thought that dissolution is inhibited because of the large activation energy that must be overcome to simultaneously remove two surface metal centers [45]. Pokrovsky et al. [46] have measured the bulk dissolution rates of brucite in the presence of eight different metals, including Ni, Co, and Pb. Brucite dissolution was found to be promoted by Pb, whereas strong inhibitory effects were observed for Ni and Co. Overall, they documented that the brucite dissolution rates in the presence of these metals correlated well with the water molecule exchange rates in the first hydration sphere of the corresponding cation.

In the present study, an inhibitory effect of Co and Ni could not be confirmed. Instead, our in situ AFM observations indicate that each of the six metals that were tested could promote brucite dissolution. The reason for this apparent discrepancy is not clear. It should, however, be noted that the etch pit spreading rate measurements are not directly comparable to the bulk dissolution rates, as in the Pokrovsky et al. [46] study. Inherent variations in the surface energies, for example, because of local differences in the number, distribution, and nature of reactive sites, make it difficult (if not impossible) to translate locally measured etch pit spreading rates into a bulk dissolution rate [47,48]. Also, during our AFM experiments, we followed dissolution only on (001) cleavage surfaces, which may be less reactive than (hk0) faces [49]. Hence, a decrease of the bulk dissolution rates could be primarily due to metal adsorption and formation of multi-dentate surface complexes at these edge surfaces. One possible explanation for our observation, that etch pit spreading rates increase in the presence of metals, could be the formation of aqueous metal–hydroxyl complexes or clusters, or the precipitation of metal hydroxide phases that consume part of the released OH$^-$ ions. In both cases, this would lower the saturation state with respect to brucite and hence increase the driving force for brucite dissolution.

3.2. Precipitation at the Dissolving Brucite Surface

The dissolution of brucite in the metal-bearing solutions was followed by the formation of new phases. At relatively high metal concentrations (i.e., ≥10 mmol/L for Ni, Co, Zn, and Cd, and ≥1 mmol/L for Cu and Pb), precipitation was observed immediately after solution injection into the AFM fluid cell (Figure 2). Initially, the precipitates nucleated as small particles (average width 20–30 nm) with rounded shapes and average heights of just a few nanometers. They tended to form preferentially at kink sites and along step edges on the dissolving brucite surface (Figure 2A–C). At kink sites, dissolution is more intense, because of the lower coordination of Mg atoms [43,50]. Consequently, the faster release of OH$^-$ ions (along with Mg^{2+}) may lead to a faster supersaturation in these areas. This means that the nucleation of the precipitates was rate-limited by the dissolution of brucite, and the dissolution of brucite is coupled with the precipitation of the new phase. Moreover, the higher amount of unsatisfied bonds at step edges and kink sites should result in a higher tendency to adsorb ions or molecules from solution, thus promoting nucleation. Precipitation also tended to be enhanced at lower pH values, where brucite dissolution is faster, again suggesting that dissolution was the rate-limiting step for nucleation. However, at pH <3, precipitation was no longer observed on the brucite surface, suggesting that the dissolution of brucite became kinetically too fast for nucleation to keep pace at the retreating surface.

In the earliest precipitation stages, the nucleated particles generally showed a weak adhesion to the brucite surface, as they were easily pushed aside by the AFM tip during scanning. Therefore, areas that had been scanned multiple times often had much fewer particles than the surrounding areas (Figure 2D). With time, the particles became more and more abundant on the brucite surfaces and they tended to aggregate to form larger particle clusters. In some cases, these particle clusters developed into aggregates of two-dimensional (2D) plates (see Section 3.3). However, the lateral spreading or growth of individual particles was limited, and no continuous precipitate layers formed within the time frame of our in situ AFM experiments (typically lasting 1–2 h).

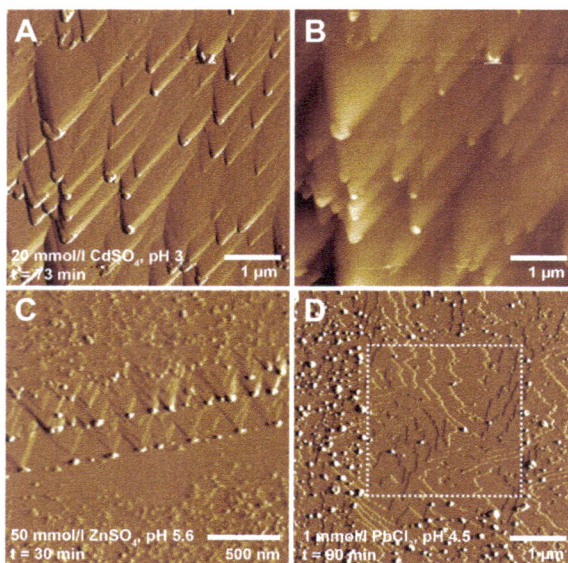

Figure 2. In situ AFM images showing the nucleation of precipitates on brucite surfaces. (**A**) Deflection and (**B**) height image (acquired in liquid) showing the nucleation of nanometer size particles at kink sites on a brucite surface after 73 min in a 20 mmol/L CdSO$_4$ (pH 3) solution; (**C**) Deflection image (acquired in liquid) of a brucite surface after 30 min in a 50 mmol/L ZnSO$_4$ (pH 5.6) solution. The image reveals two generations of particles that nucleated at kink sites along a retreating step edge; (**D**) Deflection image (acquired in liquid) showing the covering of a brucite surface with precipitates after 90 min in a 1 mmol/L PbCl$_2$ (pH 4.5) solution. The central area (outlined by a dashed square) corresponds to an area that was scanned before. This area contains far less particles than the surroundings, showing that the particles were initially weakly attached to the brucite surface, as they could be removed by the AFM tip during scanning.

Precipitation also occurred at low concentrations (0.1–1 mmol/L) for all of the six metals (Figure 3). However, in these experiments, it typically took several hours before any particles could be observed with the AFM. For example, after 18–48 h, rounded particles or particle clusters (40–100 nm in height and 100–400 nm in diameter) were observed on all of the reacted brucite surfaces, regardless of the metal used in solution (Figure 3). Again, the precipitates were concentrated near step edges or deep etch pits (i.e., at sites of higher energy and hence enhanced dissolution). Most of the particles formed at low metal concentrations had more or less rounded shapes and showed no clear evidence of crystallographic facets, possibly indicating a poorly ordered internal structure. The clustering of particles, however, tends to indicate at least some short-range order. Moreover, some particles, such as the ones that formed in Cu-bearing solutions, were slightly elongated and showed a preferred orientation (Figure 3E). This may suggest that their nucleation and growth was, to some degree, crystallographically controlled by the underlying brucite substrate.

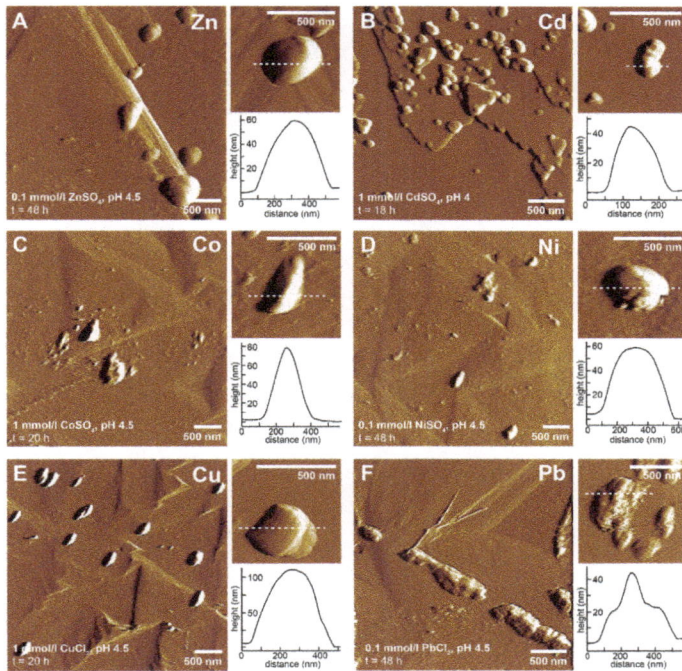

Figure 3. AFM deflection images (acquired in air) and height profiles of nanoparticles and particle clusters on brucite surfaces formed during ex situ experiments. Experimental conditions were (**A**) 0.1 mmol/L ZnSO$_4$, pH 5, 48 h; (**B**) 1 mmol/L CdSO$_4$, pH 4, 18 h; (**C**) 1 mmol/L CoSO$_4$, pH 4.5, 20 h; (**D**) 0.1 mmol/L NiSO$_4$, pH 4.5, 48 h; (**E**) 1 mmol/L CuCl$_2$, pH 4.5, 20 h; and (**F**) 0.1 mmol/L PbCl$_2$, pH 4.5, 48 h. The locations of the height profiles are indicated by dashed lines in the corresponding AFM images. Prior to the ex situ experiments, all of the surfaces were exposed to the corresponding solutions for ~2 h in the AFM fluid cell under stop-flow conditions.

3.3. Identification of the Precipitates

The particles that formed at low metal concentrations were generally too small in size (<100 nm) and number to be readily characterized by SEM-EDX or Raman spectroscopy (in both cases, the spot size is usually ~1 μm). Therefore, some brucite crystals were exposed to metal solutions at high concentrations (10 mmol/L for Ni, Zn, Cd, Co, and Cu, and 1 mmol/L for Pb) for 36 h, in order to increase the amount of precipitates. The chemical and morphological characteristics of the so formed surface precipitates are shown in Figure 4, and described in the following paragraphs.

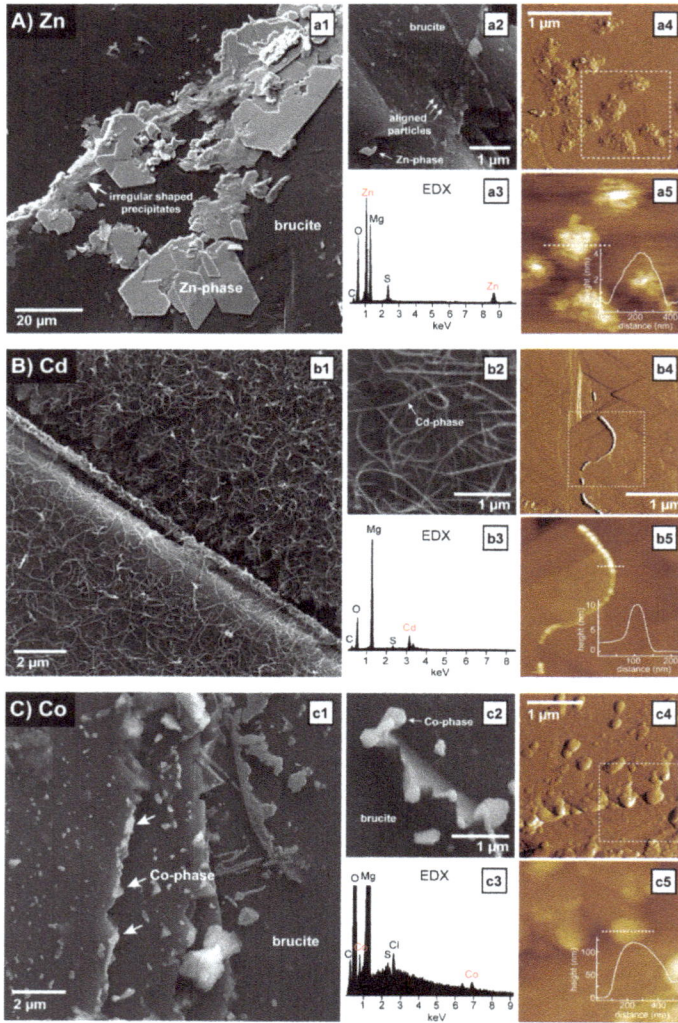

Figure 4. *Cont.*

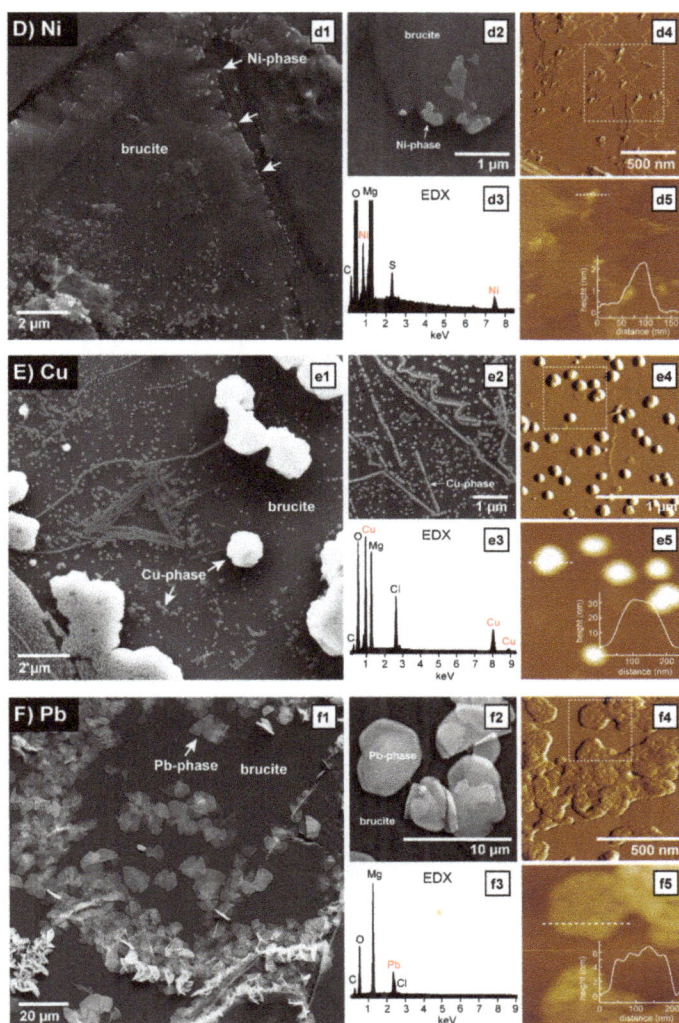

Figure 4. Chemical and morphological characterization of surface precipitates formed during reactions with solutions of (**A**) ZnSO$_4$, (**B**) CdSO$_4$, (**C**) CoSO$_4$, (**D**) NiSO$_4$, (**E**) CuCl$_2$, and (**F**) PbCl$_2$. (**a1–f1**) SEM images showing the coverage of brucite surfaces with precipitates. All of the SEM images are from brucite surfaces that reacted for 36 h with 7 mL of the respective metal solution at pH 4.5. Concentrations were 10 mmol/L for Zn, Cd, Co, Ni, and Cu (**a1–e1**), and 1 mmol/L for Pb (**f1**). The brucite surfaces were initially clean and free of any precipitates. (**a2–f2**) Higher magnification SEM images of surface precipitates from the same samples as in a1–f1. (**a3–f3**) Energy dispersive X-ray (EDX) spectra of the surface precipitates showing the incorporation of metals. The high Mg peak that is present in all of the spectra originates from the underlying brucite surface. (**a4–f4**) AFM deflection and (**a5–f5**) height images of surface precipitates highlighting some similar and additional morphological features. The location of the height images is indicated by a dashed square in the corresponding deflection image. Insets in (**a5–f5**) show the height profiles along the sections marked by the dashed lines. The experimental conditions for the AFM images were (**a4,a5**) 50 mmol/L ZnSO$_4$, pH 5.75, ~2 h in situ (flow) + 16 h ex situ (static), image acquired in air; (**b4,b5**) 20 mmol/L CdSO$_4$, pH 4.5, 40 min in situ (flow), image acquired in liquid; (**c4,c5**) 50 mmol/L CoSO$_4$, pH 5, ~2 h in situ (flow) + 95 h ex situ (static), image acquired in air; (**d4,d5**) 20 mmol/L NiSO$_4$, pH 5, 45 min in situ (flow), image acquired in liquid; (**e4,e5**) 20 mmol/L CuCl$_2$, pH 4.5, ~2 h in situ (flow) + 20 h ex situ (static), image acquired in air; and (**f4,f5**) 1 mmol/L PbCl$_2$, pH 4.5, ~2 h in situ (flow) + 20 h ex situ (static), image acquired in air.

41

Zinc: The reaction of brucite with 10 mmol/L $ZnSO_4$ (pH 4.5) resulted in the abundant surface precipitation of platy six-sided crystals, reaching diameters of more than 20 μm in some cases (Figure 4(a1)). Most of them were located near deep step edges and their basal planes were always parallel to the (001) cleavage surface of the underlying brucite. This non-random distribution and orientation of the crystals indicates that they formed in situ on the brucite surface and that they were not deposited onto the surface during drying. The newly formed crystals had well-developed shapes and straight edges and they were mostly formed on top of more irregular shaped precipitates (Figure 4(a1)). In addition to the larger crystals, clusters of smaller rounded particles, ca. 50–500 nm in size and often grouped in lines near step edges, etch pits, or kink sites, were also observed (Figure 4(a2)). The EDX analyses revealed the presence of Zn and S in both the large crystals and the particle clusters (Figure 4(a3)), thus confirming the formation of a Zn- and S-rich phase. The Raman spectroscopic analyses of the larger crystals produced spectra with several peaks, as shown in Figure 5A. The two peaks at 444 and 278 cm^{-1} are consistent with the strech/bend of Mg–O in the underlying brucite crystal [51]. The remaining peaks therefore originate from the precipitate itself. These were found to be in good agreement with the reference spectra of namuwite ($Zn_4(SO_4)(OH)_6 \cdot 4H_2O$) (Figure 5A). Namuwite is a zinc hydroxy-sulfate mineral that has been previously observed to form during the alkalinization of $ZnSO_4$ solutions [52]. The early formation stages of platy precipitates (presumably namuwite) could also be imaged with AFM (Figure 4(a4,a5)). In the presence of 50 mmol/L $ZnSO_4$ (pH 5.75), abundant precipitates were quickly formed at kink sites and along step edges. With time, these precipitates developed into flat islands that were about 100–500 nm wide and 4 nm thick, and typically consisted of several layers with irregular outlines.

Cadmium: After a reaction with 10 mmol/L $CdSO_4$ (pH 4.5), the brucite surface was covered in thread-like precipitates (Figure 4(b1,b2)). The threads were only 50–100 nm thick, but reached lengths of more than 10 μm. The EDX analyses revealed the presence of Cd (and some minor amounts of S), possibly indicating a cadmium hydroxide phase (Figure 4(b3)). The Raman spectroscopic analysis of the reacted brucite surface produced no additional bands to those belonging to brucite. This is probably due to the thread-like precipitates being so thin that the Raman laser could not sample enough of the precipitates to produce peaks in the spectra. Hence, an exact phase identification was not possible. However, previous studies investigating Cd-precipitation from supersaturated aqueous solutions reported the formation of nanowires of crystalline $Cd(OH)_2$, which are very similar in size and morphology to the thread-like Cd-precipitates observed in our experiments [53]. The Cd-hydroxide threads were not concentrated at specific locations, but were more or less equally distributed over the whole brucite surface. This may suggest that they were not directly formed at the brucite surface, but rather that they were nucleated in the overlying fluid layer, before being deposited onto the surface. However, they were not formed during the drying of the sample, as we also observed evidence for their formation during in situ AFM experiments (Figure 4(b4,b5)). Interestingly, the AFM images revealed that the threads are themselves made up of smaller aligned particles with diameters between 50 and 100 nm and heights of about 10 nm (Figure 4(b5)). This may suggest that they formed through a self-assembly mechanism involving the nucleation of particulate (either nanocrystalline or amorphous Cd-hydroxide, the latter with probably some short-range order) building blocks, followed by their continued oriented attachment to form one-dimensional (1D) chains.

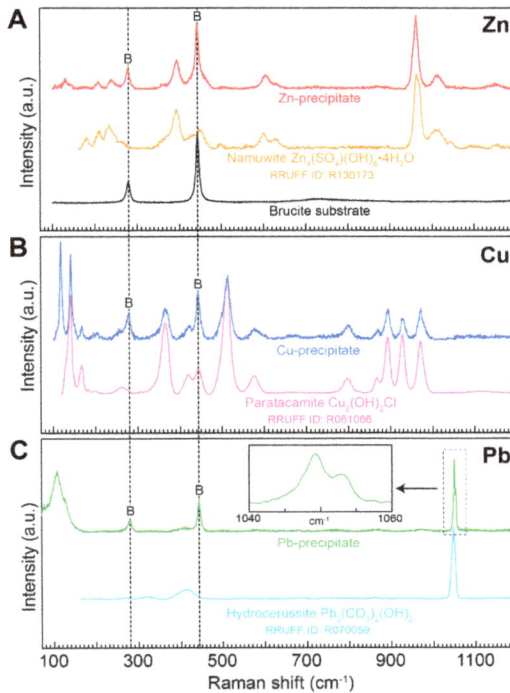

Figure 5. Raman spectra of surface precipitates formed after 36 h in (**A**) 10 mmol/L $ZnSO_4$, pH 4.5; (**B**) 10 mmol/L $CuCl_2$, pH 4.5; and (**C**) 1 mmol/L $PbCl_2$, pH 4.5. The shown spectra are representative of the micrometer sized crystals in Figure 4A,E,F, respectively. The black spectrum in (**A**) corresponds to a pristine brucite surface. The peaks belonging to the underlying brucite substrate (marked "B") are present in all of the measured spectra. The spectra of the Zn-, Cu-, and Pb-precipitates show good agreement with the reference spectra for namuwite, paratacamite, and hydrocerussite, respectively. The inset in (**C**) is a magnification of the spectral region between 1040 and 1060 cm^{-1}, and reveals a splitting of the main peak near 1050 cm^{-1}.

Cobalt: Relatively sparsely distributed clusters of particles were observed on the brucite surface after 36 h in contact with 10 mmol/L $CoSO_4$ (pH 4.5). Their diameters ranged from less than 100 nm to more than 2 μm, and they were usually more numerous and larger in areas where the dissolution was more intense, that is, near etch pits and deep step edges (Figure 4(c1)). Some particles showed platy shapes with straight edges and more or less hexagonal outlines, indicating the formation of a possible crystalline phase (Figure 4(c2)), and EDX suggests that they contained Co (along with Cl and minor S) (Figure 4(c3)). The Cl probably originates from the HCl that was added to bring the initial pH down to 4.5. Based on the elemental composition, it seems likely that the precipitate is a cobalt hydroxyl–chloride/sulfate. However, the Raman analyses showed no peaks other than those belonging to the underlying brucite, again likely due to the thinness (average <100 nm) of the precipitates. Although a higher surface coverage was observed at longer reaction times and higher Co concentrations, even after 95 h in 50 mmol/L $CoSO_4$ the maximum thickness of the precipitates barely reached 100 nm (Figure 4(c4,c5)) and thus a conclusive identification of the formed Co crystals was not possible.

Nickel: Small particles (100–500 nm in diameter) were also formed in the presence of 10 mmol/L $NiSO_4$ (pH 4.5) (Figure 4(d1)). In some cases, they formed larger aggregates reaching more than

2 μm in diameter. However, the coverage of the brucite surface was relatively sparse (i.e., similar to the cobalt experiment). Again, most of the particles were found in areas of enhanced dissolution, and kink sites generated during the dissolution at step edges were the preferred sites for nucleation (Figure 4(d1,d2)). Some of the larger particles had the form of 2D plates, indicating the formation of sheet-like crystals (Figure 4(d2)). The EDX analyses revealed the incorporation Ni and S into the precipitate, pointing to a nickel hydroxy-sulfate phase (Figure 4(d3)). Again, a conclusive identification by Raman spectroscopy was not possible because of the small particle sizes. The initial formation stages of these platy Ni-rich particles were also observed during in situ AFM imaging. For example, in the presence of 20 mmol/L $NiSO_4$ (pH 5), tiny platelets, with thicknesses of only 1–2 nm and diameters of around 50 nm, were formed after 45 min at kink sites on the dissolving brucite surface (Figure 4(d4,d5)).

Copper: The brucite surface that reacted for 36 h with a solution containing 10 mmol/L $CuCl_2$ (pH 4.5) became partly covered with patches of greenish precipitates. These were visible to the naked eye before the sample was removed from the reacting solution, and hence, they clearly are an in situ reaction product and not a drying artefact. The SEM imaging showed that the greenish patches were composed of 1–2 μm sized crystals with more or less isometric shapes (Figure 4(e1)). In addition, numerous small spherically shaped particles with diameters of ~100 nm (Figure 4(e1,e2)) were often agglomerated along step edges and steps of deep etch pits, resulting in linear structures on the brucite surface. However, they were not only observed on the brucite surface itself, but also on top of the larger, newly formed precipitate crystals, indicating that both were formed contemporaneously. The EDX analyses showed that both types of precipitates contain Cu and Cl (Figure 4(e3)). The Raman spectroscopic analysis of the larger crystals yielded spectra that agree well with the reference spectra of paratacamite ($Cu_2(OH)_3Cl$) (Figure 5B), a mineral that is commonly found as a corrosion product of copper in marine environments, and that was also observed to form in experiments where calcite was reacted with $CuCl_2$ solutions [54,55]. Spherical Cu precipitates similar to those observed in the SEM could also be imaged with AFM on a brucite surface that was left for 20 h in a 20 mmol/L $CuCl_2$ solution (Figure 4(e4,e5)). In the AFM experiments, these precipitates were between 100 and 200 nm in diameter and around 30 nm in height, and showed no evidence of crystallographic facets. Again, the small size of the spherical precipitates did not enable identification by Raman spectroscopy.

Lead: When brucite was reacted with 1 mmol/L $PbCl_2$ (pH 4.5), the formation of numerous platy crystals with approximately hexagonal shapes and diameters of around 10 μm was observed (Figure 4(f1,f2)). The basal planes of these crystals were in most cases parallel to the exposed brucite (001) surface. However, crystals protruding from the surface were also observed, particularly in the most densely covered areas. Pb (and possibly minor Cl) were confirmed for these precipitates by EDX analyses (Figure 4(f3)). The Raman spectroscopic analysis of the precipitate crystals yielded spectra with an intense and sharp peak near 1050 cm^{-1} (Figure 5C). This peak is consistent with the CO_3 symmetric stretching vibration (v_1) in hydrocerussite ($Pb_3(CO_3)_2(OH)_2$) [56], a basic lead carbonate that is a common weathering product in lead ore deposits and is often found in lead contaminated soil [57,58]. Hydrocerussite has a sheet-like structure and forms hexagonally shaped crystals (i.e., consistent with our SEM observations). Therefore, it seems reasonable to assume that the precipitated phase is hydrocerussite. This would imply that carbonate ions were present in the reacting solution. These must have resulted from CO_2 gas exchange with the laboratory air, as no carbonate was added to the solution. In the measured Raman spectrum, the v_1 band shows a clear splitting into two bands centered at 149 and 152 cm^{-1} (see inset in Figure 5C). Such a splitting has also been previously observed for hydrocerussite, and is thought to be due to the carbonate ions occupying at least two different sites in the crystal lattice [56]. In addition to the v_1 band, the measured spectrum showed another broader peak at 107 cm^{-1}, which probably relates to the lattice modes in hydrocerussite [56]. Agglomerates of the precipitates with approximately hexagonal shapes (presumably hydrocerussite) could also be imaged with AFM on a brucite surface that was left for 20 h in 1 mmol/L $PbCl_2$ (pH 4.5) (Figure 4(f4,f5)). Interestingly, the surface of these ~6 nm thick precipitates was not flat, but showed a nanoglobular morphology, possibly suggesting that they formed through the aggregation of nanoparticles.

3.4. The Importance of the Fluid Boundary Layer

Our observations show that the dissolution of brucite in the presence of dissolved metals is coupled to the precipitation of metal hydroxide phases. Precipitation was observed for each of the six metals, and occurred at all of the concentrations tested (0.1–50 mmol/L). For selected ex situ experiments, we measured the pH of the metal solution after the reaction with brucite, in order to enable solution speciation calculations with PHREEQC. Our data show that the pH increased slightly from 4.5 to values between 5 and 6 after one to three days of interaction between a 3 mm × 3 mm × 0.2 mm brucite crystal and 7 mL of 0.1–10 mmol/L metal solutions (Table 2). The thermodynamic calculations were performed using the measured pH values and initial metal concentrations (before reaction) as input parameters. Our calculations indicate that the solutions were still undersaturated with respect to all of the relevant metal phases, including various oxides, hydroxides, carbonates, and mixed hydroxy-sulfates/chlorides (Figure 6). The only exception was the 10 mmol/L $CuCl_2$ solution that was slightly supersaturated, with respect to atacamite ($Cu_2(OH)_3Cl$) (note: atacamite is a polymorph of the actually observed paratacamite and was used here as a proxy, as the thermodynamic data for paratacamite are not available in the database). It should be noted, however, that the calculated saturation indices are likely overestimates, because the actual metal concentrations (after reaction) were probably lower than those used for our calculations (due to precipitation and possibly adsorption at the brucite surface). The fact that we indeed observed precipitates on all of the reacted surfaces, even though the bulk solutions were undersaturated, must therefore mean that precipitation occurred within a thin fluid layer at the brucite–water interface, which differed in composition and pH from the bulk solution. A similar conclusion was also made for the reaction of brucite in the presence of dissolved carbonate, phosphate, and organic phosphorus [43,50,59], as well as for a wide range of other mineral–fluid systems [60–65]. Recently, the application of real-time phase-shift interferometry and ion-specific microelectrodes has indeed provided direct evidence that the fluid boundary layers at solid–fluid interfaces can become supersaturated with respect to new phases when the bulk solution is undersaturated [66]. The consequence of this is often a close spatial and temporal coupling between dissolution and precipitation reactions. This reaction mechanism is therefore known as the interface-coupled dissolution–precipitation mechanism, and it is thought to be a universal mechanism, whenever minerals are in contact with aqueous solutions with which they are out of equilibrium [67,68].

Table 2. Bulk pH values of metal solutions after reaction with brucite crystals.

Metal Concentration	Initial pH	Solution Volume	Reaction Time	pH after Reaction with Brucite Crystal (ca. 3 mm × 3 mm × 0.2 mm)					
(mmol/L)		(mL)	(days)	$ZnSO_4$	$CdSO_4$	$CoSO_4$	$NiSO_4$	$CuCl_2$	$PbCl_2$
0.1	4.5	7	3	5.96	5.84	6.86	6.6	5.56	5.31
1	4.5	7	1	5.2	5.43	5.21	5.12	5.07	4.99
10	4.5	7	3	5.54	5.56	5.65	5.77	4.66	-

Eventually, the diffusion of ions from the fluid boundary layer may also lead to supersaturation of the bulk solution. The time it takes to supersaturate the bulk solution likely depends strongly on the initial liquid-to-solid ratio. In the case of our ex situ experiments, the liquid-to-solid ratio was relatively high (~1600 mL/g), resulting in a slow concentration built-up in the bulk fluid. On the other hand, if the amount of reactive brucite surface is high relative to the volume of fluid, one might expect a much faster saturation of the bulk fluid.

The coupled dissolution–precipitation reactions at the brucite surface are schematically illustrated in Figure 7, and can be summarized as follows:

Upon contact with a metal-bearing solution, the brucite surface begins to dissolve, releasing Mg^{2+} and OH^- into the surface solution layer at the brucite–water interface (Figure 7, arrow (1)), as follows:

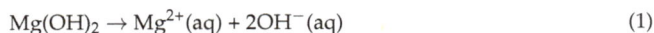

$$Mg(OH)_2 \rightarrow Mg^{2+}(aq) + 2OH^-(aq) \tag{1}$$

The release of OH$^-$ leads to a pH increase in the fluid boundary layer. At the sites where dissolution is enhanced (e.g., etch pits and kink sites), the pH will more rapidly reach values that promote the precipitation of metal hydroxide phases, as follows:

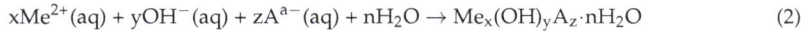

$$xMe^{2+}(aq) + yOH^-(aq) + zA^{a-}(aq) + nH_2O \rightarrow Me_x(OH)_yA_z \cdot nH_2O \qquad (2)$$

Note that in Equation (2), Me^{2+} represents the divalent metal cation (Zn^{2+}, Cd^{2+}, Co^{2+}, Ni^{2+}, Cu^{2+}, or Pb^{2+}), whereas A^{a-} refers to the corresponding counter anion (Cl$^-$ or SO$_4$$^{2-}$).

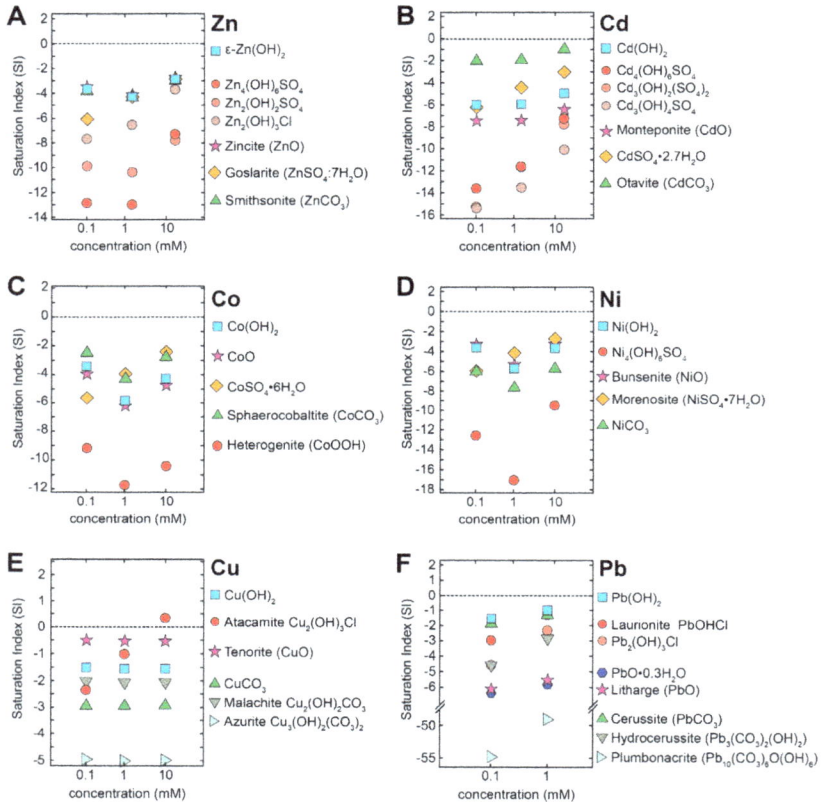

Figure 6. Bulk solution saturation indices of the relevant metal phases calculated using PHREEQC, with the minteq.v4 database. The calculations are based on the initial metal concentrations and bulk solution pH measurements (Table 2) made after the reaction of a ca. 3 mm × 3 mm × 0.2 mm large brucite crystal with 7 mL of 0.1–10 mmol/L metal solutions. All of the solutions had an initial pH of 4.5. The saturation indices of the carbonate phases were calculated under the assumption of equilibrium with atmospheric CO$_2$ (logP$_{CO_2}$ = −3.38).

Our observations indicate that precipitation within the brucite fluid boundary layer follows a range of different pathways, involving both classical ion-by-ion and non-classical particle-mediated nucleation and growth mechanisms [69,70] (Figure 7, arrows (2–8)). The appearance of small particles (few nm) at the dissolving brucite surface and their subsequent aggregation and fusion with adjacent particles (Figures 2 and 3) suggests a heterogeneous nucleation and growth process. From our in situ AFM observations, however, it is not clear whether the initial particles nucleated directly on the brucite surface (Figure 7, arrow (2)) or if they are formed within the fluid boundary layer before attaching to the brucite surface (Figure 7, arrow (3)). The latter scenario seems likely, because the first observable particles already had sizes of a few nanometers. Thus, ionic species will likely associate within the fluid boundary layer to form nanoparticles that will attach to the surface at high energy sites. The subsequent growth of precipitates occurs then via diverse pathways, as follows: (a) Particle agglomerates at specific surface sites, such as step edges and etch pits (e.g., Figures 3 and 4A,C–E), tend to indicate an aggregation process, where the successive addition of new particles occurred directly at the brucite surface (Figure 7, arrow (4)). (b) A more random distribution and arrangement of the Cd-hydroxide threads (Figure 4B) suggests that, in this case, the individual nanoparticles had already assembled within the fluid boundary layer (Figure 7, arrow (5)) before being deposited onto the brucite surface. Whether the early-formed particles were amorphous or nanocrystalline cannot be unequivocally determined from our AFM observations. Their rounded shapes and lack of clear crystallographic facets point to an amorphous or poorly ordered structure. Nevertheless, their tendency to form clusters suggests at least some short-range order. With time, some of these clusters developed into flat islands (e.g., Figure 4(a4)), possibly involving a structural rearrangement or transformation (Figure 7, arrow (6)). The further growth of these islands into microscopic, smooth-surfaced crystals (e.g., Figure 4(a1)) could have followed a classical mechanism that would require the addition of ionic species to step edges or kink sites (Figure 7, arrow (7)). However, it is also conceivable that the growth proceeded by the continued oriented attachment of nanoparticles (Figure 7, arrow (8)). In case of the Pb precipitate, the nanoglobular morphologies observed in AFM (Figure 4(f4,f5)) indeed suggest such a particle-mediated growth process. The relationships between the nm-sized particle clusters and the μm-sized crystals are, however, not always clear from our AFM and SEM observations. For the Cu-precipitates (Figure 4E), for example, it is not clear whether the large paratacamite crystals evolved from the smaller spherically shaped clusters or were formed by a separate nucleation and growth process.

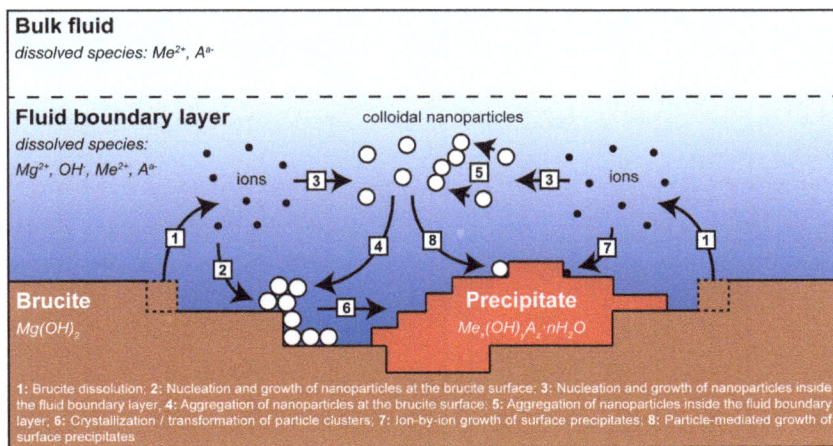

Figure 7. Schematic illustration of coupled dissolution–precipitation pathways at the brucite–water interface (see main text for more detailed explanations).

4. Conclusions

Our study shows that dissolved metals such as Zn^{2+}, Cd^{2+}, Co^{2+}, Ni^{2+}, Cu^{2+}, and Pb^{2+} can be sequestered as solid hydroxide phases on brucite surfaces by coupled dissolution–precipitation reactions. The dissolution of brucite in acidic metal solutions (pH 3–6) leads to the release of OH^- ions, and the accompanying increase in pH then enables the precipitation of the metal hydroxides.

The present results demonstrate that the precipitation of these metal hydroxides is possible even if the bulk solution is highly undersaturated. This is because supersaturation is reached only locally within a thin fluid boundary layer above the dissolving brucite surface. Thus, the dissolution of just a few monolayers of brucite may be enough to trigger the precipitation of metals. It is therefore reasonable to suggest that brucite is an effective reactant for the remediation of metal contaminated soils and waters. Moreover, our results emphasize the general importance of coupled dissolution–precipitation reactions at mineral surfaces in controlling the mobility of chemical species and toxic elements in the environment. In this context, our study may also contribute to an improved mechanistic understanding of the sorption reactions of dissolved metals to other sparingly soluble hydroxide surfaces, in particular brucite-like minerals such as layered double hydroxides or clays. These minerals are widely distributed in natural environments and of particular environmental significance, because their high ion exchange capacities and surface areas can greatly affect the fate of metals in the soil-water system [35,36].

Author Contributions: J.H. and C.V.P. conceived and designed the experiments; J.H. and C.V.P. performed the experiments; J.H. analyzed the data and wrote the paper; L.G.B. and C.V.P. helped with the discussion and interpretation of the results and edited the manuscript.

Funding: J.H. and L.G.B. acknowledge financial support by the European Union's Horizon 2020 Marie Skłodowska-Curie Innovative Training Network Metal-Aid (Project No. 675219) and the Helmholtz Recruiting Initiative (Award No. I-044-16-01). C.V.P. acknowledges funding received through the European Union's seventh Framework Marie Skłodowska-Curie Initial Training Networks CO2-React (Project No. 317235) and FlowTrans (Project No. 316889).

Acknowledgments: We thank Judith Schicks for her help with the Raman analyses.

Conflicts of Interest: The authors declare no conflict of interest.

References

1. Rinklebe, J.; Knox, A.S.; Paller, M. *Trace Elements in Waterlogged Soils and Sediments*; CRC Press, Taylor & Francis Group: New York, NY, USA, 2016.
2. Hooda, P.S. *Trace Elements in Soils*, 1st ed.; John Wiley & Sons: Chichester, UK, 2010.
3. Kabata-Pendias, A.; Mukherjee, A.B. *Trace Elements from Soil to Human*; Springer: Berlin, Germany, 2007.
4. Fu, F.; Wang, Q. Removal of heavy metal ions from wastewaters: A review. *J. Environ. Manag.* **2011**, *92*, 407–418. [CrossRef] [PubMed]
5. World Health Organisation. *Trace Elements in Human Nutrition and Health*; World Health Organization: Geneva, Switzerland, 1996.
6. Institute of Medicine. *Dietary Reference Intakes for Vitamin A, Vitamin K, Arsenic, Boron, Chromium, Copper, Iodine, Iron, Manganese, Molybdenum, Nickel, Silicon, Vanadium, and Zinc*; The National Academies Press: Washington, DC, USA, 2001.
7. Denkhaus, E.; Salnikow, K. Nickel essentiality, toxicity, and carcinogenicity. *Crit. Rev. Oncol. Hematol.* **2002**, *42*, 35–56. [CrossRef]
8. Tchounwou, P.B.; Yedjou, C.G.; Patlolla, A.K.; Sutton, D.J. Heavy metals toxicity and the environment. *Mol. Clin. Environ. Toxicol.* **2012**, *101*, 133–164. [CrossRef]
9. Papanikolaou, N.C.; Hatzidaki, E.G.; Belivanis, S.; Tzanakakis, G.N.; Tsatsakis, A.M. Lead toxicity update. A brief review. *Med. Sci. Monit.* **2005**, *11*, RA329–RA336. [PubMed]
10. Hashim, M.A.; Mukhopadhyay, S.; Sahu, J.N.; Sengupta, B. Remediation technologies for heavy metal contaminated groundwater. *J. Environ. Manag.* **2011**, *92*, 2355–2388. [CrossRef] [PubMed]
11. Barakat, M.A. New trends in removing heavy metals from industrial wastewater. *Arab. J. Chem.* **2011**, *4*, 361–377. [CrossRef]
12. Akcil, A.; Koldas, S. Acid Mine Drainage (AMD): Causes, treatment and case studies. *J. Clean. Prod.* **2006**, *14*, 1139–1145. [CrossRef]
13. Sheoran, A.S.; Sheoran, V. Heavy metal removal mechanism of acid mine drainage in wetlands: A critical review. *Miner. Eng.* **2006**, *19*, 105–116. [CrossRef]
14. Rötting, T.S.; Ayora, C.; Carrera, J. Improved passive treatment of high Zn and Mn concentrations using Caustic Magnesia (MgO): Particle size effects. *Environ. Sci. Technol.* **2008**, *42*, 9370–9377. [CrossRef] [PubMed]

15. Di Bonito, M.; Lofts, S.; Groenenberg, J.E. Models of geochemical speciation: Structure and applications. In *Environmental Geochemistry*; De Vivo, B., Belkin, H.E., Lima, A., Eds.; Elsevier: Amsterdam, The Netherlands, 2018; pp. 237–305.

16. Aziz, H.A.; Adlan, M.N.; Ariffin, K.S. Heavy metals (Cd, Pb, Zn, Ni, Cu and Cr(III)) removal from water in Malaysia: Post treatment by high quality limestone. *Bioresour. Technol.* **2008**, *99*, 1578–1583. [CrossRef] [PubMed]

17. Mirbagheri, S.A.; Hosseini, S.N. Pilot plant investigation on petrochemical wastewater treatment for the removal of copper and chromium with the objective of reuse. *Desalination* **2005**, *171*, 85–93. [CrossRef]

18. Cortina, J.L.; Lagreca, I.; De Pablo, J. Passive in situ remediation of metal-polluted water with Caustic Magnesia: Evidence from column experiments. *Environ. Sci. Technol.* **2003**, *37*, 1971–1977. [CrossRef] [PubMed]

19. Rötting, T.S.; Cama, J.; Ayora, C. Use of caustic magnesia to remove cadmium, nickel, and cobalt from water in passive treatment systems: Column experiments. *Environ. Sci. Technol.* **2006**, *40*, 6438–6443. [CrossRef] [PubMed]

20. Caraballo, M.A.; Rötting, T.S.; Macías, F.; Nieto, J.M.; Ayora, C. Field multi-step limestone and MgO passive system to treat acid mine drainage with high metal concentrations. *Appl. Geochem.* **2009**, *24*, 2301–2311. [CrossRef]

21. Macías, F.; Caraballo, M.A.; Rötting, T.S.; Pérez-López, R.; Nieto, J.M.; Ayora, C. From highly polluted Zn-rich acid mine drainage to non-metallic waters: Implementation of a multi-step alkaline passive treatment system to remediate metal pollution. *Sci. Total Environ.* **2012**, *433*, 323–330. [CrossRef] [PubMed]

22. Dong, J.; Li, B.; Bao, Q. In situ reactive zone with modified Mg(OH)$_2$ for remediation of heavy metal polluted groundwater: Immobilization and interaction of Cr(III), Pb(II) and Cd(II). *J. Contam. Hydrol.* **2017**, *199*, 50–57. [CrossRef] [PubMed]

23. Garcia, M.A.; Chimenos, J.M.; Fernandez, A.I.; Miralles, L.; Segarra, M.; Espiell, F. Low-grade MgO used to stabilize heavy metals in highly contaminated soils. *Chemosphere* **2004**, *56*, 481–491. [CrossRef] [PubMed]

24. Navarro, A.; Chimenos, J.M.; Muntaner, D.; Fernández, A.I. Permeable reactive barriers for the removal of heavy metals: Lab-scale experiments with low-grade magnesium oxide. *Gr. Water Monit. Remediat.* **2006**, *26*, 142–152. [CrossRef]

25. Shao, L.; Zhou, Y.; Chen, J.F.; Wu, W.; Lu, S.C. Buffer behavior of brucite in removing copper from acidic solution. *Miner. Eng.* **2005**, *18*, 639–641. [CrossRef]

26. Lin, X.; Burns, R.C.; Lawrance, G.A. Heavy metals in wastewater: The effect of electrolyte composition on the precipitation of cadmium(II) using lime and magnesia. *Water Air Soil Pollut.* **2005**, *165*, 131–152. [CrossRef]

27. Pilarska, A.A.; Klapiszewski, Ł.; Jesionowski, T. Recent development in the synthesis, modification and application of Mg(OH)$_2$ and MgO: A review. *Powder Technol.* **2017**, *319*, 373–407. [CrossRef]

28. Balducci, G.; Bravo Diaz, L.; Gregory, D.H. Recent progress in the synthesis of nanostructured magnesium hydroxide. *CrystEngComm* **2017**, *19*, 6067–6084. [CrossRef]

29. Mahdavi, S.; Jalali, M.; Afkhami, A. Heavy metals removal from aqueous solutions using TiO$_2$, MgO, and Al$_2$O$_3$ nanoparticles. *Chem. Eng. Commun.* **2013**, *200*, 448–470. [CrossRef]

30. Hua, M.; Zhang, S.; Pan, B.; Zhang, W.; Lv, L.; Zhang, Q. Heavy metal removal from water/wastewater by nanosized metal oxides: A review. *J. Hazard. Mater.* **2012**, *211–212*, 317–331. [CrossRef] [PubMed]

31. Gao, C.; Zhang, W.; Li, H.; Lang, L.; Xu, Z. Controllable fabrication of mesoporous MgO with various morphologies and their absorption performance for toxic pollutants in water. *Cryst. Growth Des.* **2008**, *8*, 3785–3790. [CrossRef]

32. Zhang, S.; Cheng, F.; Tao, Z.; Gao, F.; Chen, J. Removal of nickel ions from wastewater by Mg(OH)$_2$/MgO nanostructures embedded in Al$_2$O$_3$ membranes. *J. Alloys Compd.* **2006**, *426*, 281–285. [CrossRef]

33. Feng, J.; Gao, M.; Zhang, Z.; Liu, S.; Zhao, X.; Ren, Y.; Lv, Y.; Fan, Z. Fabrication of mesoporous magnesium oxide nanosheets using magnesium powder and their excellent adsorption of Ni (II). *J. Colloid Interface Sci.* **2018**, *510*, 69–76. [CrossRef] [PubMed]

34. Xiong, C.; Wang, W.; Tan, F.; Luo, F.; Chen, J.; Qiao, X. Investigation on the efficiency and mechanism of Cd(II) and Pb(II) removal from aqueous solutions using MgO nanoparticles. *J. Hazard. Mater.* **2015**, *299*, 664–674. [CrossRef] [PubMed]

35. Liang, X.; Zang, Y.; Xu, Y.; Tan, X.; Hou, W.; Wang, L.; Sun, Y. Sorption of metal cations on layered double hydroxides. *Colloids Surfaces A Physicochem. Eng. Asp.* **2013**, *433*, 122–131. [CrossRef]

36. Uddin, M.K. A review on the adsorption of heavy metals by clay minerals, with special focus on the past decade. *Chem. Eng. J.* **2017**, *308*, 438–462. [CrossRef]

37. Ruiz-Agudo, E.; Kowacz, M.; Putnis, C.V.; Putnis, A. The role of background electrolytes on the kinetics and mechanism of calcite dissolution. *Geochim. Cosmochim. Acta* **2010**, *74*, 1256–1267. [CrossRef]

38. Park, N.S.; Kim, M.W.; Langford, S.C.; Dickinson, J.T. Atomic layer wear of single-crystal calcite in aqueous solution scanning force microscopy. *J. Appl. Phys.* **1996**, *80*, 2680–2686. [CrossRef]

39. Lafuente, B.; Downs, R.T.; Yang, H.; Stone, N. The Power of Databases: The RRUFF Project. In *Highlights in Mineralogical Crystallography*; Walter de Gruyter GmbH: Berlin, Germany, 2016.

40. Parkhurst, D.L.; Appelo, C.A.J. *User's Guide to PHREEQC (Version 2): A Computer Program for Speciation, Batch-Reaction, Onedimensional Transport, and Inverse Geochemical Calculations*; Water-Resources Investigations Report 99-4259; U.S. Geological Survey: Reston, WV, USA, 1999.

41. Chivot, J.; Mendoza, L.; Mansour, C.; Pauporté, T.; Cassir, M. New insight in the behaviour of Co-H_2O system at 25–150 °C, based on revised Pourbaix diagrams. *Corros. Sci.* **2008**, *50*, 62–69. [CrossRef]

42. Kudoh, Y.; Kameda, J.; Kogure, T. Dissolution of brucite and the (001) surface at neutral pH: In situ atomic force microscopy observations. *Clays Clay Miner.* **2006**, *54*, 598–604. [CrossRef]

43. Hövelmann, J.; Putnis, C.V.; Ruiz-Agudo, E.; Austrheim, H. Direct nanoscale observations of CO_2 sequestration during brucite [$Mg(OH)_2$] dissolution. *Environ. Sci. Technol.* **2012**, *46*, 5253–5260. [CrossRef] [PubMed]

44. Ruiz-Agudo, E.; Putnis, C.V. Direct observations of mineral fluid reactions using atomic force microscopy: The specific example of calcite. *Miner. Mag.* **2012**, *76*, 227–253. [CrossRef]

45. Stumm, W. Reactivity at the mineral-water interface: Dissolution and inhibition. *Colloids Surfaces A Physicochem. Eng. Asp.* **1997**, *120*, 143–166. [CrossRef]

46. Pokrovsky, O.S.; Schott, J.; Castillo, A. Kinetics of brucite dissolution at 25 °C in the presence of organic and inorganic ligands and divalent metals. *Geochim. Cosmochim. Acta* **2005**, *69*, 905–918. [CrossRef]

47. Fischer, C.; Arvidson, R.S.; Lüttge, A. How predictable are dissolution rates of crystalline material? *Geochim. Cosmochim. Acta* **2012**, *98*, 177–185. [CrossRef]

48. Fischer, C.; Kurganskaya, I.; Schäfer, T.; Lüttge, A. Variability of crystal surface reactivity: What do we know? *Appl. Geochem.* **2014**, *43*, 132–157. [CrossRef]

49. Jordan, G.; Rammensee, W. Dissolution rates and activation energy for dissolution of brucite (001): A new method based on the microtopography of crystal surfaces. *Geochim. Cosmochim. Acta* **1996**, *60*, 5055–5062. [CrossRef]

50. Hövelmann, J.; Putnis, C.V. In Situ nanoscale imaging of struvite formation during the dissolution of natural brucite: Implications for phosphorus recovery from wastewaters. *Environ. Sci. Technol.* **2016**, *50*, 13032–13041. [CrossRef] [PubMed]

51. Dawson, P.; Hadfield, C.D.; Wilkinson, G.R. The polarized infra-red and Raman spectra of $Mg(OH)_2$ and $Ca(OH)_2$. *J. Phys. Chem. Solid* **1973**, *34*, 1217–1225. [CrossRef]

52. Tandon, K.; John, M.; Heuss-aßbichler, S.; Schaller, V. Influence of Salinity and Pb on the Precipitation of Zn in a Model System. *Minerals.* **2018**, *8*, 43. [CrossRef]

53. Shinde, V.R.; Shim, H.S.; Gujar, T.P.; Kim, H.J.; Kim, W.B. A solution chemistry approach for the selective formation of ultralong nanowire bundles of crystalline $Cd(OH)_2$ on substrates. *Adv. Mater.* **2008**, *20*, 1008–1012. [CrossRef]

54. Veleva, L.; Quintana, P.; Ramanauskas, R.; Pomes, R.; Maldonado, L. Mechanism of copper patina formation in marine environments. *Electrochim. Acta* **1996**, *41*, 1641–1646. [CrossRef]

55. Gibert, O.; De Pablo, J.; Cortina, J.L.; Ayora, C. Municipal compost-based mixture for acid mine drainage bioremediation: Metal retention mechanisms. *Appl. Geochem.* **2005**, *20*, 1648–1657. [CrossRef]

56. Brooker, M.H.; Sunder, S.; Taylor, P.; Lopata, V.J. Infrared and Raman spectra and X-ray diffraction studies of solid lead(II) carbonates. *Can. J. Chem.* **1983**, *61*, 494–502. [CrossRef]

57. Hardison, D.W.; Ma, L.Q.; Luongo, T.; Harris, W.G. Lead contamination in shooting range soils from abrasion of lead bullets and subsequent weathering. *Sci. Total Environ.* **2004**, *328*, 175–183. [CrossRef] [PubMed]

58. Sangameshwar, S.R.; Barnes, H.L. Supergene processes in zinc-lead–silver sulfide ores in carbonates. *Econ. Geol.* **1983**, *78*, 1379–1397. [CrossRef]

59. Wang, L.; Putnis, C.V.; King, H.E.; Hövelmann, J.; Ruiz-Agudo, E.; Putnis, A. Imaging organophosphate and pyrophosphate sequestration on brucite by in situ atomic force microscopy. *Environ. Sci. Technol.* **2017**, *51*, 328–336. [CrossRef] [PubMed]

60. Putnis, C.V.; Renard, F.; King, H.E.; Montes-Hernandez, G.; Ruiz-Agudo, E. Sequestration of selenium on calcite surfaces revealed by nanoscale imaging. *Environ. Sci. Technol.* **2013**, *47*, 13469–13476. [CrossRef] [PubMed]

61. Renard, F.; Putnis, C.V.; Montes-Hernandez, G.; Ruiz-Agudo, E.; Hovelmann, J.; Sarret, G. Interactions of arsenic with calcite surfaces revealed by in situ nanoscale imaging. *Geochim. Cosmochim. Acta* **2015**, *159*, 61–79. [CrossRef]

62. Wang, L.; Putnis, C.V.; Ruiz-Agudo, E.; Hövelmann, J.; Putnis, A. In situ imaging of interfacial precipitation of phosphate on goethite. *Environ. Sci. Technol.* **2015**, *49*, 4184–4192. [CrossRef] [PubMed]

63. Wang, L.; Putnis, C.V.; Ruiz-Agudo, E.; King, H.E.; Putnis, A. Coupled dissolution and precipitation at the cerussite-phosphate solution interface: Implications for immobilization of lead in soils. *Environ. Sci. Technol.* **2013**, *47*, 13502–13510. [CrossRef] [PubMed]

64. Ruiz-Agudo, E.; Putnis, C.V.; Putnis, A. Coupled dissolution and precipitation at mineral–fluid interfaces. *Chem. Geol.* **2014**, *383*, 132–146. [CrossRef]

65. Putnis, C.V.; Ruiz-Agudo, E.; Hövelmann, J. Coupled fluctuations in element release during dolomite dissolution. *Miner. Mag.* **2014**, *78*, 1355–1362. [CrossRef]

66. Ruiz-Agudo, E.; King, H.E.; Patiño-Ĺpez, L.D.; Putnis, C.V.; Geisler, T.; Rodriguez-Navarro, C.; Putnis, A. Control of silicate weathering by interface-coupled dissolution–precipitation processes at the mineral-solution interface. *Geology* **2016**, *44*, 567–570. [CrossRef]

67. Putnis, A.; Putnis, C. V The mechanism of reequilibration of solids in the presence of a fluid phase. *J. Solid State Chem.* **2007**, *180*, 1783–1786. [CrossRef]

68. Putnis, A. Mineral Replacement Reactions. *Rev. Miner. Geochem.* **2009**, *70*, 87–124. [CrossRef]

69. De Yoreo, J.J.; Gilbert, P.U.P.A.; Sommerdijk, N.A.J.M.; Penn, R.L.; Whitelam, S.; Joester, D.; Zhang, H.; Rimer, J.D.; Navrotsky, A.; Banfield, J.F.; et al. Crystallization by particle attachment in synthetic, biogenic, and geologic environments. *Science* **2015**, *349*, aaa6760. [CrossRef] [PubMed]

70. Teng, H. How ions and molecules organize to form crystals. *Elements* **2013**, *9*, 189–194. [CrossRef]

© 2018 by the authors. Licensee MDPI, Basel, Switzerland. This article is an open access article distributed under the terms and conditions of the Creative Commons Attribution (CC BY) license (http://creativecommons.org/licenses/by/4.0/).

minerals

MDPI

Article

Water Structure, Dynamics and Ion Adsorption at the Aqueous {010} Brushite Surface

Natalya A. Garcia [1], Paolo Raiteri [1], Elias Vlieg [2] and Julian D. Gale [1,*]

[1] Curtin Institute for Computation/The Institute for Geoscience Research (TIGeR), School of Molecular and Life Sciences, Curtin University, P.O. Box U1987, Perth, WA 6845, Australia; natalya.garcia@postgrad.curtin.edu.au (N.A.G.); p.raiteri@curtin.edu.au (P.R.)

[2] Institute for Molecules and Materials, Radboud University, Heyendaalseweg 135, 6525 AJ Nijmegen, The Netherlands; e.vlieg@science.ru.nl

* Correspondence: J.Gale@curtin.edu.au

Received: 29 June 2018; Accepted: 1 August 2018; Published: 3 August 2018

check for
updates

Abstract: Understanding the growth processes of calcium phosphate minerals in aqueous environments has implications for both health and geology. Brushite, in particular, is a component of certain kidney stones and is used as a bone implant coating. Understanding the water–brushite interface at the molecular scale will help inform the control of its growth. Liquid-ordering and the rates of water exchange at the brushite–solution interface have been examined through the use of molecular dynamics simulation and the results compared to surface X-ray diffraction data. This comparison highlights discrepancies between the two sets of results, regardless of whether force field or first principles methods are used in the simulations, or the extent of water coverage. In order to probe other possible reasons for this difference, the free energies for the adsorption of several ions on brushite were computed. Given the exothermic nature found in some cases, it is possible that the discrepancy in the surface electron density may be caused by adsorption of excess ions.

Keywords: brushite; mineral growth; calcium phosphate; adsorption; simulation; metadynamics

1. Introduction

Calcium phosphate phases are ubiquitous both as biominerals and geologically. The mineral apatite $(Ca_5(PO_4)_3X$, where X is an anion such as F^-, Cl^- or OH^-) is the most stable of the calcium phosphate phases over a large pH range. However, in acidic aqueous conditions brushite ($CaHPO_4 \cdot 2H_2O$, dicalcium phosphate dihydrate, DCPD) is often formed. Brushite crystallises to hydroxyapatite at neutral pH and is thought to be a precursor phase [1]. Brushite forms a particularly aggressive kidney stone variant [2], and while a correlation between pH and calcium content has been identified there is not yet a full understanding of this undesirable calcification [3]. Additionally, of the calcium phosphate minerals, brushite in particular has been heavily researched for use in biological applications, such as drug delivery, orthopaedics, craniofacial surgery, cancer therapy, and biosensors [4]. Brushite is a favoured choice because of its high solubility, but there is a limited understanding in regard to its successful resorption into the body [5,6]. Understanding these growth and dissolution processes, particularly in a biological, geological [7] and industrial [8,9] context relies on understanding brushite's structure and especially its interaction with water at the morphologically important surfaces.

As stated above, brushite has the chemical formula $CaHPO_4 \cdot 2H_2O$, making it a structural analogue to gypsum ($CaSO_4 \cdot 2H_2O$). Like gypsum [10], brushite has alternating bilayers of water molecules and calcium phosphate (or sulfate, in gypsum's case) rows, and each bilayer is offset. Unlike gypsum, the space group of brushite is I a (non-standard setting of Cc) as opposed to I 2/a (C2/c) for gypsum. Indeed the epitaxial relationship between brushite and gypsum has been subject of much study, it was first studied in depth by Heijnen and Hartman by periodic bond chain analysis [11], and later through microscale observations [12] and in situ AFM [13]. In the as-synthesised morphology, the {010} face is most commonly exposed [14,15]. When imaged via atomic force microscopy (AFM) in water the {010} brushite surface typically exhibits steps of height 7.6 Å [15,16], which correspond to one bilayer of calcium phosphate and one bilayer of water molecules. The dissolution of brushite is then observed to proceed via the formation of triangular etch pits with [001], [201] and [101] steps [17,18] (see Figure 1).

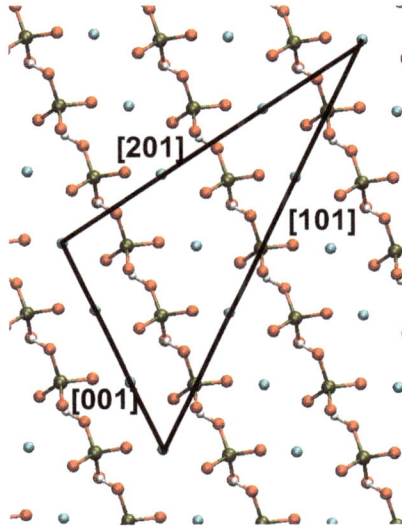

Figure 1. One of the two Brushite {010} surfaces viewed from above with crystallographic directions indicated with solid black lines. Water is omitted for clarity. H is white, O is red, P is tan, and Ca is cyan.

Understanding the structure of bulk brushite, with its intrinsic hydration, has so far not been sufficient to understand the water ordering that occurs at hydrated external surfaces. The {010} face of brushite is parallel to the brushite water layers and the natural cleavage plane lies between the two water layers. Thus no reconstruction of the surface is needed and the first layer of water can be considered as a part of the brushite crystal. Furthermore, it might be supposed that the water ordering of the second layer of water should also be particularly strong and follow that which occurs in the bulk. With this as a hypothesis, surface X-ray diffraction (SXRD) was used by Arsic et al. [16] to determine the electron density distribution of the water–brushite interface. However, less in-plane ordering was found than expected. While the first water layer was found to be ordered similarly to the bulk, the second layer appeared to be without in-plane ordering, though both exhibited strong perpendicular ordering.

A similar SXRD study investigated the water ordering of KH_2PO_4 (KDP) and found an unexpected amount of water ordering which they theorised arose from strong ionic interactions with the charged surface [19]. For KDP they found that the first two water layers exhibited strong ordering, both laterally and perpendicularly, while the third layer had reduced order in both dimensions. The fourth water layer had perpendicular order, but lacked discernible in-plane order. In comparison to this extended ordering for KDP, the lack of structuring of the water over brushite appears a surprising result.

Understanding the water ordering on brushite can also be explored through the use of atomistic simulation. Although there have been studies of the brushite-water interface, to date there has been no precise determination of the water ordering. Jiang et al. [20] studied the adsorption of aspartic acid on brushite and as a part of their study attempted to reproduce the SXRD spectrum from Arsic et al. [16]. The force field employed was based upon existing interatomic potentials (OPLS-AA [21], TIP3P water model [22]) with modifications for brushite based on Hauptmann's model [23] with partial charges from density functional theory (DFT). There was no specific parameterisation of the force field for aqueous-mineral interfaces and the thermodynamic accuracy of the model was not validated. While this model reproduces some features of the SXRD spectrum, discrepancies were also evident. For example, in the bulk structure, though the peaks were approximately at the same positions as the experimental peaks they were much broader indicating a lack of order, perhaps due to the vibrational modes of the mineral being too soft.

The main aim of the present work was to examine the water ordering and dynamics at the surface of brushite. Here a new force field [24] is used that has been specifically parameterised to reproduce a range of both structural and thermodynamic data for calcium phosphate phases, including the hydration free energy of the phosphate anions and the solubility of different phosphate mineral phases. This was complemented by the use of *ab initio* molecular dynamics to validate the results, where feasible, and a modified version of our previous force field [24] to examine the sensitivity to choices made during the parameterisation. Simulations were performed for both a flat surface and a surface with one of the possible steps, as the latter represents the most likely site for growth and dissolution. This work examined the extent to which it was possible to reproduce the experimentally acquired SXRD spectrum using simulations of the flat surface. A discrepancy between experiment and simulation led us to further consider the thermodynamics of phosphate and calcium ion adsorption as processes that might lead to disruption of the water structure. Finally, the rates of water exchange were explored based on an analysis of the residence time at the exposed calcium sites, since such data has not been previous reported, yet may be an important kinetic limitation for growth and dissolution.

2. Methods

In this study we have simulated the brushite aqueous interface using atomistic molecular dynamics based on both interatomic potentials and density functional theory. For the classical simulations, interactions were predominately described using the force field of Demichelis et al. [24] (hereafter labelled C1), which was developed with a specific focus on reproducing the thermodynamic properties of phosphate minerals. In particular, the hydration free energy of the phosphate anions and the solubility of different phosphate mineral phases were used for fitting, as these properties are important for an accurate description of the water structure and dynamics around the phosphate and calcium ions. The SPC/Fw water model [25] was chosen for consistency with the previous force field parameterisation developed for carbonate minerals [26]. At 300 K, the present force field gives deviations in the lattice parameters relative to the experimental values [27] of Δa −0.03% Δb −1.03% and Δc +7.48%. As compared to the force field reported by Jiang et al. (at 300 K: Δa 3.89% Δb +1.96% and Δc +1.32%), this force field performs more accurately in the a and b directions, but worse in c [20]. We also modified the parameters of [24] to produce a force field (C2) which more closely matched the experimental lattice parameters (See Table S1 in the Supplementary Information for force field parameters). With this revised model, the errors in the cell parameters after optimisation are Δa −2.38% Δb 0.1% and Δc +0.69%. However, for C2 the thermodynamic quantities were not targeted, and resulted in a worse performance. For example, the solvation free energy for HPO_4^{2-} had an error of −4.80% (as opposed to the unmodified force field error of +0.03%) relative to the reference value used for fitting based on quantum mechanical data.

The brushite-water interface was modelled through a periodic cell with lattice parameters of 45.999 Å, 150.978 Å, 53.545 Å, 90.00°, 114.13°, 90.00°, containing 1280 formula units of $CaHPO_4$ and 8239 water molecules (2206 of which were part of the brushite slab). The brushite slab and water layer were approximately 72 and 76 Å thick, respectively, with the mineral surface lying in the plane of the a and c lattice parameters. The same cell with a single additional ion [Ca^{2+}, PO_4^{3-}, HPO_4^{2-}, $H_2PO_4^-$]

was used for the adsorption studies. Further simulations were performed using the same cell, but with reduced water coverage, in order to investigate the influence of the experimental conditions, where only a few layers of water were present, rather than bulk solvent. The same cell with 784, 682, and 596 water molecules was used instead of bulk water for three simulations investigating the effect of the experimental water coverage. Here one side of the slab had 261 water molecules in each case and the other side whose density profile was analysed had 523, 421 and 335 water molecules, respectively.

The brushite slab was built from unit cells reported through X-ray structure data [27] with lattice parameters 5.810 Å, 15.176 Å, 6.234 Å, 90.00°, 116.41°, 90.00°. Brushite and water were pre-equilibrated in separate cells before being combined into a cell that was relaxed in the y direction prior to performing production runs in the canonical ensemble. All classical molecular dynamics simulations were run using the LAMMPS package [28] with a 1 fs time step. The temperature was kept at 300 K via a Nosé-Hoover chain of thermostats (length 5) with a relaxation time of 0.1 ps and atomic trajectories were sampled every 1 ps. The dynamics of water using force field C1 at the flat and stepped brushite surfaces were monitored during 50 and 58 ns runs, respectively. The simulations with reduced water coverage and force field C2 were analysed based on shorter 2 ns runs, as they were used only for calculating electron density profiles, rather than the dynamics of water exchange.

In previous experimental studies, the [001], [201] and [101] steps have all been observed as sites for dissolution [17,18]. A [001] or [201] step on brushite would result in either a row of Ca^{2+} or HPO_4^{2-} ions, if unreconstructed, whereas a [101] step naturally results in alternating Ca^{2+} and HPO_4^{2-} ions. Rows of a single charged species could potentially result in an unstable surface termination and thus the [101] step was chosen for study (see Figure 1). For the investigation of water dynamics at this step a larger periodic cell in the *a* direction was needed in order to have sufficient separation between steps. Hence, in this case the cell was doubled, with half of the top row of phosphate ions removed and replaced with water molecules.

Analysis of the trajectory resulting from the molecular dynamics simulations was used to calculate the density distribution, radial distribution functions and water residence times. To compare to experiment, the electron density distribution was calculated by weighting atomic positions by the number of electrons per atom. This approximation, if anything, should lead to the simulation overestimating the sharpness of peaks in the electron density distribution, though in practice this is not observed to be the case. The method of fitting the residence times has been previously described by De La Pierre et al. [29].

In order to complement the classical force field study of the hydration of the brushite surface, we have also performed ab initio molecular dynamics (AIMD). Here a reduced surface model consisting of 424 atoms was constructed with lattice parameters of 11.5, 30.196 and 13.386 Å, for *a*, *b*, and *c*, respectively, and a monoclinic angle of 114.13°. The system was first equilibrated with the force field C1 to initialise the AIMD. Here all quantum mechanical energies and forces were determined within Kohn–Sham density functional theory, as implemented in the code CP2K [30], using the Gaussian-augmented plane-wave approach within the Quickstep module [31]. The BLYP generalised gradient approximation exchange-correlation functional was used in combination with the D3 dispersion corrections of Grimme and co-workers [32]. A plane-wave cutoff of 400 Ry was set for the auxiliary basis employed for the electron density, in combination with the Goedecker-Teter-Hutter TZVP basis sets and corresponding pseudopotentials [33] for all elements except Ca, where a DZVP basis was chosen. For the AIMD, simulations were also performed in the NVT ensemble with a time step of 0.5 fs and a stochastic thermostat with a time constant of 10 fs to ensure fast equilibration given that only the water structure and not the dynamics were to be analysed. The mass of hydrogen was also set to 2 amu, corresponding to deuterium, and a slightly elevated temperature of 330 K was employed based on previous results that show that BLYP-D3 under these conditions leads to an improved description of the structure of liquid water, which can otherwise be over-structured in AIMD [34].

Multiple-walker [35], well-tempered [36], metadynamics was used to investigate the adsorption of ions [Ca^{2+}, PO_4^{3-}, HPO_4^{2-}, $H_2PO_4^-$] on brushite within the force field simulations. Again LAMMPS [28] was used, but with the inclusion of the PLUMED 2.0 [37] plug-in in order to compute the binding free energies. The collective variable chosen was the normal component of the distance between the ion (taken as the position of Ca or P, as appropriate to the species) and the middle Ca of the brushite slab. Gaussians were laid every 1 ps with a width of 0.2 Å and an initial height of $k_b T$. Using the multiple walkers algorithm, 30 parallel simulations with a bias factor of 5 were run for each energy profile. Aggregated simulation times of at least 500 ns were performed for each system to ensure well-converged sampling.

3. Results and Discussion

3.1. Electron Density Profiles Normal to the {010} Surface

The time averaged electron density was calculated for the interface between the flat {010} surface of brushite and bulk water, resulting in the density profiles normal to the surface shown in Figure 2. In the bulk brushite region (Figure 2b) both of the classical force fields (MD-C1 and MD-C2), and ab initio methods (MD-A), reproduce the SXRD electron density, whether fitted without (SXRD-1) or with (SXRD-2) bulk water. There is some discrepancy in z positions (as discussed in the Methods) and the classical force fields do not quite reproduce the peaks for the brushite Ca atoms, but overall the electron density profiles have similar heights, widths and z positions of peaks. This comparison provides some validation that the force field from Demichelis et al. [24] is appropriately describing the phosphate-water interactions, at least within the bulk brushite environment. Additionally, due to the smaller system size of the AIMD simulations it is possible that the peaks widths might be underestimated due to the truncation of the phonon spectrum in reciprocal space; however the comparison to experimental results shows the peak widths match well.

The comparison of the brushite-water interface between force fields, ab initio methods and SXRD, while qualitatively similar, shows differences in both heights, widths and position of peaks for the interfacial region (see Figure 2c). Using a fit with bulk water (SXRD-2) or without (SXRD-1) did not significantly affect the intensity or position of peaks at the interface, however the peak shape was slightly different between the two fits. In the first water layer both *ab initio* and classical force fields show a reduced intensity compared to the SXRD profiles and a trough where none exists in the experimental results. All three simulation methods show the three peaks present in the SXRD data at this reduced intensity in the 0.5 to 2 Å region, but the peak positions vary. The results of the modified Demichelis et al. force field (MD-C2) more closely resemble those of the *ab initio* molecular dynamics (MD-A) in both the first and second water layer, though at reduced intensity in the second water layer. In the second water layer the original Demichelis et al. force field (MD-C1) has a much lower intensity than the SXRD results, but does have the correct position, whereas MD-C2 and MD-A have a higher intensity but have peak positions closer to the brushite surface than the SXRD data suggests is the case. In addition, MD-A has a more pronounced minimum in the density to the right of the peak before tending to bulk water. It is important to recognise that the shorter run length that is currently possible with AIMD compared with force field approaches will partly contribute to a greater degree of structuring in the water distribution. However, despite this caveat, the AIMD results are at least suggestive of under-structuring in the force field results.

Arsic et al. estimated in [16] that there was a coverage of approximately 2–3 layers of water during the experiments as opposed to bulk water for the simulated interface, which could cause a discrepancy between simulated and measured density profiles. However, a comparison to results using other methods suggests this was an underestimate. For example, using ATR-IR Asay and Kim [38] calculated the average thickness of adsorbed water on silicon oxide and found that at 100% relative humidity there were ten monolayers of water. In another experiment, for measurements on an NaCl surface, Arsic et al. found that above 75% relative humidity a macroscopic water layer formed causing the crystal to dissolve [39]. Still, to verify whether the level of coverage affected the measured density profiles we performed additional simulations as a function of this quantity. In Figure 3 we see that calculating the electron density in the presence of bulk water or with a reduced water coverage gives essentially the same result for a distance <2.5 Å from the top layer Ca atom. While the density tails off above the surface for low coverages due to the presence of vacuum, the lack of bulk water in the simulation does not result in an increased peak height for the first two water layers using the classical force fields, but gives a small but non-negligible increase using ab initio methods. Although reducing the water coverage at the surface clearly alters the electron density away from the surface, it fails to explain the discrepancy with the experimental results for the strongly-ordered peaks close to the brushite-water interface.

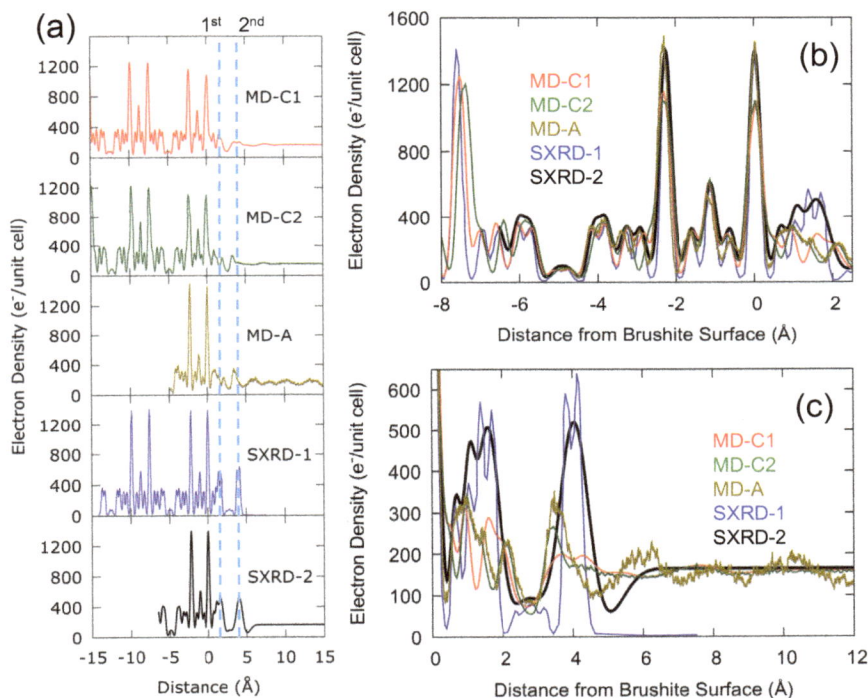

Figure 2. (**a**) Time averaged electron density profiles obtained by MD simulations (MD-C1 refers to simulations using the Demichelis et al. force field [24]; MD-C2 refers to the modified version of MD-C1; MD-A refers to the results from *ab initio* molecular dynamics) compared with the electron density profiles obtained via SXRD (SXRD-1 refers to the published data of Arsic et al. and SXRD-2 refers to the same data with a fit including bulk water). Peak positions are listed explicitly in Table S2. The origin is chosen such that the top layer Ca is at $z = 0$ Å and densities are scaled to the experimental unit cell (lattice parameters $a = 5.812$ Å, $b = 15.18$ Å, $c = 6.239$ Å, and $\beta = 114.13°$) [16]. Dashed blue lines indicate the 1st and 2nd hydration layer positions as indicated by SXRD. Overlaid density profiles for specific regions are also shown in the case of (**b**) the top two brushite layers and (**c**) the brushite–water interface.

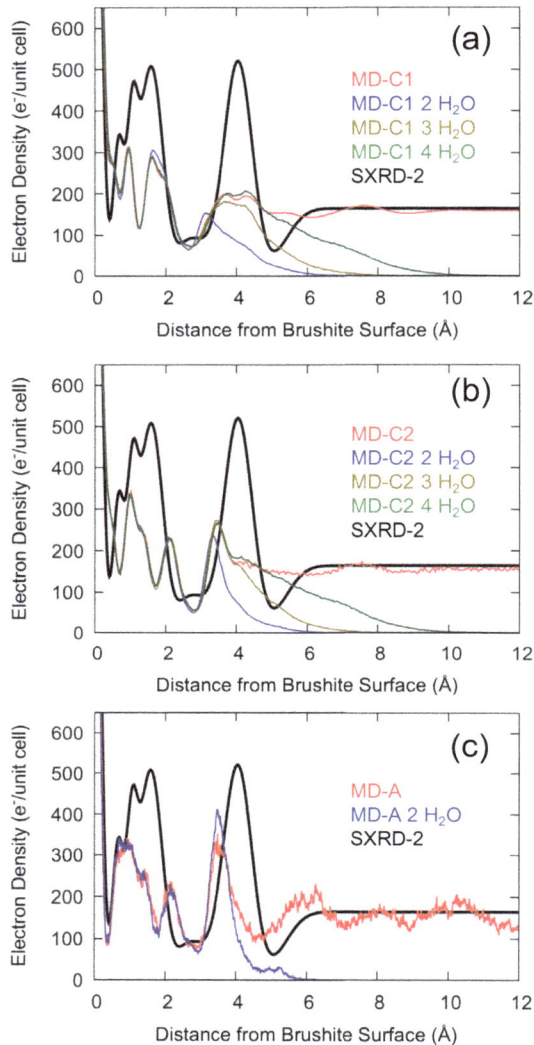

Figure 3. The interfacial water structure with 2, 3, and 4 layers of water using (**a**) the Demichelis et al. force field (MD-C1) [24], (**b**) the modified Demichelis et al. force field (MD-C2) and (**c**) *ab initio* methods (MD-A) (for this last case only the two-layer coverage was simulated).

3.2. Atomic Density Maps for Water at the {010} Surface

While acquisition of a long *ab initio* trajectory was not feasible, we performed extended simulations using the force field C1 [24] to better understand the water structure at this brushite surface in all directions, as opposed to just the surface normal projection of the electron density. Figure 4 shows the calculated 3D density isosurfaces for the different types of atom, which can be compared to Figure 1 from Arsic et al. [16]. This shows that the hydrogen bonding pattern within the bulk region, where water hydrogen bonds to oxygens of the hydrogen phosphate anions, matches between the simulated and experimental structure. We verified that the density patterns did not change greatly between force fields using a shorter simulation with C2 (see Supplementary Information Figure S1).

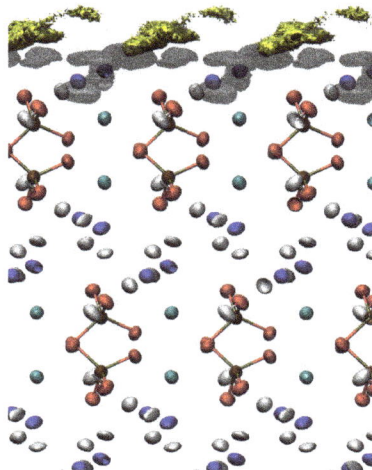

Figure 4. The atom density isosurfaces for brushite and water averaged over the molecular dynamics simulation (MD-C1). The {010} surface of $CaHPO_4 \cdot 2H_2O$ is shown through the [201] plane. For the calcium and phosphate ions, H is white, O is red, P is tan, and Ca is cyan, while for the water density isosurfaces the bulk water is blue and H is white (all isosurfaces at 2 atoms/$Å^3$). Oxygens of water from beyond the first layer are shown with yellow isosurfaces and H is shown with transparent grey in the bulk region due to its diffuse distribution (both 0.05 atoms/$Å^3$).

The first water layer in the simulations also follows this bonding pattern with some deviations. By investigating the 3D maps we can postulate H-bonding patterns to help explain the observed oxygen positions. We see in the 3D map that the distribution of hydrogens for the lower water layer at the surface (i.e., first layer above the $CaHPO_4$ layer) follows the bonding pattern of the bulk. In contrast, the upper water hydrogens (shown in dark grey in Figure 4) are much more diffuse, which indicates that the second water layer will also be less well-ordered due to the lack of a well-defined hydrogen bonding pattern. Experimental results suggest no in-plane ordering in the second water layer, but a high degree of perpendicular order, whereas the calculated distribution has a lower level of perpendicular order but a greater amount of in-plane ordering.

The suggested H-bonding pattern of Arsic et al. appears to disagree with the results of the force field simulation. As seen in Figure 4, the hydrogens of the first water layer, though somewhat diffuse, follow a similar pattern as the bulk. We confirm via the 3D map of the water density viewed down the surface normal (see Figure 5) that the oxygen of the first water layer is coordinated to calcium, whereas the oxygen of the second layer lies above phosphate. The proposed loss of H-bonding suggested by Arsic et al. is seen as an increase in the diffuse nature of the distribution for hydrogen where it rotates between multiple sites. The third layer appears to be largely diffuse, giving rise only to small localised regions at low values for the isosurface density. In terms of position, these slight indications of a third layer occur at a site that is almost directly above the first water layer. These results suggest that the brushite surface is affecting the water structure at a height of as much as 6 Å away from the surface.

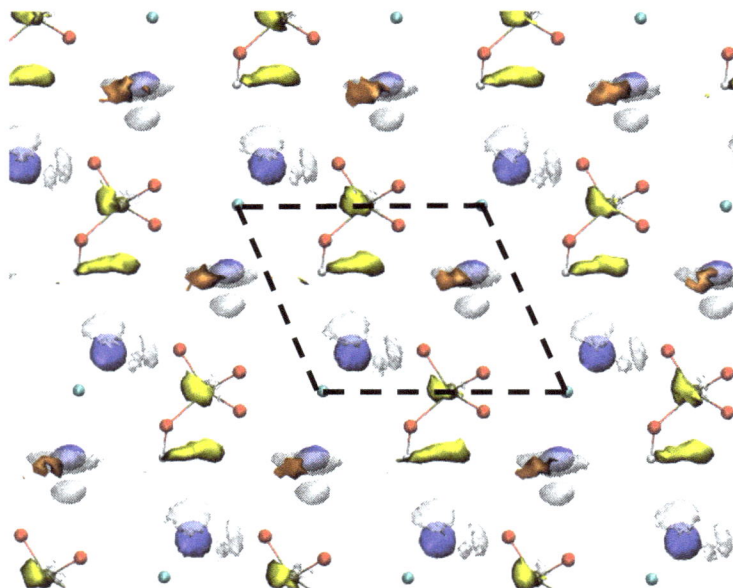

Figure 5. Top view of the 3D atomic structure of the flat brushite-water interface for the {010} plane. Average positions of the crystal are represented with a ball-and-stick model with only the top layer of the phosphate bilayer shown and calculated density isosurfaces for the first three water layers. For average positions, H is white, O is red, P is tan, and Ca is cyan. In the water density isosurfaces the first layer H is white and the first layer O is blue (both isosurfaces set to 2 atoms/$Å^3$), the second layer O is yellow (0.2 atoms/$Å^3$) and the third layer O is orange (0.05 atoms/$Å^3$). The surface unit cell is indicated with a dashed black line.

3.3. Water Density at the [101] Steps

The creation of a step corresponds to a disruption in the water structure and is frequently the site of growth and dissolution. We investigated two obtuse [101] steps; one where hydrogens of the phosphate anion were oriented outward towards the bulk water and another where they were oriented inward towards the mineral. Though the step where hydrogens are pointed inward towards the bulk does not appear in nature, since triangular islands/etch pits bounded by 3 different crystallographic directions are formed, the simulated model required the presence of this step and was thus included for analysis. We therefore expect the step with hydrogens oriented inward to be less stable than the naturally occurring one, though within the timescale of the simulations neither showed any tendency toward dissolution.

In our analysis of the water distribution at the step edges, shown in Figure 6, phosphate/calcium rows of the underlying surface are referred to by position in terms of surface height, being in the upper (U) or lower (L) terrace, and by row number where an increasing value represents a greater distance away from the step edge (example: U2 refers to the second row relative to the upper terrace step edge). The step where hydrogen is pointed inward towards the bulk is denoted with a prime e.g., U'2.

Based on the computed isosurfaces, the first water layer appears unchanged on the lower terraces for both steps as compared to the flat surface, except for the now exposed L1/L'1 sites. The second water layer has a clustering of water molecules where the step occurs, one of which appears to be coordinated to the newly exposed upper site (U1/U'1). Analysis of the radial distribution functions confirms that the U1 Ca sites have two water oxygens coordinated and the other calcium sites have only one water oxygen coordinated (See Figure S2 in the Supplementary Information). By positions

L3/U3 and L'3/U'3, the density appears to have approximately returned to the same distribution as above a surface Ca. From the top view we see more clearly the disruption of the water ordering due to the presence of both steps (see Figure S3).

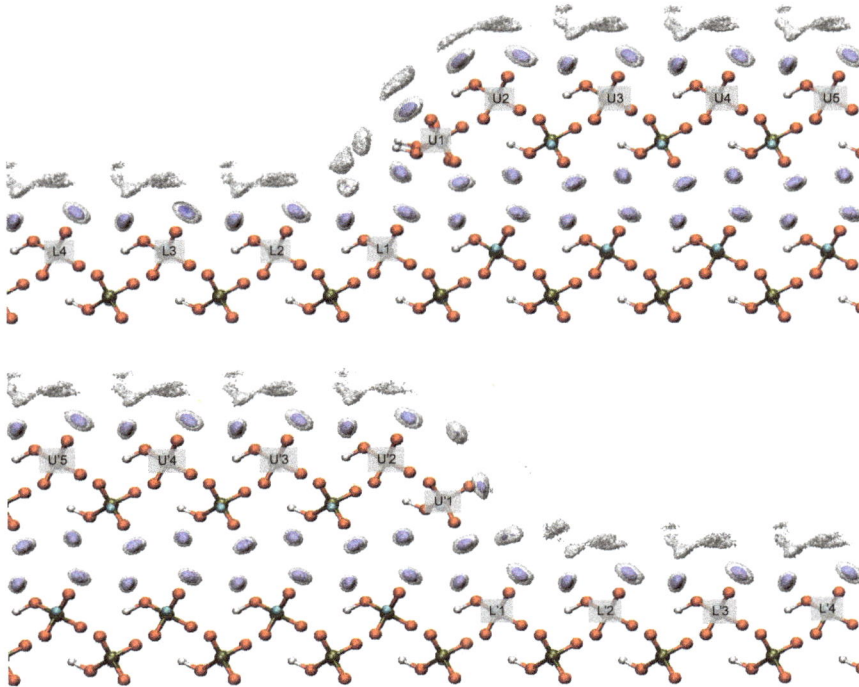

Figure 6. Side view of the 3D density maps of the [101] steps with (**Top**) hydrogens facing outward towards water (L or U) and (**Bottom**) hydrogens facing inward towards the bulk (L' or U') on the {010} face. Average positions of the phosphate and calcium ions are represented with a ball and stick model where H is white, O is red, P is tan, and Ca is cyan (all isosurfaces set to 2 atoms/$Å^3$). For water, the isosurfaces shown are in transparent colours where the darker shade is used for the higher density (blue: 2 atoms/$Å^3$ and grey 0.05 atoms/$Å^3$). Labels indicate the atomic row position (Upper or Lower) and its proximity to the step (1, 2, 3, 4 or 5) using the same notation established for calcite [29].

3.4. Water Residence Times

Experimental evidence suggests that bulk brushite water behaves in a manner that is "ice-like", where the extrapolated brushite water density based on the water bilayer is 4% higher than normal ice [16]. At the interface, water is suggested to be crystalline but not "ice-like," with only partial water ordering [16]. One way of characterising the nature of the water is through the dynamics of exchange between sites. Analysis of the residence times of water around the surface calcium sites of the flat surface indicates that they are particularly long, with no water exchange observed over the course of the 50 ns simulation. To place this in context, the exchange is much slower than for an isolated calcium ion in pure water or for calcium at the basal surface of calcite, which have been previously reported using the same force field as ≈0.2 and ≈2 ns, respectively [29].

In contrast to the flat surface, most of the calcium sites around the step edges did exhibit significant water exchange. Figures S4 and S5 show the survival function, P(t), as a function of time on a logarithmic scale for the upper and lower rows of the step. Table 1 shows the corresponding residence times of water around the calcium ions near the steps. Most residence times were ≈20–50 times longer than the residence time of an isolated calcium ion in pure water. The U1 calcium has one of the lowest residence times and thus is likely to be a reactive centre.

Table 1. Residence times of waters around sites with the terminology of sites specified in Figure 6 using fits shown in Figures S4 and S5. The column with "Hydrogen Out" refers to the step where the hydrogens of the hydrogen phosphate are pointed outward towards the bulk water and "Hydrogen In" refers to the step where the hydrogens of the hydrogen phosphate are pointed inward towards the bulk phosphate.

Site	Time (ns)	
	Hydrogen Out	Hydrogen In
L1/L′1	20.1	31.1
L2/L′2	9.9	17.1
L3/L′3	9.9	10.5
L4/L′4	10.3	10.9
U1/U′1	2.7	3.1
U2/U′2	3.6	7.3
U3/U′3	13.7	10.4
U4/U′4	7.4	11.3
U5/U′5	12.2	4.2

We expect that as the distance increases from the step, the residence times should converge to those for the flat brushite surface, which would correspond to no observable exchange over the duration of the simulation. However, this was not found to be the case, at least within the range of 5 rows either side of the step; some degree of exchange was still observed, suggesting a long-range perturbation of the water dynamics occurs due to the presence of the step. It should be noted that as the residence times increased to the order of the simulation length, the uncertainties became large, as can be seen in Figure S5. In particular, the results for site U′5 are not well fitted, indicating a degree of uncertainty regarding its value. Sites L1 and L′1 also had long residence times (>20 ns) but were more accurately sampled than U′5 as seen from the survival functions.

3.5. Adsorption of Ions

To further investigate the discrepancy in calculated electron densities at the {010} brushite-water interface, the possibility that this was due to the adsorption of excess ions has been explored. The experimental technique used to measure the electron density distribution used a high relative humidity and a reduced X-ray intensity in order to try to balance beam damage and healing of the surface, in order to prolong the lifetime of the surface during data collection [16]. Despite this, it remains possible that there could be phosphate and calcium ions in solution produced by the X-rays and local changes in pH, and that adsorption of these ions (Ca^{2+}, PO_4^{3-}, HPO_4^{2-} and $H_2PO_4^-$) could lead to a change in the measured electron density. At neutral pH, the predominant phosphate ions are likely to be HPO_4^{2-} and $H_2PO_4^-$, but the adsorption of PO_4^{3-} was also included, as proton transfers may occur in these unusual environments, though this is unlikely.

To probe this, the free energies of adsorption of the above ions on the brushite {010} surface from bulk water were calculated. Of the ions studied, adsorption of HPO_4^{2-} and $H_2PO_4^-$ were thermodynamically favourable (see Table 2). The trend of lower free energies (i.e., more exothermic) of adsorption for ions that have a lower charge magnitude suggests that the desolvation of the ions in water dominates over the strength of adsorption to the brushite surface. As the magnitude of the charge increases, the solvation shell is more tightly bound and thus desolvation is less favourable. This is evidenced by the trend in hydration free energy of the ions reported previously with the same force field, where $H_2PO_4^-$, HPO_4^{2-}, Ca^{2+} and PO_4^{3-} are −339, −1140, −1350 to −1503 and −2495 kJ/mol, respectively [24,40]. Though Ca^{2+} and HPO_4^{2-} have the same charge magnitude, HPO_4^{2-} has a less exothermic hydration free energy. Thus it is not surprising that of the two ions HPO_4^{2-} has a more favourable adsorption, as it is more easily desolvated. Correspondingly, $H_2PO_4^-$ with both the lowest charge magnitude and the highest hydration free energy has the lowest free energy of adsorption of the ions studied.

Table 2. Free energies of adsorption for calcium and phosphate ions at the aqueous-brushite {010} interface relative to bulk solution. Heights above the surface are given relative to the first layer of Ca ions.

Ion	Height of Minimum (Å)	ΔG (kJ/mol)
Ca^{2+}	5.60	0.96
PO_4^{3-}	5.99	8.52
HPO_4^{2-}	4.82	−3.81
$H_2PO_4^-$	4.18	−13.13

The position of the minima for $H_2PO_4^-$ was also the closest to the arguably anomalous peak in the experimental spectrum, 0.14 Å away, as indicated by the dashed line in Figure 7. With a relatively strong minima located within the second water layer, there is a possibility of $H_2PO_4^-$ ions being adsorbed to the surface. Further investigation would be needed to determine whether this could account for the apparent discrepancy between experiment and simulation.

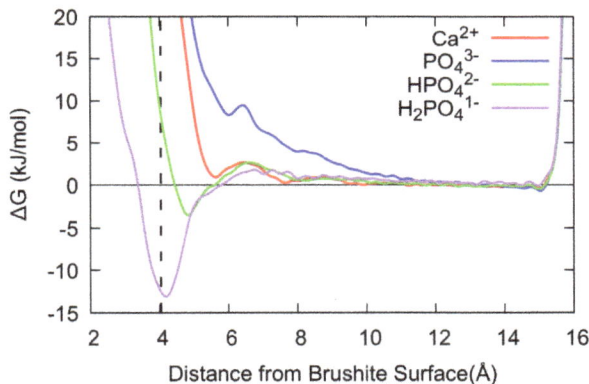

Figure 7. Free energies of adsorption of Ca^{2+}, PO_4^{3-}, HPO_4^{2-} and $H_2PO_4^-$ at the brushite {010}-aqueous interface with respect to the distance of Ca/P, as applicable, from the central surface calcium. Curves are aligned to $\Delta G = 0$ in the long-range limit prior to the constraining wall. The location of the second water layer in the experimental spectrum is indicated with a dashed line at a height above the surface of 4.04 Å.

4. Conclusions

In this study the aqueous interface at the brushite {010} surface has been simulated using both classical and *ab initio* methods. All methods reproduced the bulk brushite structure; however, there were discrepancies between the experimental electron density for the first two water layers as reported from SXRD and those obtained from the present simulations. In particular, the molecular dynamics simulations all find a reduced degree of ordering perpendicular to the surface. Repeated simulations with varying amounts of water showed that the experimental conditions, where there were only 2–3 layers of water [16], could effect the final results, but failed to explain the higher density in the second water layer for the experimental electron density at the level observed. A modified version of a published force field for aqueous calcium phosphate systems [24] better reproduced the electron density obtained using *ab initio* methods, but the differences between the distribution as a function of force field parameterisation are far smaller than those between simulation and experiment, which suggests that the difference is due to factors beyond the choice of level of theory.

Density maps obtained from our simulations are consistent with the H-bonding patterns proposed experimentally for the water within bulk brushite, while providing further insights as to how that H-bonding pattern would change for an exposed surface of aqueous brushite. Analysis of the density isosurfaces suggests that the second water layer is hydrogen bonded to phosphate ions, while the third layer, though very diffuse, does appears to show a slight preference to be located over the first layer water oxygen. While experimental results suggest that beyond the first water layer there is no lateral order, our results suggest a modified continuation of the bulk H-bonding pattern.

Analysis of water exchange dynamics at calcium ions in the surface indicates that the water is strongly bound leading to no observable exchange of water on the timescale of our simulations for the flat surface. The same analysis for [101] steps indicate that they are potentially much more reactive as water exchange is far more rapid, allowing us to determine likely active sites for growth/dissolution. Analysis of water residence times as a function of proximity to the step edge indicated that positions as far as 17 Å away may still be affected by the presence of a step, with the caveat that the uncertainty in these values increases as the exchange slows down. Investigation of the [001] and [201] steps could be used to further characterise the water dynamics on the {010} brushite surface, as the different steps could have very different residence times and coordination patterns.

Investigation of the free energies associated with adsorption of various ions suggest that there is a possibility that the experimental electron density distribution may include some contribution from the adsorption of excess phosphate ions, as their adsorption is found to be thermodynamically favourable. These ions would be present in low concentration due to equilibration with the water layers to achieve saturation, though this may be enhanced as a result of beam damage to the sample leading to the creation of defects. At this stage it is not possible to determine unambiguously the reason for the discrepancy between simulation and experiment, since both approaches suffer from limitations when studying interfacial systems, and further studies will be required in future as techniques progress.

Supplementary Materials: The following are available online at http://www.mdpi.com/2075-163X/8/8/334/s1, Section: Radial Distribution Functions. Table S1: Modified force field parameters for C2, Table S2: Electron density peak positions and intensities, Figure S1: Electron density normal to the {010} brushite-water interface from MD-C2, Figure S2: Side and top view of 3D density maps of the {010} brushite-water interface from MD-C2, Figure S3: Radial distribution functions for oxygen around Ca, Figure S4: Top view of 3D density maps of the [101] steps, Figure S5: Survival functions for the lower row of the step, Figure S6: Survival functions for the upper row of the step.

Author Contributions: E.V., J.G. and P.R. conceived the work; N.G. set up the classical simulations, with help from P.R., and then performed the molecular dynamics and analysed the results. J.G. ran the *ab initio* MD based on an initial configuration generated by N.G. E.V. was responsible for providing the SXRD data and additional fits. N.G. drafted the manuscript and all authors contributed to the revision.

Funding: This research was funded by the Australian Research Council under grant numbers FT130100463 and DP160100677.

Acknowledgments: The Australian National Computational Infrastructure (NCI) and the Pawsey Supercomputing Centre are acknowledged for the provision of computing resources.

Conflicts of Interest: The authors declare no conflict of interest.

References

1. Johnsson, M.S.A.; Nancollas, G.H. The Role of Brushite and Octacalcium Phosphate in Apatite Formation. *Crit. Res. Oral Biol. Med.* **1992**, *3*, 61–82.
2. Pramanik, R.; Asplin, J.R.; Jackson, M.E.; Williams, J.C. Protein content of human apatite and brushite kidney stones: Significant correlation with morphologic measures. *Urol. Res.* **2008**, *36*, 251–258, doi:10.1007/s00240-008-0151-7.

3. Siener, R.; Netzer, L.; Hesse, A. Determinants of Brushite Stone Formation: A Case-Control Study. *PLoS ONE* **2013**, *8*, e78996, doi:10.1371/journal.pone.0078996.

4. Tamimi, F.; Sheikh, Z.; Barralet, J. Dicalcium phosphate cements: Brushite and monetite. *Acta Biomater.* **2012**, *8*, 474–487, doi:10.1016/j.actbio.2011.08.005.

5. Cama, G.; Barberis, F.; Capurro, M.; Silvio, L.D.; Deb, S. Tailoring brushite for in situ setting bone cements. *Mater. Chem. Phys.* **2011**, *130*, 1139–1145, doi:10.1016/j.matchemphys.2011.08.047.

6. Bannerman, A.; Williams, R.L.; Cox, S.C.; Grover, L.M. Visualising phase change in a brushite-based calcium phosphate ceramic. *Sci. Rep.* **2016**, *6*, 32671, doi:10.1038/srep32671.

7. Laviano, R.; Fiore, S. Brushite, hydroxylapatite, and taranakite from Apulian caves (southern Italy): New mineralogical data. *Am. Miner.* **1991**, *76*, 1722–1727.

8. Tortet, L.; Gavarri, J.R.; Nihoul, G.; Dianoux, A.J. Proton mobilities in brushite and brushite/polymer composites. *Solid State Ion.* **1997**, *97*, 253–256, doi:10.1016/S0167-2738(97)00047-7.

9. Kumar, M.; Dasarathy, H.; Riley, C. Electrodeposition of brushite coatings and their transformation to hydroxyapatite in aqueous solutions. *J. Biomed. Mater. Res.* **1999**, *45*, 302–310, doi:10.1002/(SICI)1097-4636(19990615)45:4<302::AID-JBM4>3.0.CO;2-A.

10. Yan, C.; Nishida, J.; Yuan, R.; Fayer, M.D. Water of Hydration Dynamics in Minerals Gypsum and Bassanite: Ultrafast 2D IR Spectroscopy of Rocks. *J. Am. Chem. Soc.* **2016**, *138*, 9694–9703, doi:10.1021/jacs.6b05589.

11. Heijnen, W.M.M.; Hartman, P. Structural morphology of gypsum ($CaSO_4 \cdot 2H_2O$), brushite ($CaHPO_4 \cdot 2H_2O$) and pharmacolite ($CaHAsO_4 \cdot 2H_2O$). *J. Cryst. Growth* **1991**, *108*, 290–300, doi:10.1016/0022-0248(91)90376-G.

12. Pinto, A.; Jiménez, A.; Prieto, M. Interaction of phosphate-bearing solutions with gypsum: Epitaxy and induced twinning of brushite ($CaHPO_4 \cdot 2H_2O$) on the gypsum cleavage surface. *Am. Mineral.* **2009**, *94*, 313–322, doi:10.2138/am.2010.3557.

13. Pinto, A.J.; Ruiz-Agudo, E.; Putnis, C.V.; Putnis, A.; Jimenez, A.; Prieto, M. AFM study of the epitaxial growth of brushite ($CaHPO_4 \cdot 2H_2O$) on gypsum cleavage surfaces. *Am. Mineral.* **2010**, *95*, 1747–1757.

14. Abbona, F.; Christensson, F.; Angela, M.; Madsen, H. Crystal habit and growth conditions of brushite, $CaHPO_4 \cdot 2H_2O$. *J. Cryst. Growth* **1993**, *131*, 331–346, doi:10.1016/0022-0248(93)90183-W.

15. Kanzaki, N.; Onuma, K.; Treboux, G.; Ito, A. Dissolution kinetics of dicalcium phosphate dihydrate under pseudophysiological conditions. *J. Cryst. Growth* **2002**, *235*, 465–470, doi:10.1016/S0022-0248(01)01771-7.

16. Arsic, J.; Kaminski, D.; Poodt, P.; Vlieg, E. Liquid ordering at the Brushite-{010}-water interface. *Phys. Rev. B* **2004**, *69*, 245406, doi:10.1103/PhysRevB.69.245406.

17. Scudiero, L.; Langford, S.C.; Dickinson, J.T. Scanning force microscope observations of corrosive wear on single-crystal brushite ($CaHPO_4 \cdot 2H_2O$) in aqueous solution. *Tribol. Lett.* **1999**, *6*, 41–55, doi:10.1023/A:1019134901387.

18. Giocondi, J.L.; El-Dasher, B.S.; Nancollas, G.H.; Orme, C.A. Molecular mechanisms of crystallization impacting calcium phosphate cements. *Phil. Trans. Math. Phys. Eng. Sci.* **2010**, *368*, 1937–1961, doi:10.1098/rsta.2010.0006.

19. Reedijk, M.F.; Arsic, J.; Hollander, F.F.A.; de Vries, S.A.; Vlieg, E. Liquid Order at the Interface of KDP Crystals with Water: Evidence for Icelike Layers. *Phys. Rev. Lett.* **2003**, *90*, 066103, doi:10.1103/PhysRevLett.90.066103.

20. Jiang, W.; Pan, H.; Zhang, Z.; Qiu, S.R.; Kim, J.D.; Xu, X.; Tang, R. Switchable Chiral Selection of Aspartic Acids by Dynamic States of Brushite. *J. Am. Chem. Soc.* **2017**, *139*, 8562–8569, doi:10.1021/jacs.7b03116.

21. Jorgensen, W.L.; Maxwell, D.S.; Tirado-Rives, J. Development and Testing of the OPLS All-Atom Force Field on Conformational Energetics and Properties of Organic Liquids. *J. Am. Chem. Soc.* **1996**, *118*, 11225–11236, doi:10.1021/ja9621760.

22. Jorgensen, W.L.; Chandrasekhar, J.; Madura, J.D.; Impey, R.W.; Klein, M.L. Comparison of simple potential functions for simulating liquid water. *J. Am. Chem. Soc.* **1983**, *79*, 926–935, doi:10.1063/1.445869.

23. Hauptmann, S.; Dufner, H.; Brickmann, J.; Kast, S.M.; Berry, R.S. Potential energy function for apatites. *Phys. Chem. Chem. Phys.* **2003**, *5*, 635–639, doi:10.1039/B208209H.

24. Demichelis, R.; Garcia, N.A.; Raiteri, P.; Malini, R.I.; Freeman, C.L.; Harding, J.H.; Gale, J.D. Simulation of Calcium Phosphate Species in Aqueous Solution: Force Field Derivation. *J. Phys. Chem. B* **2017**, *122*, 1471–1483, doi:10.1021/acs.jpcb.7b10697.

25. Wu, Y.; Tepper, H.L.; Voth, G.A. Flexible simple point-charge water model with improved liquid-state properties. *J. Chem. Phys.* **2006**, *124*, 024503, doi:10.1063/1.2136877.

26. Raiteri, P.; Demichelis, R.; Gale, J.D. Thermodynamically Consistent Force Field for Molecular Dynamics Simulations of Alkaline-Earth Carbonates and Their Aqueous Speciation. *J. Phys. Chem. C* **2015**, *119*, 24447–24458, doi:10.1021/acs.jpcc.5b07532.

27. Schofield, P.F.; Knight, K.S.; van der Houwen, J.A.M.; Valsami-Jones, E. The role of hydrogen bonding in the thermal expansion and dehydration of brushite, di-calcium phosphate dihydrate. *Phys. Chem. Miner.* **2004**, *31*, 606–624, doi:10.1007/s00269-004-0419-6.

28. Plimpton, S. Fast Parallel Algorithms for Short-Range Molecular Dynamics. *J. Comput. Phys.* **1995**, *117*, 1–19, doi:10.1006/jcph.1995.1039.

29. Pierre, M.D.L.; Raiteri, P.; Gale, J.D. Structure and Dynamics of Water at Step Edges on the Calcite {1014} Surface. *Cryst. Growth Des.* **2016**, *16*, 5907–5914, doi:10.1021/acs.cgd.6b00957.

30. Vondele, J.V.; Krack, M.; Mohamed, F.; Parrinello, M.; Chassaing, T.; Hutter, J. Quickstep: Fast and accurate density functional calculations using a mixed Gaussian and plane waves approach. *Comput. Phys. Commun.* **2005**, *167*, 103–128, doi:10.1016/j.cpc.2004.12.014.

31. Lippert, G.; Hutter, J.; Parrinello, M. The Gaussian and augmented-plane-wave density functional method for ab initio molecular dynamics simulations. *Mol. Phys.* **1999**, *103*, 124–140, doi:10.1007/s002140050.

32. Grimme, S.; Antony, J.; Ehrlich, S.; Kreig, H. A consistent and accurate ab initio parametrization of density functional dispersion correction (DFT-D) for the 94 elements H-Pu. *J. Chem. Phys.* **2010**, *132*, 154104, doi:10.1063/1.3382344.

33. Goedecker, S.; Teter, M.; Hutter, J. Separable dual-space Gaussian pseudopotentials. *Phys. Rev. B* **1996**, *54*, 1703–1710, doi:10.1103/PhysRevB.54.1703.

34. Bankura, A.; Karmakar, A.; Carnevale, V.; Chandra, A.; Klein, M.L. Structure, Dynamics, and Spectral Diffusion of Water from First-Principles Molecular Dynamics. *J. Phys. Chem. C* **2014**, *118*, 29401–29411, doi:10.1021/jp506120t.

35. Raiteri, P.; Laio, A.; Gervasio, F.L.; Micheletti, C.; Parrinello, M. Efficient Reconstruction of Complex Free Energy Landscapes by Multiple Walkers Metadynamics. *J. Phys. Chem. B* **2006**, *110*, 3533–3539, doi:10.1021/jp054359r.

36. Barducci, A.; Bussi, G.; Parrinello, M. Well-Tempered Metadynamics: A Smoothly Converging and Tunable Free-Energy Method. *Phys. Rev. Lett.* **2008**, *100*, 020603, doi:10.1103/PhysRevLett.100.020603.

37. Bonomi, M.; Branduardi, D.; Bussi, G.; Camilloni, C.; Provasi, D.; Raiteri, P.; Donadio, D.; Marinelli, F.; Pietrucci, F.; Broglia, R.A.; et al. PLUMED: A portable plugin for free-energy calculations with molecular dynamics. *Comput. Phys. Commun.* **2009**, *180*, 1961–1972, doi:10.1016/j.cpc.2009.05.011.

38. Asay, D.B.; Kim, S.H. Evolution of the Adsorbed Water Layer Structure on Silicon Oxide at Room Temperature. *J. Phys. Chem. B* **2005**, *109*, 16760–16763, doi:10.1021/jp053042o.

39. Arsic, J.; Kaminski, D.M.; Radenovic, N.; Poodt, P.; Graswinckel, W.S.; Cuppen, H.M.; Vlieg, E. Thickness-dependent ordering of water layers at the NaCl(100) surface. *J. Chem. Phys.* **2004**, *120*, 9720–9724, doi:10.1063/1.1712971.

40. Raiteri, P.; Gale, J.D.; Quigley, D.; Rodger, P.M. Derivation of an Accurate Force-Field for Simulating the Growth of Calcium Carbonate from Aqueous Solution: A New Model for the Calcite-Water Interface. *J. Phys. Chem. C* **2010**, *114*, 5997–6010, doi:10.1021/jp910977a.

© 2018 by the authors. Licensee MDPI, Basel, Switzerland. This article is an open access article distributed under the terms and conditions of the Creative Commons Attribution (CC BY) license (http://creativecommons.org/licenses/by/4.0/).

![minerals logo] *minerals*

MDPI

Article

The Carbonation of Wollastonite: A Model Reaction to Test Natural and Biomimetic Catalysts for Enhanced CO$_2$ Sequestration

Fulvio Di Lorenzo [1], Cristina Ruiz-Agudo [2], Aurelia Ibañez-Velasco [1], Rodrigo Gil-San Millán [3], Jorge A. R. Navarro [3], Encarnacion Ruiz-Agudo [1] and Carlos Rodriguez-Navarro [1,*]

[1] Department of Mineralogy and Petrology, University of Granada, 18071 Granada, Spain; fulvio@ugr.es (F.D.L.); aureliaiv@ugr.es (A.I.-V.); encaruiz@ugr.es (E.R.-A.)

[2] Department of Physical Chemistry, University of Konstanz, 78457 Konstanz, Germany; cristina.ruiz-agudo@uni-konstanz.de

[3] Department of Inorganic Chemistry, University of Granada, 18071 Granada, Spain; rodrigsm@correo.ugr.es (R.G.-S.M.); jarn@ugr.es (J.A.R.N.)

* Correspondence: carlosrn@ugr.es; Tel.: +34-958-246-616

Received: 13 April 2018; Accepted: 8 May 2018; Published: 11 May 2018

check for updates

Abstract: One of the most promising strategies for the safe and permanent disposal of anthropogenic CO$_2$ is its conversion into carbonate minerals via the carbonation of calcium and magnesium silicates. However, the mechanism of such a reaction is not well constrained, and its slow kinetics is a handicap for the implementation of silicate mineral carbonation as an effective method for CO$_2$ capture and storage (CCS). Here, we studied the different steps of wollastonite (CaSiO$_3$) carbonation (silicate dissolution → carbonate precipitation) as a model CCS system for the screening of natural and biomimetic catalysts for this reaction. Tested catalysts included carbonic anhydrase (CA), a natural enzyme that catalyzes the reversible hydration of CO$_{2(aq)}$, and biomimetic metal-organic frameworks (MOFs). Our results show that dissolution is the rate-limiting step for wollastonite carbonation. The overall reaction progresses anisotropically along different [*hkl*] directions via a pseudomorphic interface-coupled dissolution–precipitation mechanism, leading to partial passivation via secondary surface precipitation of amorphous silica and calcite, which in both cases is anisotropic (i.e., (*hkl*)-specific). CA accelerates the final carbonate precipitation step but hinders the overall carbonation of wollastonite. Remarkably, one of the tested Zr-based MOFs accelerates the dissolution of the silicate. The use of MOFs for enhanced silicate dissolution alone or in combination with other natural or biomimetic catalysts for accelerated carbonation could represent a potentially effective strategy for enhanced mineral CCS.

Keywords: carbonation; wollastonite; catalysts; carbonic anhydrase; MOFs; carbon capture and storage

1. Introduction

The alarming increase in the concentration of atmospheric CO$_2$ from pre-industrial levels of ~280 ppmv to current levels of ~400 ppmv and the effects this greenhouse gas can have on the Earth system [1], have prompted extensive research aiming at reducing anthropogenic CO$_2$ emissions and its capture and storage [2,3]. Among the different solutions under investigation [4–6], the conversion of calcium and magnesium (and/or ferrous iron) silicate minerals (e.g., olivine, pyroxenes, wollastonite, amphiboles, serpentine, and Ca-plagioclase) into carbonates, the so-called Urey-type or carbonation reaction [7], i.e., MSiO$_3$ + CO$_2$ = MCO$_3$ + SiO$_2$, where M is a divalent metal cation such as Ca^{2+} and Mg^{2+} (or Fe^{2+}), has been proposed and tested for the safe and long-term storage of this greenhouse gas [2,8,9].

Carbonation of primary silicates is a natural weathering process [10] that has regulated Earth's atmospheric CO_2 concentration and climate over geologic time-scales [11–15]. Such a natural process is the basis of several current ex situ and in situ technologies for carbon capture and storage (CCS) [4,6,16,17]. Among the many ex situ methods proposed so far, involving either solid–gas or liquid–gas carbonation reactions, the most promising ones are those based on aqueous mineral carbonation [6,18]. This is the case, for instance, of the acid-promoted dissolution of primary silicates in a reactor followed by the injection of CO_2 in the resulting Ca- and/or Mg-rich solution, whose pH is increased to favor carbonate precipitation [17]. This two-stage mineral carbonation is known as the "pH-swing" process [19,20]. Ex situ silicate carbonation in solution following a one step process (i.e., direct carbonation) would be more cost-effective, but typically requires circumneutral pH conditions, which limit silicate dissolution [6,16]. Silicate activation by means of heat treatment and/or intensive grinding can result in enhanced dissolution during ex situ mineral carbonation, but these processes have a high energy penalty and are too costly [2,4], which may make such an approach impractical [18]. Alternatively, complexing agents (e.g., ethylenediaminetetraacetic acid, EDTA) and weak organic acids (e.g., acetic acid) can enhance silicate dissolution at moderately acid (pH 2–4.5) or circumneutral pH conditions [21–23], but their cost and complex recycling are strong handicaps for their widespread implementation. In situ carbonation involves the direct deep injection of CO_2 (e.g., dissolved in water as a brine) into mafic and ultramafic silicate rock formations [2,4,24]. The reactive brine (with very low pH) induces the dissolution of the alkaline-earth silicates (or silicate glass), a process which consumes protons. The subsequent pH increase ultimately favors carbonate precipitation [24]. A successful example of in situ carbon mineral sequestration via silicate carbonation is the case of the pilot test CarbFix (Iceland), which involved injection of reactive CO_2 brines into basaltic rocks [24,25]. Matter et al. [24] report that over 95% of the CO_2 injected into the CarbFix site might have been mineralized in less than two years.

Silicate minerals transform into carbonates via a combined dissolution–precipitation process. The alkaline-earth metal cations released by the dissolution of the silicate mineral act as building units for the formation of secondary carbonate phases [26–28]. The final result is the incorporation of CO_2 as carbonate structural units into the product phases. The thermodynamic driving force behind the overall carbonation process is the undersaturation with respect to the silicate minerals of the (carbon-rich) aqueous solution interacting with natural silicate rocks [26,27]. As soon as silicate minerals' dissolution starts, the aqueous phase becomes increasingly richer in alkaline-earth metal cations until a critical supersaturation with respect to the carbonate minerals is reached, triggering their precipitation [4,26,27]. The overall ΔG of the dissolution–precipitation reactions is negative, so this process takes place spontaneously [27,29]. As an example, the overall carbonation of wollastonite is exothermic, with a ΔH of -87 kJ·mol^{-1} and ΔG of -44 kJ·mol^{-1} at standard T and P conditions [9,29]. However, the kinetics of these coupled dissolution–precipitation reactions are typically very slow, which poses a strong handicap for the effective implementation of this CCS strategy [22]. The use of catalysts that can speed up both the silicate dissolution and the carbonate precipitation reactions (neither being consumed by the reaction nor needing a costly recycling) would thus represent a significant advance for the effective and widespread implementation of mineral carbonation for CCS.

The hydration of CO_2 via the overall reaction $CO_{2(aq)} + H_2O = HCO_3^- + H^+$ is the rate-limiting step of the carbonate precipitation reaction [30,31]. This reaction is significantly accelerated by carbonic anhydrase (CA) enzyme and, as a result, CA also accelerates the precipitation of alkaline-earth carbonates [32–34]. CAs are an ubiquitous family of natural metalloenzymes, its main function being the catalysis of the reversible hydration of carbon dioxide to form bicarbonate ions [35], thereby playing a critical role in various cellular functions, such as pH-regulation, CO_2 transport, and bone resorption [36], as well as in the biomineralization of $CaCO_3$ [37]. The active site of CA is formed by a Zn atom tetrahedrally coordinated to three histidine residues and a water molecule. Histidine residues polarize the electronic cloud around the metallic atom and facilitate anchoring of bonded OH$^-$ (formed after water splitting) to the electrophilic carbon of dissolved CO_2, resulting in its hydration (nucleophilic attack) to form HCO_3^-. The bicarbonate ion is then displaced from Zn by a water molecule, which again undergoes protolysis, rendering the active site ready to catalyze the hydration of another CO_2 molecule [35]. CA is one of the fastest enzymes known, with a turnover of up to $10^6 s^{-1}$ [34]. Such a biocatalyst has been found to be effective for enzymatic pre- and post-combustion (flue gas) CO_2 capture [6,38], and for accelerated ex situ CO_2 capture and storage via carbonation of industrial alkaline brines [32], $Ca(OH)_2$ solutions [39] and pastes [31], and $Mg(OH)_2$ slurries [34], as well as solutions produced during the pH-swing process [40,41].

In addition to its "hydrase" catalytic activity, CA also acts as an "esterase" enzyme, being able to hydrolyze a range of ester bonds [38]. Despite its proved catalytic effect on carbonate precipitation, it is not known if CA plays any catalytic role on the dissolution of silicate minerals subjected to carbonation, which is the rate-limiting step in most studied systems [4,16,20,22,26,27], or on the development of secondary amorphous silica layers, also known as silica altered layers, SALs [42,43], which reportedly have a detrimental effect on mineral carbonation [20,29,44–49]. Because CA can catalyze the breaking of ester bonds [38], this enzyme could potentially be able to facilitate silicate dissolution or contribute to the dissolution of passivating SALs via Si–O ester-like bond breaking. Interestingly, some sponges that are able to dissolve amorphous silica, express an enzyme called silicase that belongs to the family of CAs [50]. CA and silicase possess the same active site (i.e., the Zn atom coordinated to three histidine residues). However, while silicase can also catalyze the reversible hydration of CO_2 [50], its main biochemical role is to catalyze the breaking of Si–O ester-like bonds producing silicic acid [50,51], thereby speeding the dissolution and/or precipitation of silica. The structural similarity of silicase and CA gives some bearing to the notion that CA could accelerate the dissolution of silicate minerals as recently proposed by Xiao et al. [52]. Conversely, the proved capacity of silicase to break Si–O bonds would make it an ideal biocatalyst for the accelerated dissolution of Ca and/or Mg silicates, prior to carbonate precipitation. Unfortunately, silicase extraction (from SiO_2-biomineralizing sponges) is very complex [50], and at present this enzyme is not commercially available. In contrast, CA from bovine serum is commercially available. Hence, here we decided to explore the potential silicase-like (i.e., esterase) role of CA during the dissolution stage of wollastonite subjected to carbonation.

In any case, the limited availability of natural enzymes, and their poor stability (i.e., rapid denaturation) are strong handicaps for their widespread use in CCS [6]. Therefore, it would be of advantage to evaluate the potential of biomimetic synthetic metalloenzymes for the accelerated dissolution of silicates and the subsequent precipitation of carbonates. While several biomimetic catalysts used as CA substitutes for accelerated CO_2 capture and carbonate precipitation exist [53–57], to our knowledge, no biomimetic catalyst for the dissolution of silicates has been developed. In this context, the catalytic potential offered by the recent development of metal-organic frameworks (MOFs) [58] could represent a significant step-ahead for CCS. MOFs are extremely porous crystalline materials containing metal cations (or metal nanoclusters) bonded by multidentate organic ligands that have been shown to be effective substrates for a range of chemical processes, including pre- and post-combustion CO_2 capture [57,59]. MOFs have also been thoroughly investigated as effective and economic alternatives to natural enzymes [58]. The advantages of using MOFs as catalysts instead of enzymes are several: (i) the cost of MOFs is significant lower because of the easier synthesis and purification process; (ii) their physical/chemical resistance is extremely high in comparison with enzymes, which tend to suffer denaturation; and (iii) the catalytic properties of MOFs can be tuned through an easy variation of substrates or through tailoring synthesis reactions [58]. Their main disadvantage is the lower catalytic effect exerted by the metal clusters in MOFs compared to the active metal sites of enzymes. It should be remarked, however, that the number of metal clusters in a MOF particle can be several orders of magnitude higher than the number of active sites present in an enzyme, so MOFs can achieve an overall comparable catalytic efficiency [58]. In this context, we decided to test the catalytic effect on silicate dissolution exerted by chemically resistant Zr-based MOFs such as UiO-66 [60], MOF-808 [61], and the analogous $Mg(OH)_2$-doped MOF-808@$Mg(OH)_2$ (Figure 1) [62]. Zr-based MOFs have been demonstrated to be highly catalytically active in the detoxification of chemical-warfare agents (CWAs) [63], because they are able to induce the hydrolysis of the very strong phosphate ester bonds (P–O, P–F) due to the suitable combination of the strong Lewis acidic nature of Zr^{4+} ions and basic oxide and hydroxide bridging ligands giving rise to a phosphotriesterase-like activity [64]. We expected that Zr-based MOFs could also have the capacity to break the (very strong) Si–O ester-like bond of silicates, thereby accelerating their dissolution in a similar way to silicase. Here, the catalytic effect on silicate dissolution of Zr-based MOFs and CA are compared.

The main goal of the present study is to screen, test, and validate effective natural and biomimetic catalysts to enhance the conversion of silicates into carbonate minerals. For this task, we used wollastonite as a model alkaline-earth metal silicate mineral. Wollastonite is a single chain silicate with non-pyroxene structure (space group $P\bar{1}1$) that commonly forms during contact metamorphism [65]. Its dissolution and carbonation have been widely studied due to its availability, mineral significance, relatively high reactivity in comparison with other primary silicate minerals, and its tendency to form SALs [22,23,26,27,29,42,43,45,48,49,66–80]. These characteristics make it the perfect "standard silicate mineral" for mineral carbonation studies. We present an experimental study on the effect of CA as a catalyst during the dissolution of wollastonite and the subsequent precipitation of calcium carbonate. We also explored the catalytic effect exerted by CA and Zr-based MOFs on the initial dissolution step of the carbonation reaction. To do so, we developed a fast and reproducible experimental protocol that was validated and used to test the catalytic effect of both natural and biomimetic (MOFs) enzymes on silicate dissolution, and which can also be used to test any material aimed at enhancing silicate carbonation for CCS purposes.

2. Materials and Methods

2.1. Materials

Natural wollastonite crystals from South Africa were ground in an agate mortar and sieved to a size range of $251 < \varphi < 66$ μm. Their surface area was determined by N_2-adsorption using the Brunauer–Emmett–Teller (BET) method (1.9 ± 0.1 m^2/g). Prior to testing and characterization, powder samples were washed with 10 mL of absolute ethanol (Sigma-Aldrich, Merck, Darmstadt, Germany) and ultrasonically cleaned for 5 min. Subsequently, the liquid phase was removed and the process was repeated five times. Afterwards, the crystals were cleaned with 10 mL of H_2O to remove the excess ethanol and dried overnight at 60 °C [78]. The purity of the solids was checked by X-ray diffraction analysis showing no crystalline phase other than wollastonite. However, a few μm-sized isolated Mg-rich olivine crystals were observed during scanning electron microscopy analyses. Due to their scarcity, their influence on the results presented and discussed here is considered negligible. Experiments were conducted using ultrapure H_2O (Millipore-MilliQ®, resistivity = 18.2 MΩ·cm), 0.1 M HCl (Scharlab S.L., Barcelona, Spain), and $NaHCO_3$ (Sigma-Aldrich, Merck, Darmstadt, Germany).

The catalysts used in the present study are CA from bovine serum erythrocytes (Sigma-Aldrich, Merck, Darmstadt, Germany) and three different Zr-based MOFs—namely, UiO-66, MOF-808, and MOF-808@Mg(OH)$_2$ (Figure 1)—synthesized at the Department of Inorganic Chemistry of the University of Granada. Catalysts were tested individually and experiments were repeated three times to ensure reproducibility.

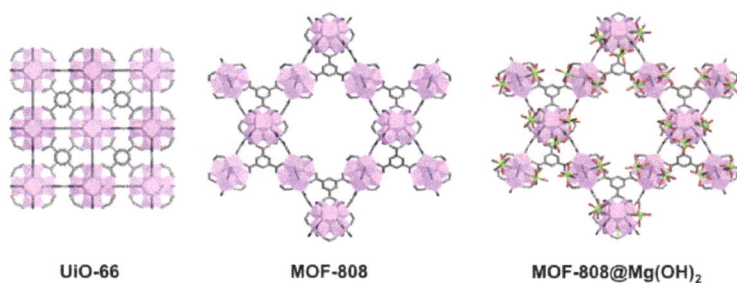

Figure 1. Structure of UiO-66, MOF-808, and MOF-808@Mg(OH)$_2$. The aromatic multidentate ligands used to construct the Zr-MOFs are benzene-1,4-dicarboxylate and benzene-1,3,5-tricarboxylate for UiO-66 and MOF-808, respectively [60–62].

2.2. Carbonation Experiments

Carbonation experiments involved the initial dissolution of wollastonite and the subsequent precipitation of calcium carbonate. To investigate such coupled dissolution–precipitation reactions taking place during the overall carbonation reaction, Falcon® tubes (V_{tot} = 15 mL) were filled with 50 mg of wollastonite crystals and 10 mL of a 0.1 M NaHCO$_3$ aqueous solution acidified to pH 7.5 (using HCl). At this pH a significant fraction of bicarbonate ions will be present as dissolved CO$_{2(aq)}$. This ensures that the addition of CA catalyst used to speed up the hydration of CO$_2$ to form bicarbonate ions is still a relevant reaction in our system. Inorganic carbon lost due to CO$_2$ degassing at this pH (standard temperature and pressure, STP) should be lower than the maximum of 5 mol % reached at equilibrium, as predicted by PHREEQC computer simulation (version 3.3.12, USGS, Reston, VA, USA) [81]. CA was added to the solutions at a concentration of 6 μM (175 ppm). The Falcon® tubes were sealed with Parafilm® and inserted into a thermostatic chamber equipped with an orbital rotator to maintain the particles in suspension. Experiments were performed at 40 °C under vigorous stirring (200 rpm). To analyze the evolution of the mineral replacement reaction, Falcon® tubes were collected at pre-set reaction time intervals and centrifuged (5 min, 2500 rpm) to separate the solids from the supernatant. Both the liquid and solid phases were subsequently analyzed (see below).

2.3. Dissolution Experiments

In order to single out the effect of the studied catalysts on the first stage of silicate carbonation (i.e., silicate dissolution), tests were performed focusing on the dissolution of wollastonite in the absence of a carbon source. Dissolution experiments were performed using PTFE (polytetrafluoroethylene) reactors. As in the previous case, plastic reactors were used to avoid potential silicon contamination from borosilicate glassware. The dissolution process was monitored using an automatic titration system (Titrando 905, Metrohm, Gallen, Switzerland) equipped with an automatic burette (V_{tot} = 2 mL, Metrohm, Gallen, Switzerland), pH and conductivity probes (Metrohm, Gallen, Switzerland), and an ion selective electrode for calcium (Ca^{2+} ISE, Mettler Toledo, Twinsburg, OH, USA). The instrument was controlled with the Tiamo software (version 2.5, Metrohm, Gallen, Switzerland) to achieve a pH-stat system where the pH increase due to silicate dissolution was compensated by the addition of the proper amount of HCl. Dissolution tests were run at 25 ± 1 °C (this *T* was selected to be able to compare dissolution rates with those previously published) and at a constant pH of 4. The reaction vessel was filled with 169.8 mL H$_2$O, the catalyst (30 ppm) and the necessary HCl to achieve pH 4 (about 0.15 mL). Such a relatively low pH was selected to speed up the dissolution of wollastonite and to minimize CO$_2$ dissolution in the reaction cell. After pH stabilization, wollastonite crystals (100 mg) were added to the solution and dissolution started. After 1 h elapsed time, the experiment was stopped and the solution was filtered under low-vacuum (Millipore®, φ < 0.2 μm) to recover the reacted wollastonite crystals. Crystals were stored in sealed plastic vials until further analysis.

2.4. Characterization of Reactant/Product Phases and Geochemical Modeling

Reactant and product solid phases were analyzed by means of X-ray diffraction (XRD, PANalytical X'pert Pro, Eindhoven, The Netherlands); Cu kα radiation (λ= 1.5405 Å), voltage 45 kV, current 40 mA, 3–80° 2θ exploration range and 0.01° 2 θ s^{-1} scanning rate). Detailed μ-XRD analyses of individual reacted wollastonite crystals were performed using a Bruker Discover8 (Bruker, Karlsruhe, Germany) microdiffractometer equipped with a Cu microfocus X-ray source and a 2D area detector, which enabled obtaining better XRD patterns in the cases of reacted wollastonite (i.e., with a higher signal to noise ratio). Textural and compositional details were analyzed using a field emission scanning electron microscope (FESEM) (FEI Qemscan 650F, Hillsboro, OR, USA and/or Zeiss Auriga, Jena, Germany) equipped with energy dispersive X-ray spectroscopy (EDS) for elemental analysis. To quantify the final ratio between residual wollastonite and newly formed $CaCO_3$ we used thermogravimetric analysis (TG/DSC1, Mettler Toledo, Twinsburg, OH, USA) according to the procedure described by Huijgen et al. [29]. The concentration of Ca and Si in the aqueous phase during carbonation experiments was analyzed by ICP-OES (Optima 8300, Perkin Elmer, Waltham, MA, USA). In the case of the dissolution-only experiments, the free Ca was determined using the ion-selective electrode and Si calculated from HCl addition using the equation

$$\frac{mL_{HCl} \cdot conc_{HCl}(= mmol_{HCl})}{2} = mmol_{H_2SiO_4^{2-}} \cong mmol_{Ca^{2+}} = conc_{Ca^{2+}} \cdot V_{tot} \tag{1}$$

where mL_{HCl} = volume of HCl (mL) automatically dosed to keep a constant pH, $conc_{HCl}$ = concentration of HCl solution (mmol·mL^{-1}), $conc_{Ca^{2+}}$ = concentration of free Ca measured by the ISE (mmol·mL^{-1}), $mmol_{H_2SiO_4^{2-}}$ = amount of silicate ions released by the mineral (mmol), $mmol_{Ca^{2+}}$ = amount of Ca ions released by the mineral (mmol), and V_{tot} = total volume of the solution (L).

Geochemical calculations and solution speciation were performed using PHREEQC computer code (version 3.3.12, USGS, Reston, VA, USA) and the minteq.dat database [81].

3. Results

3.1. Carbonation of Wollastonite in the Presence of CA

Figure 2 shows the time-evolution of the Si and Ca concentration in solution during carbonation experiments determined by ICP-OES analysis. Systems with and without CA followed the same trend. The Si concentration increased continuously with time during the carbonation experiments due to its continuous release from dissolving wollastonite. However, the Si release rate decreased with time, pointing to the precipitation of amorphous silica on wollastonite surfaces and their partial passivation—also associated with calcite precipitation (see below). In contrast, the Ca concentration reached a maximum during the first 8–24 h of the experiments and then decreased. This drop was related to calcite precipitation as demonstrated by the XRD analysis (Figure 3). Note that the molar Ca concentration measured at 24 h was already one order of magnitude lower than that of Si. However, previous studies showed that wollastonite dissolution over a range of pHs is non-stoichiometric (apparently incongruent, although related to the secondary precipitation of silica-rich layers, see References [42,43]), with the release rate of Ca being higher than that of Si [42,66,70,72,77]. Our Ca and Si analysis results thus suggest that carbonate precipitation was already taking place before 24 h. This was confirmed by SEM-EDX analyses (see below).

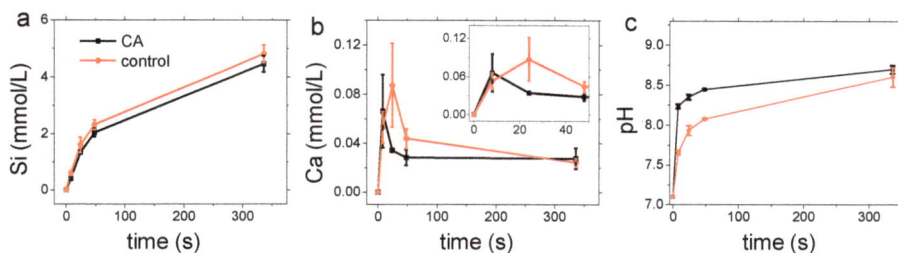

Figure 2. Evolution of the solution chemistry during wollastonite carbonation. Aqueous concentration of silicon (**a**) and calcium (**b**) in solution determined by ICP-OES. The inset in (**b**) shows the Ca evolution during the early stages of the carbonation reaction; (**c**) pH evolution (values measured at 25 °C and recalculated for a *T* of 40 °C using PHREEQC). The systems with CA (black symbols and curves) and without CA (control; red symbols and curves) are compared. Error bars show standard deviation ($\pm 1\sigma$).

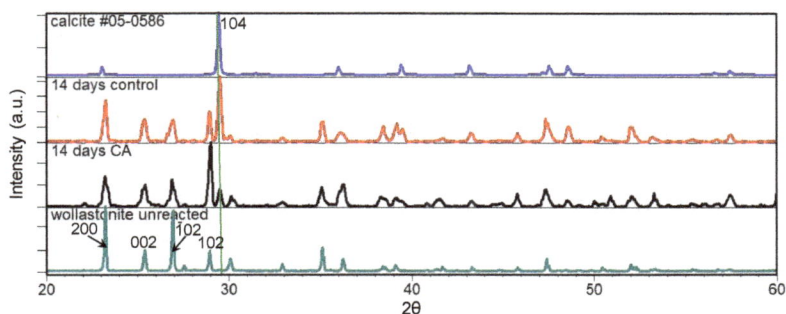

Figure 3. XRD patterns of wollastonite powders reacted for 14 days in the absence (control) and presence of CA. The XRD patterns of unreacted wollastonite and reference calcite (Joint Committee on Powder Diffraction Standards, JCPDS card # 05-0586) are included for comparison. Note the higher intensity of the main peak of calcite (marked by the green vertical line) in the control compared with the CA run, and the significant reduction in intensity ratio of wollastonite 200/002 peaks after 14 days reaction, especially in the run with CA.

The time evolution of both aqueous Si and Ca concentrations showed the effect of CA. The amount of Si released after wollastonite dissolution was slightly lower than in the control during the whole carbonation experiment, which may imply a (minor) inhibitory effect of this enzyme on wollastonite dissolution and/or a passivation-induced effect (see discussion below). In the case of Ca, its decrease in concentration occurred earlier and faster in the presence of the enzyme. This is in agreement with previous studies reporting that CA catalyzes calcite precipitation [31,39].

Figure 2c shows the pH evolution during the carbonation test. A rapid increase in pH took place during the early stages of the dissolution process, which is consistent with the reported proton consumption that takes place during dissolution of primary silicate minerals [10]. Interestingly, after 48 h, the pH continued to increase, although at a much lower rate, when calcium carbonate precipitation had already started. Note that carbonate precipitation releases one mole of H^+ per mole of $CaCO_3$ formed [27]. This would result in a net pH reduction. In contrast, the observed continuous increase in pH can only be explained if wollastonite dissolution continued over the course of the carbonation experiment, consuming two moles of H^+ per mole of wollastonite dissolved via the (overall) reaction $CaSiO_3 + 2H^+ + H_2O = Ca^{2+} + H_4SiO_4$ [67]. Figure 2c also shows a faster and higher pH increase in the presence of CA compared with the control run. This could be due to either a faster dissolution of wollastonite associated with an esterase-like catalytic effect of CA (see, however, the results of the dissolution tests, below), or (more likely) a possible CA-catalyzed dehydration of HCO_3^- resulting in (limited) CO_2 outgassing.

TG analyses enabled us to determine the extent of the replacement process (Figure 4). The amount of calcite formed was determined by the mass loss between 550–800 °C (Figure 4a), which corresponds to the loss of CO_2 from structural carbonate ions [82]. The amount of calcite formed after 14 days reaction in the presence of CA was systematically lower (6.6 ± 1.5 mol %) than in the control experiments (13.9 ± 2.2 mol %) (Figure 4b).

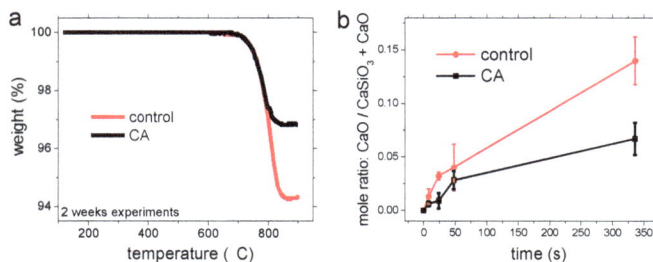

Figure 4. Thermogravimetric (TG) analysis of the reacted solid phases: (**a**) TG curves for the control and CA runs; (**b**) time-evolution of the extent of carbonation for the control and CA runs based on TG results (error bars correspond to ±1σ).

The TG results are consistent with the XRD results showing more intense calcite Bragg peaks in the control run as compared with the CA run, and the opposite in the case of wollastonite (Figure 3). TG (and XRD) results also show that only a minor fraction of wollastonite was converted into calcium carbonate (i.e., maximum conversion of ~15 mol %) at the circumneutral pH, relatively low T (40 °C) and short time-spam of the experiment (14 days). These results show that CA does not seem to be a suitable catalyst for the overall carbonation of wollastonite under our experimental conditions. Despite its observed acceleration of carbonate precipitation, the overall effect of CA was to limit the conversion of wollastonite into $CaCO_3$.

Figure 5 shows FESEM images of the very early stages of the carbonation of wollastonite crystals (i.e., 8 h reaction time). It was observed that regardless of the presence or absence of CA, abundant dissolution pits formed on the different faces of the prismatic crystals, being more abundant and better developed on {100} prismatic faces (Figure 5b,d,e). Such dissolution pits were elongated parallel to [010], with narrow sidewalls parallel to (001) planes and, in more evolved pits, deep along [100] direction. This clearly shows that the dissolution for this mineral is markedly anisotropic (i.e., faster along [010]). Note that {100} and {001} prism faces of wollastonite are easily identifiable because the later display prominent {100} cleavage planes parallel to the [010] direction, whereas {001} cleavage planes present on {100} faces are less marked (or absent) [65]. Nearly all the crystals only showed these two prismatic faces, which are the most common (stable) prismatic faces of wollastonite [65]. Any other {h0l} prismatic faces, such as {101}, which should be easily identified because it appears at an angle of ~45° with respect to the {100} and {001} faces, were not observed in our samples. This is likely due to the fact that the wollastonite crystals used here were subjected to grinding prior to testing, thereby promoting their cleavage along the {100} and {001} cleavage planes. In some cases, coalescence of deep dissolution pits led to a rough surface with prominent pyramidal hillocks, more evident in sections normal to {100} cleavage planes (Figure 5e). In all cases, abundant Si-rich (according to EDS results) nanogranular precipitates corresponding to amorphous silica [28], were observed covering areas close to dissolution pits (Figure 5b,c,f). Individual nanoparticles were ~20–30 nm in size (Figure 5c), and tended to coalesce forming SALs, as those depicted in Figure 5f, where the individual nanoparticles were still visible. The Si-rich nanoparticles making up SALs observed here were nearly identical to the amorphous silica nanoparticles observed by Béarat et al. [44] on olivine crystals subjected to carbonation and developing abundant SALs.

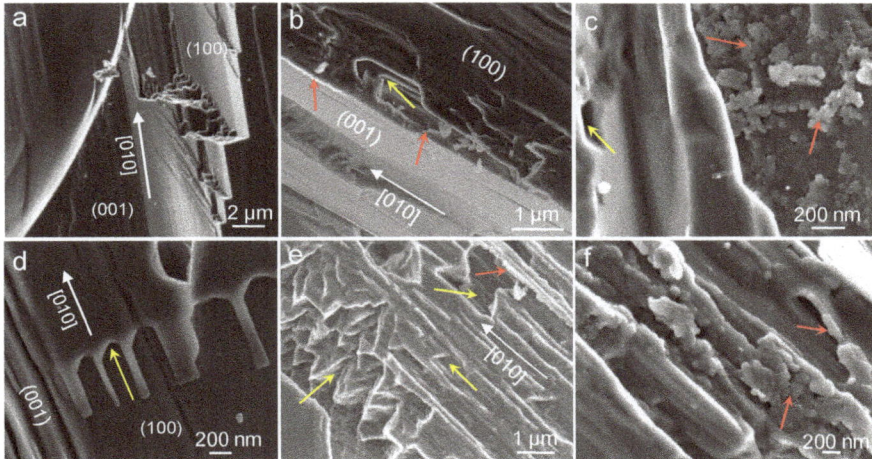

Figure 5. FESEM images of unreacted wollastonite crystals (**a**) and those reacted for 8 h showing dissolution pits (**b–e**) and surface precipitates of amorphous silica nanoparticles (**b,c,e,f**). Images (**b–d**) correspond to the control, and (**e,f**) to the CA run. Red arrows point to amorphous silica nanoparticles/surface layers, and yellow arrows point to dissolution pits (elongated parallel to <010>$_{wollastonite}$).

At such an early stage of reaction, calcite crystals (identified by their habit and Ca content in EDS maps) were already present and in some areas—such as cracks, edges of macrosteps, cleavage planes, and {010} faces—they were particularly abundant (Figure 6). Remarkably, in the presence of CA, calcite precipitates tended to coalesce forming a layer of equally oriented crystals on the {100} faces of wollastonite (Figure 6f,g).

Figures 7 and 8 show time-resolved SEM photomicrographs of samples carbonated for 1, 2, and 14 days in the absence and presence of CA, respectively. Backscattered electron images in combination with EDS elemental maps allowed us to distinguish between wollastonite (light grey), amorphous silica (grey) and calcite (dark grey) due to their compositional differences. Irrespectively of the presence or absence of CA, Si-rich layers corresponding to SALs were observed as rims in proximity of etch pits and/or lining fractures and cleavage planes on wollastonite faces. They were predominantly concentrated on {010} and {100} faces of wollastonite, as previously observed in this mineral that shows anisotropic dissolution [42].

It was remarkable the thickness (up to ~5 μm) reached in just 14 days by SALs developed on the {010} faces of wollastonite (Figures 7d and 8e). It was also noticeable how on {001} faces SALs developed along {100} cleavage planes and grew (thickened) in directions normal to these planes (i.e., <100>) and along <010> directions (Figure 8e,f). These results show that SALs preferentially grow along the <*hkl*> directions where wollastonite dissolution is apparently fastest (i.e., <010> and, to a lesser extent, <100>).

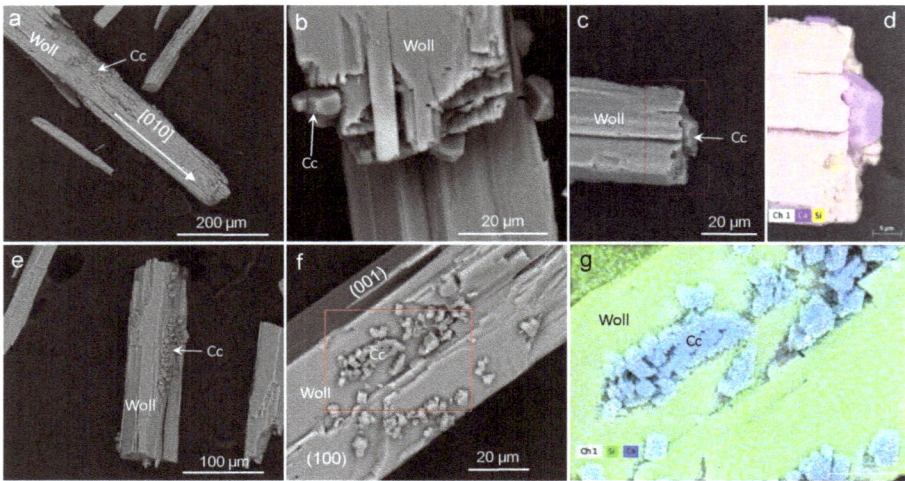

Figure 6. FESEM-EDS analysis of reactant and product phases during the early stage of wollastonite carbonation (8 h reaction time). (**a**) General overview, and (**b**,**c**) details of reacted wollastonite (Woll) in the control run showing calcite (Cc), and (**d**) EDS map of the red rectangular area in (**c**); (**e**) general overview and (**f**) detail of calcite formed on {001} faces of wollastonite in the CA-inclusive run; (**g**) shows the EDS map of the rectangular area marked in (**f**).

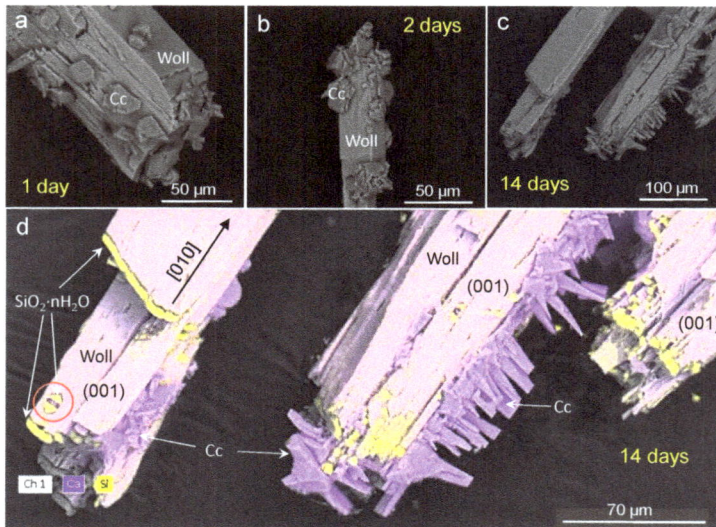

Figure 7. FESEM-EDS images of wollastonite carbonated for 1 up to 14 days in the absence of CA. (**a**) wollastonite (Woll) reacted for 1 day, (**b**) 2 days, and (**c**) 14 days; (**d**) EDS map of (**c**). Calcite crystals (Cc) tend to form dense coatings on {100} and {010} faces. SALs (amorphous $SiO_2 \cdot nH_2O$) display a remarkable thickness on the walls of dissolution pits (red circled area in (**d**)) and on {010} faces.

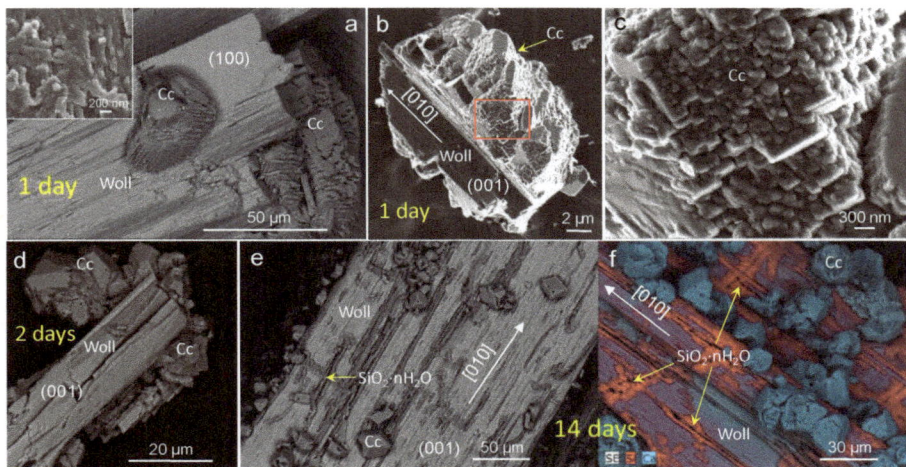

Figure 8. FESEM images and EDS map of wollastonite crystals reacted in the presence of CA. Backscattered (**a,d,e**) and secondary (**b,c**) electron images of wollastonite (Woll) subjected to carbonation for different time-periods; (**f**) EDS compositional map of wollastonite reacted for 14 days. Crystals are elongated along [010], showing marked {100} cleavage, and in some sections, less marked {001} cleavage. Note the thick calcite (Cc) coatings developed on {100} and {010} faces in (**a,b,d**). A high magnification image of the calcite aggregate in the red squared area in (**b**) is shown in (**c**). Note the preferential nucleation of calcite on the areas where amorphous silica forms at {100} cleavage planes and at fractures nearly parallel to [100] in (**e,f**). The inset in (**a**) shows a high magnification image of SALs developed on wollastonite after one day of carbonation. Yellow arrows in (**e,f**) point to amorphous silica, $SiO_2 \cdot nH_2O$ (SALs).

Irrespectively of the presence or absence of CA, most calcite crystals appeared with a poorly defined habit during the initial stages of reaction (one and two days) (Figure 7a,b and Figure 8a–d). Ageing favored morphology development of calcite crystals and their growth. Over time, calcite aggregates formed thick coatings, particularly thick on wollastonite {010} and {100} faces. In the case of the control, thick aggregates (up to ~40 µm in thickness) of elongated prismatic calcite crystals growing nearly normal to {100} faces, and almost parallel to {001} faces, were observed after 14 days (Figure 7c). Figure 7d shows the EDS compositional map overlapped on the secondary electron image of a control wollastonite reacted for 14 days: the distribution of Si and Ca helped us to identify areas where SALs and calcite crystals developed, validating the phase assignation by backscattered electron images. In the presence of CA, we observed calcite crystals growing on top of wollastonite, especially on {010} (Figure 8a) and on {100} faces (Figure 8b). The carbonate coatings showed calcite crystals with rough, and stepped faces made up of an aggregate of nanoparticles, especially during the first days of reaction (Figure 8c). Calcite crystals with a clearly defined rhombohedral habit were observed after 14 days reaction time (Figure 8e,f). The thickness of these dense calcite layers varied between ~10 and ~40 µm and, irrespectively of the presence or absence of CA, they were preferentially formed on {010} and {100} faces of wollastonite, thereby tending to concentrate where SALs were more abundant and better developed. Interestingly, one of the Bragg reflections that experienced a more significant relative intensity reduction following 14 days carbonation was that corresponding to (200) planes (Figure 3). This is consistent with the masking effect induced by massive calcite precipitation on the {100} faces observed with FESEM. Note, however, that changes in the relative intensity of different wollastonite Bragg peaks have to be interpreted with caution due to the possible preferred orientation of such fibrous/prismatic crystals. In any case, these observations suggest that calcite crystals nucleated on the areas where most active dissolution of wollastonite and surface precipitation of amorphous silica was taking place. Calcite and amorphous silica precipitates thus appear to be acting as passivating layers, precisely where most active dissolution of wollastonite was taking place. This can help explain why the Si release rate and the rate of pH rise showed a continuous decrease over the time-span of the experiment, and why the conversion of wollastonite into calcite was so limited.

3.2. Dissolution Experiments

A fast protocol was designed in this work to test the effects of a range of potential catalysts on silicate dissolution kinetics. It allowed the online recording of free Ca^{2+} concentration measured by a Ca-ISE and the dosed amount of HCl necessary to maintain a constant pH of 4. The time evolution of these two experimental values enabled us to calculate the amount (mmol) of $H_2SiO_4^{2-}$ and Ca^{2+} released during the dissolution process (Figure 9a,b).

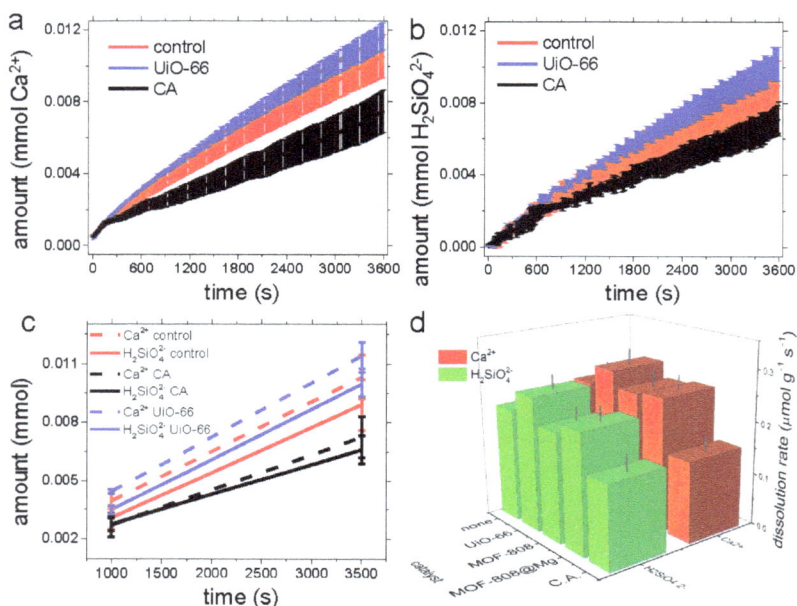

Figure 9. Dissolution of wollastonite in the presence of CA and MOFs: (**a**) Time-evolution of Ca^{2+} released by wollastonite dissolution and continuously recorded by the ISE probe; (**b**) amount (mmol) of $H_2SiO_4^{2-}$ calculated from the amount (mL) of HCl added to maintain a constant pH = 4 during the titration experiments. Shaded color areas show values of standard deviation ($\pm 1\sigma$); (**c**) time evolution of Ca^{2+} and $H_2SiO_4^{2-}$ concentration following dissolution of wollastonite in the presence of CA, UiO-66 and without catalyst (control); (**d**) 3D-graph summarizing the dissolution rates of wollastonite crystals measured in the presence of MOFs, carbonic anhydrase (CA), and without catalysts (control).

The total amount of ions released to the solution varied almost linearly with time after an initial stage (less than five minutes) when the dissolution rate slightly decreased. This is likely due to the progressive elimination of the most reactive areas/sites of the fresh wollastonite crystals, although a partial, limited passivation by the early formation of SALs cannot be ruled out (see Ruiz-Agudo et al. [42]). Control experiments were performed up to 35,000 s and only a very slight or negligible change in the slope of Ca release curves was observed after this initial rapid reduction in the dissolution rate. The very minor divergence from a straight line that we observed was due to the fact that as soon as dissolution starts, ion concentration raises and the related change in the ionic strength induces a very minor difference between the free Ca^{2+} measured by ISE and the total Ca released by the mineral. These results are in agreement with those of Bailey and Reesman [66] who showed that, at a constant pH of 4, the concentration of Ca released after dissolution of wollastonite (at 25 °C, in the absence of CO_2) increases linearly with time. Our results also suggest that the SALs that reportedly form at this pH [28,42,72] do not seem to act as an effective passivating layer, at least during the relatively short time-span of our dissolution experiments.

Figure 9a,b shows that UiO-66 MOF induced a faster release of Ca^{2+} and $H_2SiO_4^{2-}$ than the control. The effect of CA was the opposite: it decreased the release rate of these ionic species as compared with the control. The two other tested MOFs did not show any significant difference as compared with the control (data not plotted in Figure 9a,b for the sake of clarity).

From these results, the actual dissolution rates of wollastonite were calculated considering the 1000–3500 s interval (Figure 9c). The rate constants for wollastonite dissolution in the absence and presence of the different catalysts tested were calculated considering the average BET surface area of unreacted wollastonite crystals and are reported in Figure 9d and Table 1. UiO-66 achieved a moderate catalytic effect on wollastonite dissolution. The other MOFs (MOF-808 and MOF-808@Mg(OH)$_2$) showed no significant catalytic effect. Dissolution experiments confirmed that despite its structural and chemical similarity to silicase, CA had no catalytic effect on this process. In fact, CA showed a weak inhibition effect confirming that the observed catalysis of the carbonate precipitation reaction achieved by using this enzyme (see above), appears to be the results of its ability to increase the availability of carbonate ions, but not because it can enhance silicate dissolution. Figure 9c,d also shows that dissolution rates calculated independently using free Ca^{2+} and dosed HCl (i.e., for neutralization of $H_2SiO_4^{2-}$ to form H_4SiO_4) were identical within experimental error.

Table 1. Dissolution rates of wollastonite ($mol \cdot m^{-2} \cdot s^{-1}$) in the absence and presence of catalysts.

	Control	CA	UiO-66	MOF-808	MOF-808@Mg(OH)$_2$
$R_{Ca^{2+}}$	$1.28 \pm 0.14 \times 10^{-8}$	$0.96 \pm 0.11 \times 10^{-8}$	$1.47 \pm 0.11 \times 10^{-8}$	$1.34 \pm 0.16 \times 10^{-8}$	$1.31 \pm 0.13 \times 10^{-8}$
$R_{H_2SiO_4^{2-}}$	$1.19 \pm 0.11 \times 10^{-8}$	$0.86 \pm 0.09 \times 10^{-8}$	$1.32 \pm 0.12 \times 10^{-8}$	$1.19 \pm 0.15 \times 10^{-8}$	$1.19 \pm 0.12 \times 10^{-8}$

4. Discussion

4.1. Carbonation of Wollastonite

The time evolution of the aqueous Ca and Si concentrations are fundamental parameter to understand the chemistry of our mineral–solution system. In the case of calcium, it was determined by the relative rates of Ca supply via wollastonite dissolution and Ca consumption via calcite precipitation. Figure 10a shows calculations using PHREEQC, which demonstrate that in our system, the Ca concentration necessary to achieve equilibrium with respect to wollastonite (black line) was more than three orders of magnitude larger than that at equilibrium with calcite in a 0.1 M NaHCO$_3$ solution at 40 °C (red line). This is reflected in the calculated values of $SI_{wollastnite}$ (saturation index, $SI = \lg(IAP/K_{sp})$, where IAP is the ion activity product and K_{sp} is the solubility product of a particular phase), which are below 0 for the whole duration of the experiment (Figure 10a,b). This is also reflected by the affinity values (i.e., free energy of reaction, $\Delta G = -R \cdot T \cdot \ln(IAP/K_{sp})$, where R is the gas constant and $T = 313$ K) presented in Table 2. Such a degree of undersaturation (i.e., very high negative ΔG values) favored wollastonite dissolution throughout the whole time-span of the carbonation experiment and favored the precipitation of calcite. The competition between silicate dissolution and carbonate precipitation reactions implies that the actual time-evolution of Ca concentration depends on the relative rates of these two processes. During the initial stages of the reaction the most reactive sites of wollastonite crystals (point defects, kinks, and step edges) would significantly contribute to the initial high rates of calcium and silicon release [42]. However, after 8 h, the system was already supersaturated with respect to calcite in the presence ($SI_{calcite} = 0.59$; $\Delta G = 3.3 \pm 1.3$ kJ·mol^{-1}) and in the absence of CA ($SI_{calcite}$ = 0.03; ΔG = 0.1 ± 0.7 kJ·mol^{-1}). The fact that in the CA-bearing system the $SI_{calcite}$ (and ΔG) was significantly higher at an earlier point in time than in the control (compare Figure 10a,c; see also Table 2) demonstrates that CA acted as a catalyst for calcite precipitation. Massive calcite precipitation, especially after 48 h, led to the observed Ca depletion in the solution (Figure 10a) and the reduction in the supersaturation with respect to calcite (Figure 10c,d). This behavior demonstrates that the rate-controlling step in the overall carbonation reaction after calcite starts forming is the dissolution of the silicate. This is consistent with previous studies on the carbonation of wollastonite [27]. Nonetheless, it should be pointed out that the pool of aqueous Ca was not fully consumed during the experiment. Indeed, the solution remained supersaturated with respect to calcite, irrespectively of the presence/absence of CA. This means that equilibrium with respect to wollastonite is unachievable under these conditions (after two weeks, $SI_{wollastonite}$ = −2.96 and ΔG = −17.8 ± 1.2 kJ·mol^{-1} for the control and $SI_{wollastonite}$ = −2.80 and ΔG = −16.9 ± 0.3 kJ·mol^{-1} in presence of CA). As a result, wollastonite continued to dissolve throughout the whole time-span of the experiment, although at a lower rate than during the early stages. It could be argued that this is due to a reduction in the SI (and ΔG) associated with the observed increase in pH taking place over the course of the experiments. However, as indicated above, the precipitation of amorphous silica and calcite kept the system undersaturated with respect to wollastonite at a very high value, consistent with the above reported SI and ΔG values. This should enable wollastonite dissolution and carbonation to progress at higher rates than those observed. A partial passivation of the silicate surface due to both the secondary precipitation of an amorphous silica layer or SAL, especially on the most reactive sites (defects, steps, and cleavage zones) of the most reactive (*hkl*) faces of wollastonite [42,77], and the heterogeneous precipitation of calcite on the silicate surface and/or SALs [26,27] can explain the observed reduction in dissolution and carbonation rates. A passivation effect can also explain why under very similar SI and ΔG conditions, the carbonation rate (and carbonate yield) is smaller in the case of the CA run as compared with the control run. Note that although it has been reported that SALs developed on wollastonite are not necessarily "passivating" layers [27,66], over time they can undergo ageing by condensation/repolymerization [69], forming a more dense and less porous layer that may actually act as a diffusional barrier [29,49], especially if secondary precipitates such as Ca-phyllosilicates and Ca-carbonates form within the SALs [26,27]. Recent results by Daval et al. [77] show that while on some faces of wollastonite (e.g., (010)) SALs do not act as diffusional barriers, on other faces (e.g., (001)) they can actually form passivating layers because in this later case SALs densify with time, leading to an overall (although minor) reduction in the dissolution rates on such specific faces, typically the less reactive ones showing the slowest initial dissolution rate.

Table 2. Time evolution of free energy (ΔG) values (kJ·mol^{-1}) for the dissolution of wollastonite and the precipitation of calcite and amorphous silica in the absence and presence of CA.

Time	$\Delta G_{\text{dissolution wollastonite}}$		$\Delta G_{\text{precipitation calcite}}$		$\Delta G_{\text{precipitation Sio}_2 \text{ nH}_2\text{O}}$	
	No CA	CA	No CA	CA	No CA	CA
8 h	−30.7 ± 1.1	−25.1 ± 1.1	0.1 ± 0.7	3.3 ± 1.3	−3.9 ± 0.3	−5.0 ± 0.3
24 h	−23.9 ± 0.7	−22.4 ± 0.1	2.7 ± 0.9	2.2 ± 0.1	−1.4 ± 0.4	−1.9 ± 0.1
48 h	−23.1 ± 0.5	−21.0 ± 0.3	1.7 ± 0.5	2.0 ± 0.5	−0.4 ± 0.1	−0.9 ± 0.2
336 h	−17.8 ± 1.2	−16.9 ± 0.3	2.1 ± 0.5	2.7 ± 0.7	1.2 ± 0.1	1.0 ± 0.1

In our system a steady-state "equilibrium" between wollastonite dissolution and calcite precipitation was achieved. At this steady state, the precipitation of calcite and its growth continued, but at a relatively slower rate due to the lower supersaturation of the system ($SI_{calcite}$ = 0.36 for the control and $SI_{calcite}$ = 0.46 in the presence of CA). Note, that it has been experimentally shown that there is a positive relationship between $SI_{calcite}$ and calcite growth rates [83–85]. This means that during the latter stages of the carbonation process, and after the rapid initial precipitation and growth of calcite at a relatively higher $SI_{calcite}$ (i.e., maximum $SI_{calcite}$ values of 0.48 and 0.59 for the control and CA systems) further growth of this carbonate will be limited by the supply of Ca released after the dissolution of wollastonite, whose complete passivation is not achieved during the experimental time-span.

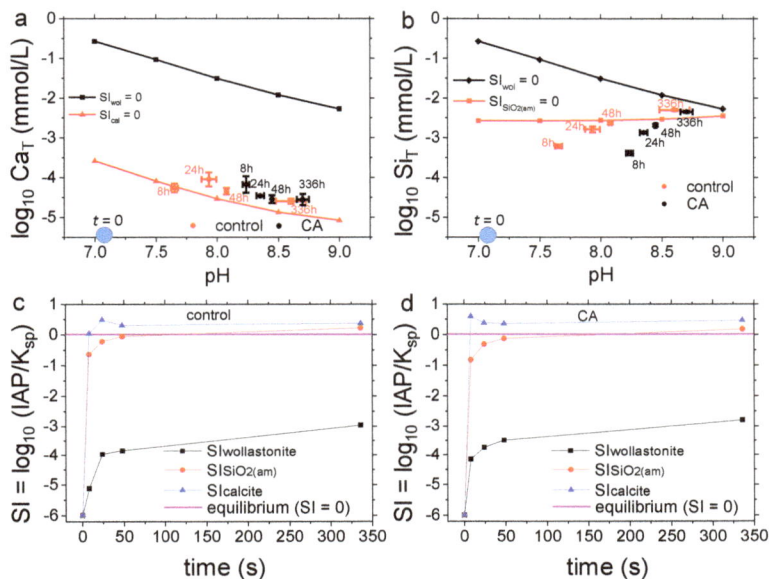

Figure 10. Geochemical evolution of the control and CA-inclusive carbonation systems. Variation of Ca (**a**) and Si (**b**) concentration vs. pH over the time course of the experiments with respect to the equilibrium solubility curves of wollastonite, calcite, and amorphous silica (continuous lines calculated using PHREEQC); time evolution of the saturation index (SI) for the control (**c**) and CA-inclusive (**d**) systems. The blue circles in (**a,b**) mark the pH of the starting solution ($t = 0$).

The system was undersaturated with respect to amorphous silica during the early stages of the carbonation experiment (up to 48 h), being supersaturated after 14 days (Figure 10b–d; and Table 2). However, the formation of SALs at specific areas of the reacted wollastonite was already observed after 8 h from the start of the experiment. This is fully consistent with previous studies showing that amorphous silica can form SALs on dissolving wollastonite in systems where the bulk fluid is undersaturated with respect to this phase [42]. This led some researchers to propose that the formation of such SALs occurred by diffusion of Ca ions out of the wollastonite structure via an exchange with H^+ coming from the solution, leaving behind a Ca-depleted amorphous silica layer, the so-called "leached layer", leading to an apparent incongruent (non-stoichiometric) dissolution of this silicate [69,79]. However, it has been experimentally shown that dissolution of wollastonite over a range of pH conditions is in fact congruent and takes place via an interface-coupled dissolution–precipitation mechanism [42,43]. Dissolution occurs via etch pit formation and spreading, resulting in the actual stoichiometric release of Ca and Si. Such dissolution pits have been detected in our FESEM analysis (Figures 5–8), spread during the course of the experiments and were blanketed by a layer (SAL) of silica nanoparticles deposited on their stepped faces. Precipitation of silica nanoparticles forming SALs on wollastonite, as shown here, can take place at the mineral–fluid interface were strong compositional gradients have been measured [43]. Such strong compositional gradients can lead to local supersaturation with respect to amorphous silica and its precipitation forming SALs, despite the fact that the bulk fluid is undersaturated with respect to this phase [42,43]. This is fully consistent with our experimental results and observations. Remarkably, such a compositional gradient also exists for the case of released Ca, so it was not unexpected that calcite precipitation also occurred preferentially on the areas where a more active dissolution is taking place, and where SALs are preferentially formed, as observed here. It follows that, as occurs in the case of SALs [42,43], calcite precipitation after wollastonite dissolution occurs via an interface-coupled dissolution–precipitation mechanism.

Importantly, thicker SALs can form in batch experiments and under acidic conditions, compared to those formed in flow-through experiments and under circum-neutral conditions [10,26,27]. Thus, it seems that parameters of the solution such as the composition or flow regime could ultimately dictate the formation or not of silica altered layers. While Schott et al. [79] reported that SALs do not form on wollastonite in mixed-flow reactors at 25 °C for pH > 4 (at a very high undersaturation with respect to amorphous silica), in our specific case, working at circumneutral conditions and using batch reactors, thick SALs formed after 14 days at 40 °C. In parallel, we observed a continuous reduction in the Si release rate, which demonstrates that a passivation effect was taking place. As indicated above, this effect cannot be only ascribed to the formation of SALs but also to the formation of thick carbonate coatings, specifically on the most reactive sites/surfaces of wollastonite. Our results are thus consistent with those of Daval et al. [27] who observed that passivation was due to the formation of SAL plus carbonate coatings. These authors indicated that the sole formation of SALs was not sufficient to induce significant passivation.

Remarkably, the progress of dissolution, degree of SALs development, and amount of heterogeneous carbonate precipitation on wollastonite were clearly face- and direction-specific. This is fully consistent with the fact that wollastonite has an anisotropic structure (Figure 11).

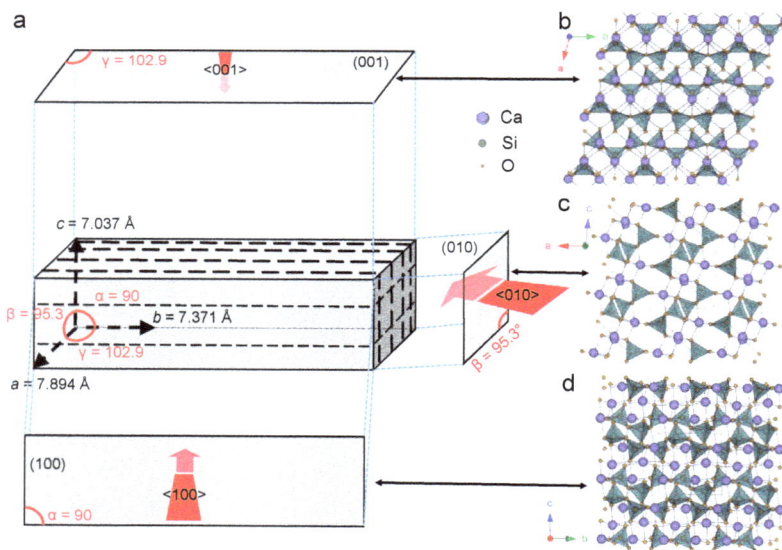

Figure 11. Crystallographic features and structure of wollastonite. (**a**) prism of wollastonite showing the main (*hkl*) faces and cleavage planes (dashed black lines). The (010), (100), and (001) planes are separately outlined with red arrows indicating main <*hkl*> directions for the progress of dissolution. Their size qualitatively represents the rate of dissolution along these directions; Structure of wollastonite projected along [001] (**b**), [010] (**c**), and [100] (**d**).

It has long been known that the anisotropic structure of many primary silicates strongly affects their dissolution rate along different crystallographic directions [86,87]. Acidic dissolution of silicates takes place via H-promoted attack to Si–O–Si bonds linking silica tetrahedra and M–O bonds of metal cations in octahedral coordination directly facing the solution. Protons adsorbed on the mineral surface can polarize the metal–oxygen bonds, thereby weakening the bonding with the underlying lattice [10]. As a result, metal-oxygen bonds are hydrolyzed, being the breaking of the strong Si–O bonds the rate controlling step for the dissolution of multicomponent silicate minerals [10].

Hydrolysis (i.e., proton-promoted Si–O bond breaking) of chain silicates would preferentially occur at the ends of tetrahedral chains where Q1 tetrahedra (i.e., SiO_4 tetrahedra sharing one oxygen) and associated metal octahedra directly face the solution. In the case of wollastonite, such structural units are concentrated on the {010} faces (Figure 11c). In these faces, hydrolysis of one single oxygen bond linking two terminal tetrahedra would lead to the release of one Si (i.e., the terminal Q1 tetrahedron). In contrast, in the case of prismatic faces (Figure 11b,d), at least two oxygen bonds linking two Q2 tetrahedra (i.e., tetrahedra sharing two oxygens) have to be broken to release one Si tetrahedron, leaving behind two Q1 tetrahedra for dissolution to progress in opposite directions along the (now broken) tetrahedral chain. This explains why the {010} faces of wollastonite show the highest dissolution rate of all faces in this mineral [77]. This also helps explaining why dissolution progresses faster along <010> in wollastonite, as shown here by the elongated shape of dissolution pits formed on {100}, or along <001> in chain silicates such as diopside [88]. The prismatic faces of wollastonite reportedly show a lower dissolution rate as compared with the basal {010} faces [77]. Of the two main prismatic faces, {100} faces dissolve faster than {001} faces [77]. The higher reactivity of {100} faces as compared with that of {001} ones is consistent with the presence of a higher density of Q2 tetrahedra with Si–O–Si bonds directly facing the solution (compare Figure 11c,d). Those can be preferentially attacked by protons, leading to their hydrolysis. Daval et al. [77] suggested that an additional effect that can help explaining the faster dissolution along [010] and [100] is the higher density of outcropping dislocations on (010) and (100) faces, which would facilitate the progress of dissolution via a classic step-wave mechanism [89]. However, under far-from-equilibrium conditions such as those taking place in most reported wollastonite dissolution and carbonation experiments, dissolution rates would be controlled by etch pits nucleated at point defects, not by dislocations [90]. Indeed, our previous AFM studies of wollastonite dissolution in acidic solutions showed that dissolution of {100} faces progressed via the nucleation of abundant shallow etch pits and the subsequent retreat of their step edges (not by unwinding of steps at emerging dislocations) [42]. It follows that the most likely and straightforward explanation for the anisotropic dissolution of wollastonite is the distinctive structural differences along different [*hkl*] directions. Overall, the structural anisotropy of wollastonite is reflected by the highly anisotropic dissolution of this chain-silicate. This is a prominent feature of this mineral that has been previously underlined [42,43,77]. Because in the case of wollastonite, as well as several other silicates—such as olivine, pyroxenes, amphiboles, and feldspars where dissolution results in the formation of SALs—the dissolution step is tightly coupled at the mineral–fluid interface with the precipitation of a secondary amorphous silica phase (i.e., the nanoparticles observed here) and not to a diffusional leaching of Ca, the thickness and density of the SALs is directly related to the reactivity (dissolution rate) of the faces on which SALs develop, as shown by Daval et al. [77]. The authors observed that the dissolution rate $R_{(hkl)}$ of the main faces of wollastonite followed the trend $R_{(010)} > R_{(100)} > R_{(001)}$, as did the thickness of SALs. The relation between the structural features of silicates and their anisotropic dissolution and SALs development appears to be general, as also demonstrated by the anisotropic dissolution of olivine [46], diopside [88], and K-feldspar [91]. Overall, these results are fully consistent with our observations of the anisotropic progress of dissolution and SALs formation during wollastonite carbonation.

More significantly, carbonate precipitation on wollastonite surfaces was also anisotropic. Calcite crystals were concentrated primarily on {010} and {100} wollastonite faces, forming thick coatings. This appears to be directly related to the higher rate of dissolution of these specific faces, which can lead to a faster release of Ca and the subsequent build-up of a high supersaturation with respect to calcite in the solution at the mineral–fluid interface. This can result in the coupling of both the formation of SALs and the subsequent precipitation of calcite via an interface-coupled dissolution–precipitation mechanism. This is consistent with results reported by Daval et al. [26,27], showing that at circumneutral conditions calcite tends to form on and within SALs developed on wollastonite subjected to carbonation, thereby acting as a passivating coating. In our case, the formation of SALs plus calcite passivating coatings on the most reactive {010} and {100} faces helps to explain the observed drastic reduction in the Si release rates and the very limited carbonate yield.

Regarding the effect of CA, our results show that its main role during carbonation of wollastonite is the acceleration of calcium carbonate precipitation. CA also seems to affect the morphology of the precipitates due to its adsorption on specific calcite (*hkl*) faces. The latter results in rhombohedral calcite crystals with stepped, rounded edges and an overall nanogranular surface structure. This is likely due to adsorption of this protein via interaction of its deprotonated carboxylic groups (at the alkaline pH conditions reached during carbonate precipitation) on terraces and growth steps of calcite. CA can thus induce step pinning, poisoning crystal growth along specific [*hkl*] directions, and thereby altering the overall morphology of calcite crystals [85]. Moreover, the nanogranular surface structure is consistent with non-classical aggregation-based growth mechanisms (possibly via an amorphous calcium carbonate precursor) promoted by the presence of this organic additive [84,85]. In contrast, we observed no silicase-like catalytic effect of CA on wollastonite dissolution, which was the rate-limiting step of the overall carbonation process. Actually, the overall CA effect was to limit the extent of carbonation. This appears to be due to an enhanced passivation effect associated with the very fast, early precipitation of calcite on the most reactive sites and faces of wollastonite. Passivation was more effective than in the case of the control because: (i) CA induced the formation of calcite coatings on specific faces of wollastonite such as the reactive {100} faces, as shown by our SEM observations. This is likely due to a template effect exerted by adsorbed CA on such specific faces that favors the heterogeneous and oriented nucleation (possibly epitaxial) of calcite forming a continuous surface layer, as described for the case of other minerals such as sulfates growing epitaxially on calcite following the formation of an organic template on the calcite substrate [92], and (ii) the higher supersaturation reached in the presence of CA at the early stages of carbonate formation, compared to the control, leading to a higher nucleation density. Such a higher density of calcite crystals nucleated on specific faces of wollastonite (i.e., the most reactive ones) favored the development of a passivating layer. As discussed above, the precipitation of abundant calcite along with SALs on wollastonite has a strong passivation effect [26,27].

4.2. Dissolution Experiments

The dissolution of wollastonite ($CaSiO_3$) could be described by Equations (2) and (3),

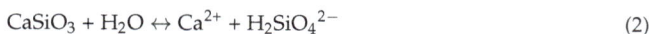

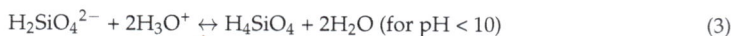

$$CaSiO_3 + H_2O \leftrightarrow Ca^{2+} + H_2SiO_4^{2-} \tag{2}$$

$$H_2SiO_4^{2-} + 2H_3O^+ \leftrightarrow H_4SiO_4 + 2H_2O \text{ (for pH < 10)} \tag{3}$$

these reactions are strongly pH-dependent, as shown by Figure 12a. Figure 12a shows a PHREEQC simulation of the amount of wollastonite dissolved at different pH values after 10 mol of wollastonite were added in 1 L of water. It is shown that at pH > 9 almost no wollastonite is dissolved. Conversely, at pH < 9 a significant fraction of the dosed wollastonite is dissolved. For instance, at pH 4 about 50% of dosed wollastonite is dissolved. During dissolution, the release of SiO_3^{2-}, that instantaneously hydrates in water to form $H_2SiO_4^{2-}$, corresponds to the addition of a strong base to the solution (i.e., Equation (2)). The resulting pH increase would hinder further dissolution (Figure 12a). However, in our dissolution experiments, the increase in pH induced by the dissolution of $CaSiO_3$ was compensated by the addition of HCl to keep a constant pH of 4. This enabled dissolution at a constant rate, thereby avoiding the inhibition effect induced by the increase in pH. On the one hand, by measuring the free Ca^{2+} concentration using the ISE electrode, we could measure the Ca release rate and the (Ca-based) dissolution rate of wollastonite (R_{Ca}). Note that at pH 4, the total concentration of calcium can be reasonably approximated by the free ion concentration [Ca^{2+}] recorded with the probe due to the weakness of the ion pairing constants for Ca^{2+} in acidic media (Figure 12b). On the other hand, the amount (mol) of HCl added during the experiments corresponded to half of the SiO_3^{2-} dissolved from the crystals because, at the experimental pH, the most stable species of silicate ion is the orthosilicic acid, H_4SiO_4 (Figure 12c) that forms according to Equation (3). Overall, two protons per mole of Si released to the solution were consumed. From the dosed amount of HCl, the dissolution rate of wollastonite $R_{H_2SiO_4^{2-}}$ was thus also determined.

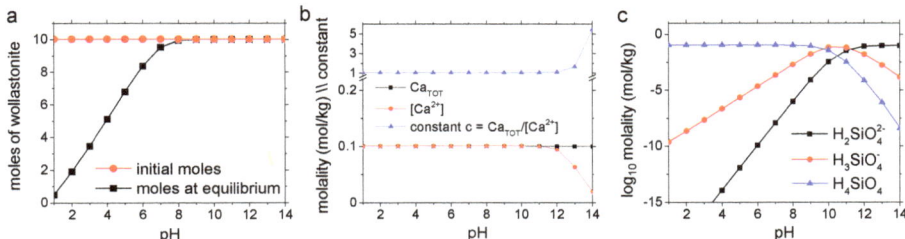

Figure 12. Results of PHREEQC modeling of: (**a**) Extent of $CaSiO_3$ dissolution vs. pH for an initial 10 mol of wollastonite dosed to 1 L H_2O. The moles at equilibrium with this aqueous solution strongly depend on pH and acidic conditions are required to reach significant dissolution; (**b**) the variation of the total (Ca_{TOT}) and free Ca (i.e., $[Ca^{2+}]$) in solution, as well as their ratio (the constant denoted, c), with pH. Note that for pH < 11, the ratio is constant and equal to 1; (**c**) the speciation of Si in water as a function of pH, which is useful to realize that the release of $H_2SiO_4^{2-}$ induces a strong pH shift toward more basic values and the fact that at pH 4 the most abundant species is the orthosilicic acid.

We have chosen to represent the dissolution of wollastonite according to Equations (2) and (3) for the sake of clarity and to underline that proton consumption is associated with the release of Si species. Note, however, that from a mechanistic point of view, Equation (2) might imply that dissolution of wollastonite is not proton-promoted, but rather occurs via direct nucleophilic attack by water followed by proton consumption according to Equation (3) once a Si species is released. This is not what Equations (2) and (3) are intended to represent, because both should be coupled into one single event in order to enable the proton-promoted stoichiometric dissolution of wollastonite, as we have shown to occur previously [42,43]. A more accurate description of the overall proton-promoted dissolution of wollastonite integrating Equations (2) and (3) would thus be the following equation,

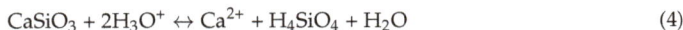

$$CaSiO_3 + 2H_3O^+ \leftrightarrow Ca^{2+} + H_4SiO_4 + H_2O \tag{4}$$

regardless whether we considered the coupled Equations (2) and (3) or the overall Equation (4) to describe wollastonite dissolution, we want to stress that proton consumption is related to the release of highly basic Si species and their neutralization to form H_4SiO_4.

It is worth noting that the process of silicate carbonation relies on the combination of a H^+ consuming process (silicate dissolution) and a H^+ releasing process (carbonate precipitation). This coupling prevents both reactions from self-inhibition. The latter explains why in our carbonation experiments equilibrium was never reached, despite the observed reduction in the rates of wollastonite dissolution and carbonate precipitation.

The dissolution rates R_{Ca} ($1.28 \pm 0.14 \times 10^{-8}$ $mol·m^{-2}·s^{-1}$) and $R_{H_2SiO_4^{2-}}$ ($1.19 \pm 0.11 \times 10^{-8}$ $mol·m^{-2}·s^{-1}$) of wollastonite at pH 4 obtained using our experimental set up are in good agreement with the R_{Ca} values of ~0.9–3.2 $\times 10^{-8}$ $mol·m^{-2}·s^{-1}$ (based on Ca release rate and normalized to the initial surface area) reported by Weissbart and Rimstidt [72] and Schott et al. [79] for experiments conducted at pH 4 and 25 °C, as well as the R_{Ca} value of ~0.5 $\times 10^{-8}$ $mol·m^{-2}·s^{-1}$ reported by Ruiz-Agudo et al. [42] (R_{Ca} value interpolated here to pH 4 from the results corresponding to the pH range 1.5–5.9). This gives us confidence about the reliability of our method to test silicate dissolution kinetics. Our values are, however, about one order of magnitude lower than those reported by Daval et al. [77] for (100) and (010) faces dissolved in batch reactors at pH 4 and 22 °C, but they approach the values reported by the authors for the case of (001) face (i.e., the less reactive face of wollastonite, as also shown here). As pointed out by Daval et al. [77], our lower rates and the lower rates reported in the literature could be simply due to the preponderance of the {001} faces in powdered wollastonite. Hence, R_{Ca} reported for powder samples would tend to approach the R_{Ca} of the (001) face.

The very good agreement between the dissolution rates measured using free Ca and dosed HCl shows that the dissolution of wollastonite under our experimental conditions is stoichiometric, despite the fact that SALs reportedly form during wollastonite dissolution at pH 4 [77]. Note that amorphous silica precipitation from dissolved H_4SiO_4, does not affect pH; therefore, its precipitation and the formation of SALs would not affect the dosed HCl. The observation of stoichiometric wollastonite dissolution is fully consistent with our previous studies demonstrating that the acid-promoted dissolution of this silicate is congruent and therefore stoichiometric [42,43]. The apparent incongruent dissolution of wollastonite, resulting in the reported non-stoichiometric release of Ca and Si [42,72,77], is due to the above-mentioned SAL formation via an interface-coupled dissolution–precipitation mechanism, leading to the reduction in the amount of Si in solution compared with Ca. This is a side-effect due to amorphous silica precipitation, not a consequence of a hypothetical incongruent and non-stoichiometric dissolution process. We are, however, aware of the fact that dissolution of wollastonite via the ion interdiffusion mechanism described above [69], would also lead to the same results using our experimental dissolution protocol. Nonetheless, an interdiffusion mechanism would not explain all the textural features of wollastonite dissolution and SALs formation shown here and elsewhere [42,43,77], such as the formation of dissolution pits, the advancement of dissolution along specific [*hkl*] direction (at different rates), or the formation of amorphous silica nanogranular precipitates and their coalescence to form SALs.

Because no carbonate precipitation took place during the dissolution experiments, the dissolution rate was nearly constant over the time spam of the experiment, despite our previous observations that at such pH, SALs formed [42]. These results demonstrate that under our experimental conditions SALs did not act as effective passivating layers, although they seem to play a role in the slight reduction of the dissolution rate observed at the very early stages of the dissolution experiments. Note, however, that under different experimental conditions SALs formed on wollastonite during acid dissolution can act as a passivating layer [42]. Similarly, SALs formed during dissolution of other silicates such as olivine, reportedly act as passivating layers [44,46,88]. It follows that the passivating (or not passivating) effect of SALs is experimental and mineral-specific.

Our dissolution experiments disclosed that CA has no catalytic "silicase-like" effect on the dissolution of wollastonite. Actually, CA slightly reduced the dissolution rate of this silicate. This could be explained considering that the backbone of this protein includes numerous carboxylic groups that may likely favor its adsorption (for instance, via H-bonding) to the surface of wollastonite, as indicated above for the case of calcite precipitation, thereby acting as a passivating organic surface coating. However, adsorption is likely not a general phenomenon due to non-favorable (negative) charging of the protein functional groups and the wollastonite surface [79] at the experimental acid pH. This may explain why the inhibitory effect is so reduced. In any case, it is surprising to find out that despite the compositional and structural similarities between CA and silicase, the former does not enhance silicate dissolution whereas the later does [50]. It seems that there are some subtle structural differences between these two enzymes that determine their specific catalytic activity. Future studies should focus on disclosing such structural differences, which may help to explain why silicase can catalyze Si–O bond breaking and CA cannot.

Regarding the dissolution-enhancement observed in the case of UiO-66 MOF, it is likely that the Lewis acidic Zr and basic hydroxide sites present on the surface of this compound [60] directly interacted with the surface of wollastonite favoring the hydrolysis of Si–O–Si bonds, thereby increasing the dissolution rate of the silicate by ~10%. This increase is, however, modest, most probably because the crystalline (insoluble) MOF particles used here, with a relatively large size (intergrown cubic crystals with size ~1–2 µm) [60], were not efficiently interacting with the surfaces of wollastonite crystals. We are currently working on ways to bypass this handicap, preparing nanosized MOFs, with higher amounts of metal nanoclusters, to foster reactivity.

Finally, we have to state that these results demonstrate that the experimental protocol presented here for the determination of dissolution rates of silicates and used to screen the effect of natural and biomimetic catalysts on the dissolution of wollastonite is simple and highly reliable, and could be used as a routine protocol for the screening of other (bio)catalysts.

5. Conclusions

The present study reports two types of experimental studies aimed at observing the possible catalytic effect of CA and Zr-based MOFs toward the reaction of replacement of a silicate mineral (wollastonite) into a carbonate (calcite). The first set of experiments involved the whole carbonation process. Wollastonite crystals were put in contact with concentrated bicarbonate solutions, in the presence and absence of CA. Our analysis of the evolution of the carbonation reaction shows that the dissolution step progresses anisotropically along the different faces of wollastonite exposed to the solution. Dissolution was coupled to the precipitation of amorphous silica nanoparticles which formed thick SALs on specific faces of the silicate, i.e., mainly on the {010} and {100} faces. Those faces were the most reactive ones. Similarly, precipitation of calcite on reacted wollastonite preferentially occurred on the areas where dissolution progressed at a faster rate. The thick SALs and carbonate coatings formed on the most reactive surfaces of wollastonite acted as passivating layers leading to a significant reduction in the dissolution and carbonation rates over the time-course of the experiment. Such a passivating effect helps explaining why the conversion of wollastonite into calcite was so limited (up to 14 mol %) under our experimental conditions. Our study also demonstrates that CA is not effective at catalyzing such a reaction probably because its main catalytic effect involves the hydration of carbon dioxide while the rate-determining step of the process is calcium release to solution by wollastonite dissolution. For this reason, we decided to search for more effective biomimetic catalysts to speed up the dissolution of the silicate mineral. To screen their effectiveness, we designed and validated a new test to study the dissolution rate of wollastonite crystals in water using an automatic titration system. The protocol was used to evaluate the catalytic effect exerted by three different Zr-based MOFs. The results were compared with the uncatalyzed system and with CA. Our experiments revealed that only UiO-66 MOF induced an increase of the wollastonite dissolution rate. It is important to note that under the experimental conditions tested ($T = 25\ °C$ and pH = 4), CA did not demonstrate any catalytic effect on wollastonite dissolution (whose rate decreased in the presence of this enzyme) an observation that could seem unexpected considering the biochemical similarity between CA and silicase. Further studies will be performed to test the catalytic effect of novel materials on silicate dissolution using the fast and reliable protocol designed and validated here. Finally, the study of the different stages that constitute the process of silicate replacement by carbonate offers the possibility to discover catalysts active on a specific reaction step and the possibility to use combination of catalysts with a synergic effect for a more effective CCS.

Author Contributions: E.R.-A., J.A.R.N., and C.R.-N. conceived and designed the experiments; F.D.L., C.R.-A., and A.I.-V. performed the experiments; E.R.A., F.D.L., and C.R.-N. analyzed the data; J.A.R.N. and R.G.-S.M. designed and synthesized MOFs; C.R.-N. and F.D.L. wrote the paper with contribution by all authors.

Acknowledgments: This research was funded by the Spanish Government (grants CGL2015-70642-R, CGL2015-73103-EXP, CTQ2017-84692-R), EU FEDER funding, the University of Granada ("Unidad Científica de Excelencia" UCE-PP2016-05) and the Junta de Andalucía (grant P11-RNM-7550 and Research Group RNM-179). We thank the personnel of the Centro de Instrumentación Científica (CIC) of the University of Granada for their help during TG-DSC, FESEM, µ-XRD, and ICP-OES analyses.

Conflicts of Interest: The authors declare no conflict of interest.

References

1. Falkowski, P.; Scholes, R.J.; Boyle, E.E.A.; Canadell, J.; Canfield, D.; Elser, J.; Gruber, N.; Hibbard, K.; Högberg, P.; Linder, S.; et al. The global carbon cycle: A test of our knowledge of Earth as a system. *Science* **2000**, *290*, 291–296. [CrossRef] [PubMed]
2. Lackner, K.S. A guide for CO_2 sequestration. *Science* **2003**, *300*, 1677–1678. [CrossRef] [PubMed]
3. IPCC. *Climate Change 2014. Synthesis Report: Contribution of Working Groups I, II and III to the Fifth Assessment Report of the Intergovernmental Panel on Climate Change*; Pachauri, R.K., Meyer, L.A., Eds.; IPCC: Geneva, Switzerland, 2014; 151p.
4. Oelkers, E.H.; Gislason, S.R.; Matter, J. Mineral carbonation of CO_2. *Elements* **2008**, *4*, 333–337. [CrossRef]
5. Cole, D.R.; Oelkers, E.H. Carbon dioxide sequestration. *Elements* **2008**, *4*, 287–362.
6. Sanna, A.; Uibu, M.; Caramanna, G.; Kuusik, R.; Maroto-Valer, M.M. A review of mineral carbonation technologies to sequester CO_2. *Chem. Soc. Rev.* **2014**, *43*, 8049–8080. [CrossRef] [PubMed]
7. Urey, H.C. *The Planets: Their Origin and Development*; Yale University Press: New Haven, CT, USA, 1952; 245p.
8. Siefritz, W. CO_2 disposal by means of silicates. *Nature* **1990**, *345*, 486. [CrossRef]
9. Lackner, K.S.; Wendt, C.H.; Butt, D.P.; Joyce, E.L.; Sharp, D.H. Carbon dioxide disposal in carbonate minerals. *Energy* **1995**, *20*, 1153–1170. [CrossRef]
10. Brantley, S.L. Kinetics of mineral dissolution. In *Kinetics of Water-Rock Interaction*; Brantley, S.L., Kubicki, J.D., White, A.F., Eds.; Springer: New York, NY, USA, 2008; pp. 151–210.
11. Walker, J.C.G.; Hays, P.B.; Kasting, J.F. A negative feedback mechanism for the long-term stabilization of Earth's surface temperature. *J. Geophys. Res.* **1981**, *86*, 9776–9782. [CrossRef]
12. Berner, R.A.; Lasaga, A.; Garrels, R.M. The carbonate-silicate geochemical cycle and its effect on atmospheric carbon dioxide over the past 100 million years. *Am. J. Sci.* **1983**, *283*, 641–683. [CrossRef]
13. Brady, P.V. The effect of silicate weathering on global temperature and atmospheric CO_2. *J. Geophys. Res. Solid Earth* **1991**, *96*, 18101–18106. [CrossRef]
14. Berner, R.A. The long-term carbon cycle, fossil fuels and atmospheric composition. *Nature* **2003**, *426*, 323–326. [CrossRef] [PubMed]
15. Houghton, R.A. Balancing the global carbon budget. *Annu. Rev. Earth Planet. Sci.* **2007**, *35*, 313–347. [CrossRef]
16. Power, I.M.; Harrison, A.L.; Dipple, G.M.; Wilson, S.A.; Kelemen, P.B.; Hitch, M.; Southam, G. Carbon mineralization: From natural analogues to engineered systems. *Rev. Miner. Geochem.* **2013**, *77*, 305–360. [CrossRef]
17. Azdarpour, A.; Asadullah, M.; Mohammadian, E.; Hamidi, H.; Junin, R.; Karaei, M.A. A review on carbon dioxide mineral carbonation through pH-swing process. *Chem. Eng. J.* **2015**, *279*, 615–630. [CrossRef]
18. Gerdemann, S.J.; O'Connor, W.K.; Dahlin, D.C.; Penner, L.R.; Rush, H. Ex situ aqueous mineral carbonation. *Environ. Sci. Technol.* **2007**, *41*, 2587–2593. [CrossRef] [PubMed]
19. Park, A.H.A.; Jadhav, R.; Fan, L.S. CO_2 mineral sequestration: Chemically enhanced aqueous carbonation of serpentine. *Can. J. Chem. Eng.* **2003**, *81*, 885–890. [CrossRef]
20. Park, A.A.; Fan, L. CO_2 mineral sequestration: Physically activated dissolution of serpentine and pH swing process. *Chem. Eng. Sci.* **2004**, *59*, 5241–5247. [CrossRef]
21. Krevor, S.C.; Lackner, K.S. Enhancing process kinetics for mineral carbon sequestration. *Energ. Procedia* **2009**, *1*, 4867–4871. [CrossRef]
22. Zhao, H.; Park, Y.; Lee, D.H.; Park, A.H.A. Tuning the dissolution kinetics of wollastonite via chelating agents for CO_2 sequestration with integrated synthesis of precipitated calcium carbonates. *Phys. Chem. Chem. Phys.* **2013**, *15*, 15185–15192. [CrossRef] [PubMed]
23. Ghoorah, M.; Dlugogorski, B.Z.; Balucan, R.D.; Kennedy, E.M. Selection of acid for weak acid processing of wollastonite for mineralisation of CO_2. *Fuel* **2014**, *122*, 277–286. [CrossRef]
24. Matter, J.M.; Stute, M.; Snæbjörnsdottir, S.Ó.; Oelkers, E.H.; Gislason, S.R.; Aradottir, E.S.; Sigfusson, B.; Gunnarsson, I.; Sigurdardottir, H.; Gunnlaugsson, E.; et al. Rapid carbon mineralization for permanent disposal of anthropogenic carbon dioxide emissions. *Science* **2016**, *352*, 1312–1314. [CrossRef] [PubMed]
25. Gislason, S.R.; Oelkers, E.H. Carbon storage in basalt. *Science* **2014**, *344*, 373–374. [CrossRef] [PubMed]

26. Daval, D.; Martinez, I.; Guigner, J.M.; Hellmann, R.; Corvisier, J.; Findling, N.; Dominici, C.; Goffé, B.; Guyot, F. Mechanism of wollastonite carbonation deduced from micro-to nanometer length scale observations. *Am. Miner.* **2009**, *94*, 1707–1726. [CrossRef]

27. Daval, D.; Martinez, I.; Corvisier, J.; Findling, N.; Goffé, B.; Guyot, F. Carbonation of Ca-bearing silicates, the case of wollastonite: Experimental investigations and kinetic modeling. *Chem. Geol.* **2009**, *265*, 63–78. [CrossRef]

28. Ruiz-Agudo, E.; Putnis, C.V.; Di Lorenzo, F.; Ruiz-Agudo, C.; Rodriguez-Navarro, C. Interfacial and surface controls on wollastonite carbonation: Insights from texture observations and geochemical modeling. *Am. J. Sci.* under review.

29. Huijgen, W.J.; Witkamp, G.J.; Comans, R.N. Mechanisms of aqueous wollastonite carbonation as a possible CO_2 sequestration process. *Chem. Eng. Sci.* **2006**, *61*, 4242–4251. [CrossRef]

30. Dreybrodt, W.; Eisenlohr, L.; Madry, B.; Ringer, S. Precipitation kinetics of calcite in the system $CaCO_3$-H_2O-CO_2: The conversion to CO_2 by the slow process $H^+ + HCO_3^- \rightarrow CO_2 + H_2O$ as a rate limiting step. *Geochim. Cosmochim. Acta* **1997**, *61*, 3897–3904. [CrossRef]

31. Cizer, Ö.; Ruiz-Agudo, E.; Rodriguez-Navarro, C. Kinetic effect of carbonic anhydrase enzyme on the carbonation reaction of lime mortar. *Int. J. Archit. Heritage* **2018**. [CrossRef]

32. Bond, G.M.; Stringer, J.; Brandvold, D.K.; Simsek, F.A.; Medina, M.G.; Egeland, G. Development of integrated system for biomimetic CO_2 sequestration using the enzyme carbonic anhydrase. *Energy Fuels* **2001**, *15*, 309–316. [CrossRef]

33. Vinoba, M.; Bhagiyalakshmi, M.; Jeong, S.K.; Nam, S.C.; Yoon, Y. Carbonic anhydrase immobilized on encapsulated magnetic nanoparticles for CO_2 sequestration. *Chem. Eur. J.* **2012**, *18*, 12028–12034. [CrossRef] [PubMed]

34. Power, I.M.; Harrison, A.L.; Dipple, G.M. Accelerating mineral carbonation using carbonic anhydrase. *Environ. Sci. Technol.* **2016**, *50*, 2610–2618. [CrossRef] [PubMed]

35. Silverman, D.N.; Lindskog, S. The catalytic mechanism of carbonic anhydrase: Implications of a rate-limiting protolysis of water. *Acc. Chem. Res.* **1988**, *21*, 30–36. [CrossRef]

36. Frost, S.C.; McKenna, R. (Eds.) *Carbonic Anhydrase: Mechanism, Regulation, Links to Disease and Industrial Applications*; Springer: London, UK, 2014; Volume 75, 429p.

37. Miyamoto, H.; Miyashita, T.; Okushima, M.; Nakano, S.; Morita, T.; Matsushiro, A. A carbonic anhydrase from the nacreous layer in oyster pearls. *Proc. Natl. Acad. Sci. USA* **1996**, *93*, 9657–9660. [CrossRef] [PubMed]

38. Pierre, A.C. Enzymatic carbon dioxide capture. *ISRN Chem. Eng.* **2012**, *2012*, 1–22. [CrossRef]

39. Molva, M.; Kilic, S.; Ozdemir, E. Effect of carbonic anhydrase on $CaCO_3$ crystallization in alkaline solution. *Energy Fuels* **2016**, *30*, 10686–10695. [CrossRef]

40. Patel, T.N.; Park, A.H.A.; Banta, S. Periplasmic expression of carbonic anhydrase in Escherichia coli: A new biocatalyst for CO_2 hydration. *Biotechnol. Bioeng.* **2013**, *110*, 1865–1873. [CrossRef] [PubMed]

41. Gadikota, G.; Park, A.H.A. Accelerated carbonation of Ca-and Mg-bearing minerals and industrial wastes using CO_2. In *Carbon Dioxide Utilisation*; Styring, P., Quadrelli, E.A., Armstrong, K., Eds.; Elsevier: Amsterdam, The Netherlands, 2015; pp. 15–137.

42. Ruiz-Agudo, E.; Putnis, C.V.; Rodriguez-Navarro, C.; Putnis, A. The mechanism of leached layer formation during chemical weathering of silicate minerals. *Geology* **2012**, *40*, 947–950. [CrossRef]

43. Ruiz-Agudo, E.; King, H.E.; Patiño-López, L.D.; Putnis, C.V.; Geisler, T.; Rodriguez-Navarro, C.; Putnis, A. Control of silicate weathering by interface-coupled dissolution–precipitation processes at the mineral-solution interface. *Geology* **2016**, *44*, 567–570. [CrossRef]

44. Béarat, H.; McKelvy, M.J.; Chizmeshya, A.V.; Gormley, D.; Nunez, R.; Carpenter, R.W.; Squires, K.; Wolf, G.H. Carbon sequestration via aqueous olivine mineral carbonation: Role of passivating layer formation. *Environ. Sci. Technol.* **2006**, *40*, 4802–4808. [CrossRef] [PubMed]

45. Whitfield, P.S.; Mitchell, L.D. In situ laboratory X-ray powder diffraction study of wollastonite carbonation using a high-pressure stage. *Appl. Geochem.* **2009**, *24*, 1635–1639. [CrossRef]

46. King, H.E.; Plümper, O.; Putnis, A. Effect of secondary phase formation on the carbonation of olivine. *Environ. Sci. Technol.* **2010**, *44*, 6503–6509. [CrossRef] [PubMed]

47. Daval, D.; Sissmann, O.; Menguy, N.; Saldi, G.D.; Guyot, F.; Martinez, I.; Corvisier, J.; García, B.; Machouk, I.; Knauss, K.G.; Hellmann, R. Influence of amorphous silica layer formation on the dissolution rate of olivine at 90 °C and elevated pCO2. *Chem. Geol.* **2011**, *284*, 193–209. [CrossRef]

48. Miller, Q.R.S.; Thompson, C.J.; Loring, J.S.; Windisch, C.F.; Bowden, M.E.; Hoyt, D.W.; Hu, J.Z.; Arey, B.W.; Rosso, K.M.; Schaef, H.T. Insights into silicate carbonation processes in water-bearing supercritical CO_2 fluids. *Int. J. Greenhouse Gas Control* **2013**, *15*, 104–118. [CrossRef]

49. Min, Y.; Li, Q.; Voltolini, M.; Kneafsey, T.; Jun, Y.S. Wollastonite carbonation in water-bearing supercritical CO_2: Effects of particle size. *Environ. Sci. Technol.* **2017**, *51*, 13044–13053. [CrossRef] [PubMed]

50. Schröder, H.C.; Krasko, A.; Le Pennec, G.; Adell, T.; Wiens, M.; Hassanein, H.; Müller, I.M.; Müller, W.E.G. *Silicon Biomineralization*; Springer: Berlin, Germany, 2003; 341p.

51. Ehrlich, H.; Demadis, K.D.; Pokrovsky, O.S.; Koutsoukos, P.G. Modern views on desilicification: Biosilica and abiotic silica dissolution in natural and artificial environments. *Chem. Rev.* **2010**, *110*, 4656–4689. [CrossRef] [PubMed]

52. Xiao, L.; Lian, B.; Hao, J.; Liu, C.; Wang, S. Effect of carbonic anhydrase on silicate weathering and carbonate formation at present day CO_2 concentrations compared to primordial values. *Sci. Rep.* **2015**, *5*, 7733. [CrossRef] [PubMed]

53. Nakata, K.; Shimomura, N.; Shiina, N.; Izumi, M.; Ichikawa, K.; Shiro, M. Kinetic study of catalytic CO_2 hydration by water-soluble model compound of carbonic anhydrase and anion inhibition effect on CO_2 hydration. *J. Biochem.* **2002**, *89*, 255–266. [CrossRef]

54. Ibrahim, M.M.; Shaban, S.Y.; Ichikawa, K. A promising structural zinc enzyme model for CO_2 fixation and calcification. *Tetrahedron Lett.* **2008**, *49*, 7303–7306. [CrossRef]

55. Zastrow, M.L.; Peacock, A.F.; Stuckey, J.A.; Pecoraro, V.L. Hydrolytic catalysis and structural stabilization in a designed metalloprotein. *Nat. Chem.* **2012**, *4*, 118–123. [CrossRef] [PubMed]

56. Bhaduri, G.A.; Šiller, L. Nickel nanoparticles catalyse reversible hydration of carbon dioxide for mineralization carbon capture and storage. *Catal. Sci. Technol.* **2013**, *3*, 1234–1239. [CrossRef]

57. Sabouni, R.; Kazemian, H.; Rohani, S. Carbon dioxide capturing technologies: A review focusing on metal organic framework materials (MOFs). *Environ. Sci. Pollut. Res.* **2014**, *21*, 5427–5449. [CrossRef] [PubMed]

58. Nath, I.; Chakraborty, J.; Verpoort, F. Metal organic frameworks mimicking natural enzymes: A structural and functional analogy. *Chem. Soc. Rev.* **2016**, *45*, 4127–4170. [CrossRef] [PubMed]

59. D'Alessandro, D.M.; Smit, B.; Long, J.R. Carbon dioxide capture: Prospects for new materials. *Angew. Chem. Int. Ed.* **2010**, *49*, 6058–6082. [CrossRef] [PubMed]

60. Cavka, J.H.; Jakobsen, S.; Olsbye, U.; Guillou, N.; Lamberti, C.; Bordiga, S.; Lillerud, K.P. A new zirconium inorganic building brick forming metal organic frameworks with exceptional stability. *J. Am. Chem. Soc.* **2008**, *130*, 13850–13851. [CrossRef] [PubMed]

61. Furukawa, H.; Gándara, F.; Zhang, Y.B.; Jiang, J.; Queen, W.L.; Hudson, M.R.; Yaghi, O.M. Water adsorption in porous metal–organic frameworks and related materials. *J. Am. Chem. Soc.* **2014**, *136*, 4369–4381. [CrossRef] [PubMed]

62. Gil-San Millan, R.; Lopez-Maya, E.; Ryo, S.G.; Kim, M.K.; Navarro, J.A.R. Improved soman and VX nerve agents degradation by magnesium hydroxide doped metal-organic frameworks. *J. Mater. Chem. A* under review.

63. López-Maya, E.; Montoro, C.; Rodríguez-Albelo, L.M.; Aznar Cervantes, S.D.; Lozano-Pérez, A.A.; Cenís, J.L.; Barea, E.; Navarro, J.A.R. Textile/metal–organic-framework composites as self-detoxifying filters for chemical-warfare agents. *Angew. Chem. Int. Ed.* **2015**, *54*, 6790–6794. [CrossRef] [PubMed]

64. Valenzano, L.; Civalleri, B.; Chavan, S.; Bordiga, S.; Nilsen, M.H.; Jakobsen, S.; Lillerud, K.P.; Lamberti, C. Disclosing the complex structure of UiO-66 metal organic framework: A synergic combination of experiment and theory. *Chem. Mater.* **2011**, *23*, 1700–1718. [CrossRef]

65. Deer, W.A.; Howie, R.A.; Zussman, J. *Rock-Forming Minerals: Single-Chain Silicates*; The Geological Society: London, UK, 1997; 668p.

66. Bailey, A.; Reesman, A.L. A survey study of the kinetics of wollastonite dissolution in H_2O-CO_2 and buffered systems at 25 degrees C. *Am. J. Sci.* **1971**, *271*, 464–472. [CrossRef]

67. Rimstidt, J.D.; Dove, P.M. Mineral/solution reaction rates in a mixed flow reactor: Wollastonite hydrolysis. *Geochim. Cosmochim. Acta* **1986**, *50*, 2509–2516. [CrossRef]

68. Murphy, W.M.; Helgeson, H.C. Thermodynamic and kinetic constraints on reaction rates among minerals and aqueous solutions. III. Activated complexes and the pH-dependence of the rates of feldspar, pyroxene, wollastonite, and olivine hydrolysis. *Geochim. Cosmochim. Acta* **1987**, *51*, 3137–3153. [CrossRef]

69. Casey, W.H.; Westrich, H.R.; Banfield, J.F.; Ferruzzi, G.; Arnold, G.W. Leaching and reconstruction at the surfaces of dissolving chain-silicate minerals. *Nature* **1993**, *366*, 253–256. [CrossRef]

70. Xie, Z.; Walther, J.V. Dissolution stoichiometry and adsorption of alkali and alkaline earth elements to the acid-reacted wollastonite surface at 25 °C. *Geochim. Cosmochim. Acta* **1994**, *58*, 2587–2598. [CrossRef]

71. Kojima, T.; Nagamine, A.; Ueno, N.; Uemiya, S. Absorption and fixation of carbon dioxide by rock weathering. *Energy Convers. Manag.* **1997**, *38*, S461–S466. [CrossRef]

72. Weissbart, E.J.; Rimstidt, J.D. Wollastonite: Incongruent dissolution and leached layer formation. *Geochim. Cosmochim. Acta* **2000**, *64*, 4007–4016. [CrossRef]

73. Wu, J.C.S.; Sheen, J.D.; Chen, S.Y.; Fan, Y.C. Feasibility of CO_2 fixation via artificial rock weathering. *Ind. Eng. Chem. Res.* **2001**, *40*, 3902–3905. [CrossRef]

74. Golubev, S.V.; Pokrovsky, O.S.; Schott, J. Experimental determination of the effect of dissolved CO_2 on the dissolution kinetics of Mg and Ca silicates at 25 °C. *Chem. Geol.* **2005**, *217*, 227–238. [CrossRef]

75. Green, E.; Luttge, A. Incongruent dissolution of wollastonite measured with vertical scanning interferometry. *Am. Miner.* **2006**, *91*, 430–434. [CrossRef]

76. Tai, C.Y.; Chen, W.R.; Shih, S.M. Factors affecting wollastonite carbonation under CO_2 supercritical conditions. *AIChE J.* **2006**, *52*, 292–299. [CrossRef]

77. Daval, D.; Bernard, S.; Rémusat, L.; Wild, B.; Guyot, F.; Micha, J.S.; Rieutord, F.; Magnin, V.; Fernandez-Martinez, A. Dynamics of altered surface layer formation on dissolving silicates. *Geochim. Cosmochim. Acta* **2017**, *209*, 51–69. [CrossRef]

78. Pokrovsky, O.S.; Shirokova, L.S.; Bénézeth, P.; Schott, J.; Golubev, S.V. Effect of organic ligands and heterotrophic bacteria on wollastonite dissolution kinetics. *Am. J. Sci.* **2009**, *309*, 731–772. [CrossRef]

79. Schott, J.; Pokrovsky, O.S.; Spalla, O.; Devreux, F.; Gloter, A.; Mielczarski, J.A. Formation, growth and transformation of leached layers during silicate minerals dissolution: The example of wollastonite. *Geochim. Cosmochim. Acta* **2012**, *98*, 259–281. [CrossRef]

80. Ding, W.; Fu, L.; Ouyang, J.; Yang, H. CO_2 mineral sequestration by wollastonite carbonation. *Phys. Chem. Miner.* **2014**, *41*, 489–496. [CrossRef]

81. Parkhurst, D.L.; Appelo, C.A.J. *Description of Input and Examples for Phreeqc Version 3. A Computer Program for Speciation, Batch-Reaction, One-Dimensional Transport, and Inverse Geochemical Calculations*; U.S. Geological Survey: Denver, VA, USA, 2013. Available online: http://pubs.usgs.gov/tm/06/a43 (accessed on 9 May 2018).

82. Rodriguez-Navarro, C.; Ruiz-Agudo, E.; Luque, A.; Rodriguez-Navarro, A.B.; Ortega-Huertas, M. Thermal decomposition of calcite: Mechanisms of formation and textural evolution of CaO nanocrystals. *Am. Miner.* **2009**, *94*, 578–593. [CrossRef]

83. Teng, H.H.; Dove, P.M.; De Yoreo, J.J. Kinetics of calcite growth: Surface processes and relationships to macroscopic rate laws. *Geochim. Cosmochim. Acta* **2000**, *64*, 2255–2266. [CrossRef]

84. Rodriguez-Navarro, C.; Burgos Cara, A.; Elert, K.; Putnis, C.V.; Ruiz-Agudo, E. Direct nanoscale imaging reveals the growth of calcite crystals via amorphous nanoparticles. *Cryst. Growth Des.* **2016**, *16*, 1850–1860. [CrossRef]

85. Rodriguez-Navarro, C.; Ruiz-Agudo, E.; Harris, J.; Wolf, S.E. Nonclassical crystallization in vivo et in vitro (II): Nanogranular features in biomimetic minerals disclose a general colloid-mediated crystal growth mechanism. *J. Struct. Biol.* **2016**, *196*, 260–287. [CrossRef] [PubMed]

86. Turpault, M.P.; Trotignon, L. The dissolution of biotite single crystals in dilute HNO_3 at 24 °C: Evidence of an anisotropic corrosion process of micas in acidic solutions. *Geochim. Cosmochim. Acta* **1994**, *58*, 2761–2775. [CrossRef]

87. Awad, A.; Van Groos, A.K.; Guggenheim, S. Forsteritic olivine: Effect of crystallographic direction on dissolution kinetics. *Geochim. Cosmochim. Acta* **2000**, *64*, 1765–1772. [CrossRef]

88. Daval, D.; Hellmann, R.; Saldi, G.D.; Wirth, R.; Knauss, K.G. Linking nm-scale measurements of the anisotropy of silicate surface reactivity to macroscopic dissolution rate laws: New insights based on diopside. *Geochim. Cosmochim. Acta* **2013**, *107*, 121–134. [CrossRef]

89. Lasaga, A.C.; Luttge, A. Variation of crystal dissolution rate based on a dissolution stepwave model. *Science* **2001**, *291*, 2400–2404. [CrossRef] [PubMed]

90. Dove, P.M.; Han, N.; De Yoreo, J.J. Mechanisms of classical crystal growth theory explain quartz and silicate dissolution behavior. *Proc. Natl. Acad. Sci. USA* **2005**, *102*, 15357–15362. [CrossRef] [PubMed]

91. Pollet-Villard, M.; Daval, D.; Ackerer, P.; Saldi, G.D.; Wild, B.; Knauss, K.G.; Fritz, B. Does crystallographic anisotropy prevent the conventional treatment of aqueous mineral reactivity? A case study based on K-feldspar dissolution kinetics. *Geochim. Cosmochim. Acta* **2016**, *190*, 294–308. [CrossRef]

92. Ruiz-Agudo, E.; Putnis, C.V.; Pel, L.; Rodriguez-Navarro, C. Template-assisted crystallization of sulfates onto calcite: Implications for the prevention of salt damage. *Cryst. Growth Des.* **2012**, *13*, 40–51. [CrossRef]

© 2018 by the authors. Licensee MDPI, Basel, Switzerland. This article is an open access article distributed under the terms and conditions of the Creative Commons Attribution (CC BY) license (http://creativecommons.org/licenses/by/4.0/).

Article

Interfacial Precipitation of Phosphate on Hematite and Goethite

Lijun Wang [1],*, Christine V. Putnis [2,3,*], Jörn Hövelmann [4] and Andrew Putnis [2,5]

[1] College of Resources and Environment, Huazhong Agricultural University, Wuhan 430070, China
[2] Institut für Mineralogie, University of Münster, 48149 Münster, Germany
[3] Department of Chemistry, Curtin University, Perth, WA 6845, Australia
[4] GFZ German Research Centre for Geosciences, 14473 Potsdam, Germany; jhoevelm@gfz-potsdam.de
[5] The Institute for Geoscience Research (TIGeR), Curtin University, Perth, WA 6102, Australia;
 andrew.putnis@curtin.edu.au
* Correspondence: ljwang@hzau.edu.cn (L.W.); putnisc@uni-muenster.de (C.V.P.);
 Tel.: +86-27-87288382 (L.W.); +49-251-8333454 (C.V.P.)

Received: 19 March 2018; Accepted: 9 May 2018; Published: 10 May 2018

check for updates

Abstract: Adsorption and subsequent precipitation of dissolved phosphates on iron oxides, such as hematite and goethite, is of considerable importance in predicting the bioavailability of phosphates. We used in situ atomic force microscopy (AFM) to image the kinetic processes of phosphate-bearing solutions interacting with hematite or goethite surfaces. The nucleation of nanoparticles (1.0–4.0 nm in height) of iron phosphate (Fe(III)-P) phases, possibly an amorphous phase at the initial stages, was observed during the dissolution of both hematite and goethite at the earliest crystallization stages. This was followed by a subsequent aggregation stage where larger particles and layered precipitates are formed under different pH values, ionic strengths, and organic additives. Kinetic analysis of the surface nucleation of Fe-P phases in 50 mM $NH_4H_2PO_4$ at pH 4.5 showed the nucleation rate was greater on goethite than hematite. Enhanced goethite and hematite dissolution in the presence of 10 mM $AlCl_3$ resulted in a rapid increase in Fe-P nucleation rates. A low concentration of citrate promoted the nucleation, whereas nucleation was inhibited at higher concentrations of citrate. By modeling using PHREEQC, calculated saturation indices (SI) showed that the three Fe(III)-P phases of cacoxenite, tinticite, and strengite may be supersaturated in the reacted solutions. Cacoxenite is predicted to be more thermodynamically favorable in all the phosphate solutions if equilibrium is reached with respect to hematite or goethite, although possibly only amorphous precipitates were observed at the earliest stages. These direct observations at the nanoscale may improve our understanding of phosphate immobilization in iron oxide-rich acid soils.

Keywords: interfacial precipitation; phosphate; hematite; goethite; dissolution-precipitation; citrate

1. Introduction

Phosphorus (P) is a limiting nutrient for plant growth [1,2] due to both its low P bioavailability in soils and the increasingly limited mineable resources of P-containing rocks [3]. The mineral interface-mediated P adsorption and precipitation are major mechanisms of P immobilization in calcite-rich alkaline soils [4] or (hydr)oxides of iron (Fe) and aluminium (Al)-rich acidic soils [5], and this leads to the lowering of its bioavailability following the application of P fertilizers. Therefore, a knowledge of the transformation and mobility of phosphate at the mineral interfaces at the nanoscale may improve our understanding of the fate of dissolved phosphate species in soils.

Despite the widely observed bulk P adsorption, especially from modeling of both non-specific and specific adsorption [6–11], and surface precipitation [12–14] at iron oxide mineral surfaces, the nanoscale surface reaction kinetics of solutions containing phosphate with iron oxides remain limited. Therefore, the objective of this study was to observe the kinetics of hematite and goethite dissolution and the coupled precipitation of new phases in the presence of aqueous solutions of phosphate. To achieve these goals, we used in situ atomic force microscopy (AFM) to image the real-time kinetics of hematite and goethite dissolution and the subsequent formation of new Fe-P phases. In the present study, we chose hematite (Fe_2O_3) and goethite (α-FeOOH) because they have been recognized as two of the most abundant of the naturally occurring iron oxides in soils [15,16]. We directly compared the surface nucleation rates of Fe-P phases on hematite and goethite under the same solution conditions and calculated solution speciation and thermodynamics of precipitating phases using PHREEQC modeling. These in situ observations may improve the mechanistic understanding of phosphate immobilization at the mineral interface in soils.

2. Materials and Methods

In situ dissolution experiments were performed using a Bruker Nanoscope IIIa AFM equipped with a fluid cell in contact mode. Natural Fe-bearing minerals, crystalline hematite, and goethite were chosen (Jackson Mine, MI, USA) as the most common Fe-bearing mineral sources in soils. Elongated small tabular crystals of goethite were cleaved in order to expose a fresh cleavage (010) surface immediately prior to their mounting in the fluid cell and contact with reaction solutions. A massive hematite crystal was parted along (001) to present a flat surface suitable for imaging. Ammonium dihydrogen phosphate solutions (50 mM $NH_4H_2PO_4$ at pH 2.0 or 4.5, adjusted using 0.01 M HCl), in the absence and presence of 10 mM $AlCl_3$ as background electrolytes, or 1–50 µM citrate as an organic additive, were passed through the fluid cell at a rate of 2 ml per 1.5 min. AFM images were collected in contact mode using Si_3N_4 tips (Bruker, tip model NP-S10, spring constants of 0.12 N/m and 0.58 N/m) and were recorded sequentially using a scan rate of 4 Hz and a resolution of 256 pixels (apart from occasional images when this was increased to 512 for some images chosen for reproduction). This imaging rate allows an image to be acquired approximately every 1 min 20 s and was used for all AFM experiments reported here. The scan angle is chosen to give the best trace and retrace output fits indicating a good imaging quality. This varies with each experiment and is dependent on the characteristic mineral surface topography. Images were analyzed using the NanoScope software (Version 5.31r1, Bruker).

Nucleation rates were measured from data collected shortly after the onset of precipitate formation [5]. All data with their mean values ± standard deviation (SD) of three independent sets of experiments are presented [5]. Ex situ dissolution experiments were performed following AFM experiments, and the sample was then removed from the AFM fluid cell and placed in a closed container filled with ca. 10 mL of the same reaction solutions for 3 days at room temperature [5]. The reacted crystal surfaces were coated with carbon and observed in a scanning electron microscope (SEM, JEOL JSM 6460 LV) equipped with an energy dispersive X-ray (EDX) detector (Oxford Instruments, UK) for the elemental analyses of precipitates formed on the hematite and goethite surfaces. SEM imaging and EDX analyses were made at an acceleration voltage of 20 kV [5].

The PHREEQC geochemical modeling software [17] was used to calculate solution speciation [5]. The PHREEQC model simulated the dissolution of either hematite or goethite into a given solution composition until equilibrium with the respective iron oxide was attained. The saturation index ($SI = IAP/K_{sp}$, where IAP is the actual ion activity product and K_{sp} is the solubility product for a given Fe-P phase) of the equilibrated solution with respect to the different iron phosphate (Fe-P) phases [18], including cacoxenite ($Fe_4(PO_4)_3(OH)_3 \cdot 12H_2O$), tinticite ($Fe_3(PO_4)_2(OH)_3 \cdot 3H_2O$), strengite ($FePO_4 \cdot 2H_2O$), or an Al-P phase wavellite ($Al_3(PO_4)_2(OH)_3 \cdot 5H_2O$) [19] was calculated using the MINTEQ.v4 database. The equilibrium constants (logK) for cacoxenite, tinticite, and wavellite are not included in the database and were therefore taken from the literature [18,19].

3. Results and Discussion

3.1. Iron Phosphate Nucleation and Growth on Hematite and Goethite

The dissolution of hematite or goethite in the presence of phosphate solutions provided a source of Fe^{3+} ions, which resulted in a supersaturation of the interfacial fluid with respect to a Fe-P phase and its nucleation on the dissolving hematite or goethite surfaces. In situ AFM imaging demonstrated that the observed nanoparticles/nanoclusters were present with a height of 1.0–4.0 nm and 2.0–3.0 nm on hematite (Figure 1A–C,G) and goethite (Figure 1D,F,H), respectively, at the earliest nucleation stages, i.e., within minutes of the phosphate solution contacting the mineral surfaces. The number of Fe-P nucleated nanoparticles increased linearly with time at 50 mM $NH_4H_2PO_4$ (pH 4.5), and faster nucleation rates on goethite compared with hematite were observed (Figure 1I). However, after longer reaction times (3 days), the almost complete coverage of the reacting hematite surface by the deposition of Fe-P phases occurred (Figure 2A), whereas non-covered areas of the dissolving goethite substrate still remained (Figure 2B). The nucleation rates on both hematite (Figure 3) and goethite (Figure 4) increased when the pH was decreased to 2.0 from 4.5 in 50 mM $NH_4H_2PO_4$ solutions.

For a given $NH_4H_2PO_4$ concentration (50 mM, pH 4.5), in the presence of 10 mM $AlCl_3$, faster nucleation rates were also observed on goethite than hematite (Figure 5), and both surfaces induced faster nucleation rates in the presence of $AlCl_3$ than in the absence of $AlCl_3$ (Figure 1). After 94 and 29 min, no further growth of nucleated crystals was observed on hematite and goethite, respectively. This could be related to the fact that both the hematite and goethite surfaces were rapidly covered by a layer of fine precipitates (Figure 5C,F), effectively passivating the surface so that no further Fe^{3+} ions could be released to the interfacial fluid layer. Similar phenomena could be observed in the presence of various concentrations of NaCl (10–100 mM) [5]. In the present study, passivation seems highly likely due to the slowing down of the reaction. However, it is quite difficult to know if the complete coverage was porous and therefore permeable to allow fluid access within the original crystal. This would be unlikely because, from considerations of the crystal's structural characteristics, there is probably no epitaxy between the iron oxide surface and amorphous/polycrystalline Fe-P precipitates.

Upon the input of a 50 mM $NH_4H_2PO_4$ solution in the presence of 1.0 μM citrate (pH 4.5), greater nucleation rates on hematite and goethite were observed than in the absence of citrate (Figure 6A–D and Figure 7A–C). Faster rapid growth of the nucleated particles with different sizes (height in nm) almost fully covered the goethite substrates in a short time period (16 min) (Figure 7C), compared to hematite (Figure 6D). When the citrate concentration was raised to 10 μM, the Fe-P nucleation on hematite was completely inhibited as no precipitation was seen under in situ AFM (Figure 6E,F), whereas a similar result only occurred on goethite surfaces at an increased citrate concentration of 50 μM (Figure 7D,E).

3.2. SEM-EDX Identification of Surface Precipitates on Hematite and Goethite Faces

After exposing cleaved hematite or goethite surfaces to 50 mM $NH_4H_2PO_4$ solutions (pH 4.5) for 3 days, the hematite surfaces were almost completely covered by thicker layers (several tens of nanometers) of precipitates (Figure 8A,B), whereas thinner layers with scattered particles of precipitates were formed on goethite (Figure 9A). SEM-EDX analyses indicated these precipitates are composed of Fe, P, and O (Figures 8C and 9B). The SEM analysis showing P present in the precipitate phases is an indication that a Fe-P phase is evident.

3.3. Kinetics and Thermodynamics of Coupled Dissolution-Precipitation at the Iron Oxide-Phosphate Solution Interface

In general, iron oxides dissolve slowly for in situ AFM study, but for both minerals, goethite, and hematite, dissolution features were evident on the exposed surfaces. Our previous results showed that dissolution of goethite surfaces exposed to $NH_4H_2PO_4$ solutions (5.0 mM, pH 4.5), in the absence and presence of background electrolytes, occurs through a simple step retreat process by showing surfaces comprised of multiple straight-edged and non-straight preexisting steps without evidence of etch pit formation within the experimental time frame [5]. For hematite, we infer that dissolution in $NH_4H_2PO_4$ solutions also proceeds by simple step retreat. This was based on the results that the nucleation of Fe-P nanoparticles, which occurred preferentially along the step edges of preexisting steps on a hematite surface (Figure 6A–D). For hematite or goethite surfaces (only preexisting straight-edged steps with no evidence of pitting) under far from equilibrium conditions (as is the case for our solution conditions), the dissolution rates increased with acidic pH (Figures 3 and 4).

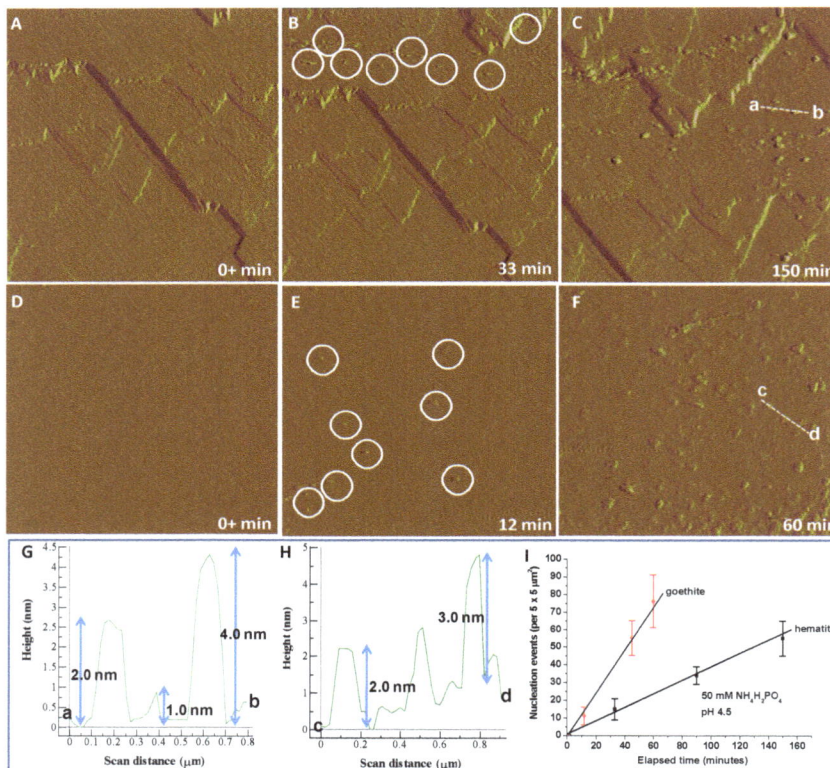

Figure 1. AFM time sequence (deflection images) showing in situ nucleation kinetics of Fe-P phases and the growth evolution of precipitates on dissolving (**A–C**) hematite and (**D–F**) goethite surfaces in 50 mM $NH_4H_2PO_4$ (pH 4.5). Images (**A–F**), 5 μm × 5 μm. Height profile of the initially formed Fe-P nanoparticles (circled for clarity in (**B** or **E**)) on a (**G**) hematite along section a → b (dashed line in (**C**)) or (**H**) goethite face along section c → d (dashed line in (**F**)). The average height of Fe-P nanoparticles is about 1.0–4.0 nm, and was almost unchanged in a short period (150 min) of reaction time. (**I**) Kinetic analysis of surface nucleation of Fe-P phases on hematite and goethite substrates in short periods of reaction times in 50 mM $NH_4H_2PO_4$ at pH 4.5, showing linear scaling of surface nucleation with time and dependence of the nucleation rate given by slopes of the lines (the solid line is a linear fit to the AFM data (the number of nucleated particles in a short time period)). Vertical error bars (the standard deviation of the mean) correspond to the variation in measured number of nucleated particles in three different areas in each experiment.

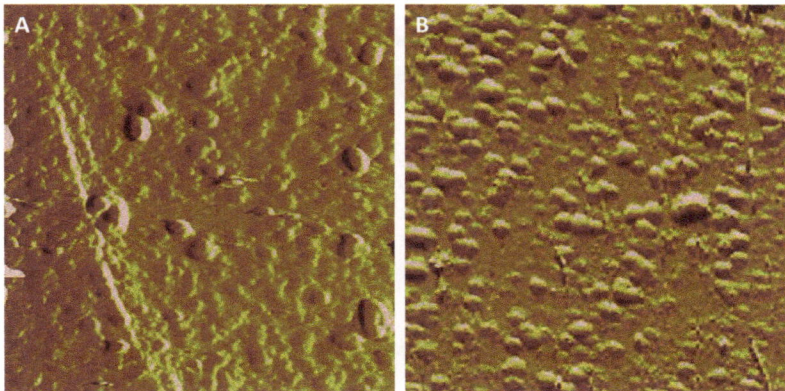

Figure 2. Ex situ AFM images of the same (**A**) hematite and (**B**) goethite substrate as the above (Figure 1), reacting in solutions (50 mM $NH_4H_2PO_4$, pH 4.5) for 3 days at room temperature.

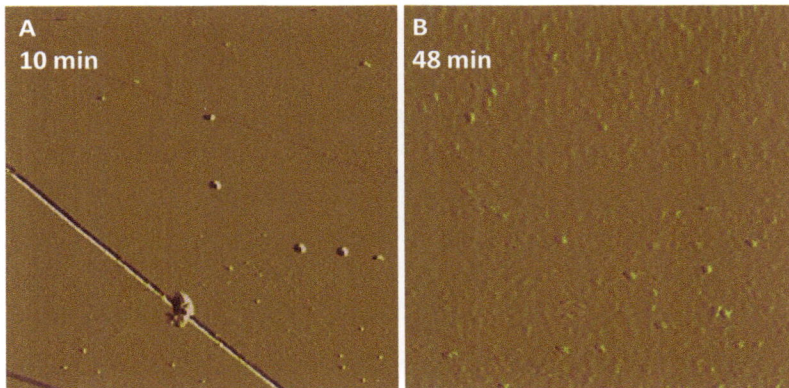

Figure 3. Following (**A**) 10 min, particles initially formed, and after (**B**) 48 min of injecting reaction solution (50 mM $NH_4H_2PO_4$, pH 2.0), the scan area (5 μm × 5 μm) on hematite was covered with a larger number of the new nucleating particles, compared with 50 mM $NH_4H_2PO_4$ reaction solution at pH 4.5 after the same reaction time in Figure 1I.

The dissolution of hematite or goethite in the presence of all phosphate solutions provided a continuous source of Fe^{3+} ions, increasing the driving force *SI* at the mineral interfacial layer with respect to several Fe-P phases, and eventually resulting in nucleation. Therefore, the dissolution of either hematite or goethite is coupled at the reaction interface with the precipitation of the new phase [20]. For nucleation to occur the solution at the mineral surface must become supersaturated with respect to the precipitating phase. In fact, some of the precipitating Fe-P phases have an extremely low solubility. For example, according to the PHREEQC calculations, you would only need to dissolve 3.3×10^{-11} moles of FeOOH into 1 L of solution to saturate it with respect to cacoxenite. Therefore, the interfacial layer will become quickly saturated and this may lead to a strong interfacial coupling between dissolution and precipitation. This occurs despite the fact that the solutions flowing over the goethite or hematite surfaces were all highly undersaturated with respect to any possible phase [21], indicating the presence of a boundary fluid layer whose composition differs from the bulk solution. In effect, the nucleation and growth rates of the newly formed phases are controlled by the thermodynamic component of the interfacial energy and kinetic barriers associated with hydration, desolvation, degree of epitaxy with the parent surfaces, attachment/detachment, and diffusion [5,22]. In the present study, the flow rate of approximately 2 mL/1.5 min has been shown in previous studies [5] to make surface-control the rate determining step in the reaction, rather than diffusion through the fluid at the interface. However, we accept that the solution used may also play a role in the evolution of fluid concentration at the interface for certain AFM experimental conditions. A net increase in the dissolution rates of preexisting steps at low pH can increase the nucleation rate due to faster dissolution and therefore reach a higher *SI* at the mineral-fluid interface (Figure 4) in a shorter time. An almost constant inter-particle distance of 218 ± 43 nm (Figure 4E,D) is present during the formation of Fe-P nanoparticles on goethite. This implies that there is a fast build-up of ions in the interfacial solution until precipitation occurs. Transport via diffusion may also play a role in the interfacial solution chemistry.

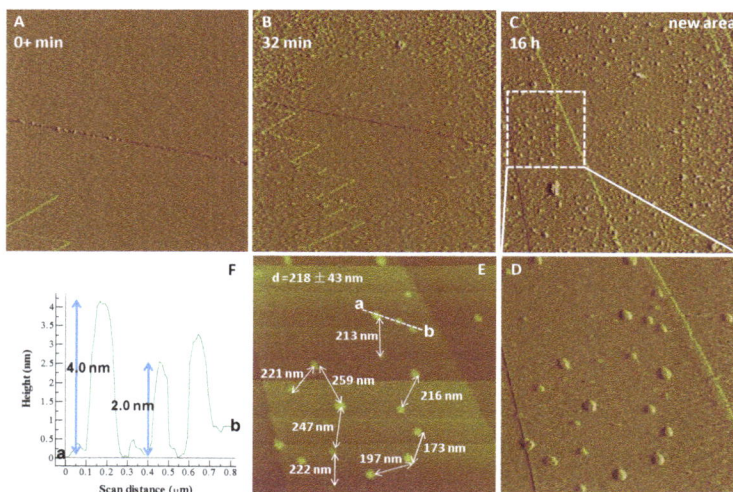

Figure 4. AFM time sequence (deflection images) showing in situ nucleation kinetics of Fe-P phases on a dissolving goethite surface in 50 mM $NH_4H_2PO_4$ (pH 2.0) at (**A**) *t* = 0 min, (**B**) 32 min, and (**C**) 16 h, respectively. (**D**) A zoom of a dotted rectangle in (**C**) showing Fe-P nanoparticles on a goethite (010) cleavage face and (**E**) inter-particle distances (218 ± 43 nm). (**F**) Height profile for Fe-P nanoparticles on a goethite (010) cleavage face along section a → b (dashed line in (**E**)). The average height of Fe-P nanoparticles is about 2.0–4.0 nm. AFM images, (**A**) 5 μm × 5 μm; (**B**,**C**) 10 μm × 10 μm; (**D**,**E**) 3 μm × 3 μm.

At the initial nucleation stages, faster nucleation rates on goethite compared to hematite were observed (Figure 1I), most likely related to the faster dissolution (higher reactivity) of goethite surfaces compared to hematite, and hence the faster release of Fe^{3+} ions in the former case. On the contrary, after longer reaction times (3 days), the almost complete coverage of the reacting hematite surface by thicker layers of precipitates of possibly amorphous Fe(III)-P phases occurred (Figures 2A and 8A,B), compared with the goethite substrate (Figure 9A). This may be due to the fact that the goethite surface was covered with a thin layer of Fe-P particles by a rapid initial reaction and was subsequently inhibited from further dissolution. We observed that goethite reacted faster than hematite in the presence of any phosphate solutions, however in the latter case perhaps a slower dissolution was kinetically favorable to form a thicker Fe-P layer. This also could be confirmed by the fact that the hematite surface was rapidly covered by a layer of fine precipitates (Supplementary Materials Figure S1) in the presence of 50 mM $NH_4H_2PO_4$ at pH 2.0, effectively passivating the surface.

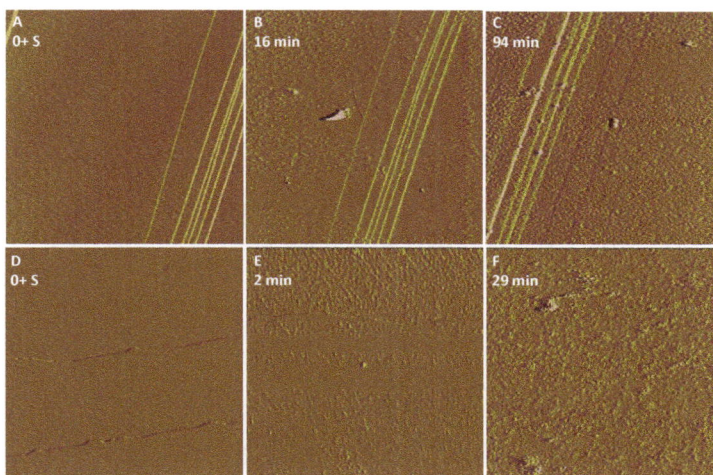

Figure 5. AFM images of fast nucleation of Fe-P particles formed on (**A–C**) hematite or (**D–F**) goethite surfaces after injecting 50 mM $NH_4H_2PO_4$ + 10 mM $AlCl_3$ (pH 4.5). The scan area was almost fully covered with the nucleating fine nanoparticles following 29-min-reaction, passivating the reacting surface of goethite. AFM images, 5 μm × 5 μm.

Using PHREEQC, calculated *SI* values show that the three most likely Fe-P phases on hematite and goethite are cacoxenite, tinticite, and strengite. Cacoxenite is predicted to be most favorable due to its higher *SI* in all the experimental solutions (Supplementary Materials Tables S1–S4), i.e., only very little hematite or goethite needs to be dissolved before the interfacial solution would become supersaturated with respect to cacoxenite. Saturation indices (*SI*) of possible Fe-P phases formed on goethite, in equilibrium with 50 mM $NH_4H_2PO_4$ solution (pH 4.5), are 1.4, 19.8, and 28.2, in relation to strengite, tinticite, and cacoxenite, respectively, and for hematite, the *SI* are 0.2, 16.2, 23.4, respectively, under the same solution conditions (Table S1). As pH was decreased to 2.0 in 50 mM $NH_4H_2PO_4$ solutions, the *SI* with respect to strengite, tinticite, or cacoxenite, formed on hematite or goethite, was significantly increased (Table S1), although all the observed precipitates may be amorphous at the initial stages. This suggests that the effect of pH is more favorable for both hematite and goethite dissolution, and subsequent Fe-P phase formation, than for the effect of ionic strength. As shown in Table S2, following the addition of 10 mM $AlCl_3$ into a 50 mM $NH_4H_2PO_4$ solution (pH 4.5), *SI* with respect to strengite, tinticite, and cacoxenite formed on hematite, was slightly raised to 0.6, 16.9, and 24.4, respectively. This suggests that the presence of background electrolytes enhances the mineral dissolution rate [5,23] because the electrolytes possibly modify water structure dynamics, as well as solute and surface hydration [5,24]. In addition, a new phase of wavellite $(Al_3(PO_4)_2(OH)_3 \cdot 5H_2O)$ may become stable in the presence of Al^{3+} ions (Table S2).

Furthermore, calculated *SI* also predicts that low concentrations of citrate can promote the formation of strengite, tinticite, or cacoxenite formed on hematite or goethite [5] due to elevated *SI* (Tables S3 and S4). Organic acids in the soil solution typically range from 0.1 μM to 0.1 mM [5,25,26]. Our experiments also show that higher concentrations of citric acid retard Fe-P precipitation formation. Therefore, the presence of relatively high concentrations of citrate may increase P availability by suppressing the precipitation of Fe-P phases, for example, cacoxenite [5]. The role of citrate is very interesting in that small concentrations seem not to affect or even to enhance the sequestration of PO_4, but high concentrations inhibit Fe-phosphate formation, i.e., prevent PO_4 sequestration. This phenomenon is highly relevant to P-depleted soils when a higher citrate concentration will inhibit P-phosphate formation and hence allow for the availability of P for plant uptake. The citrate ion forms complexes with metallic cations, such as Fe. The stability constants for the formation of these complexes are quite large because of the chelate effect. Consequently, it forms complexes even with alkali metal cations. However, when a chelate complex is formed using all three carboxylate groups, the chelate rings have 7 and 8 members, which are generally less stable thermodynamically than smaller chelate rings. As a consequence, the hydroxyl group can be deprotonated, forming part of a more stable 5-membered ring, as in ammonium ferric citrate, denoted as $(NH_4)_5Fe(C_6H_4O_7)_2 \cdot 2H_2O$ [27]. This is most likely to be the form present in our experiments [27].

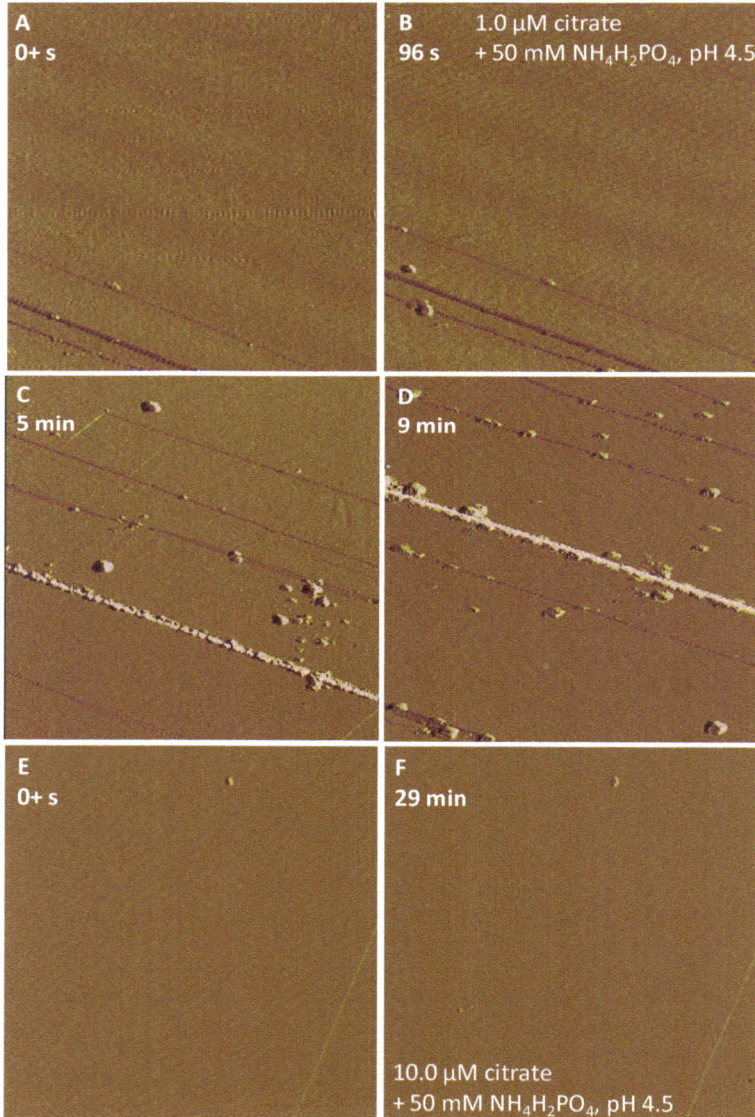

Figure 6. Nucleation kinetics of Fe-P phases on hematite in the presence of (**A–D**) 1.0 µM citrate and (**E,F**) 10.0 µM citrate with 50 mM $NH_4H_2PO_4$ at pH 4.5. AFM images, 5 µm × 5 µm.

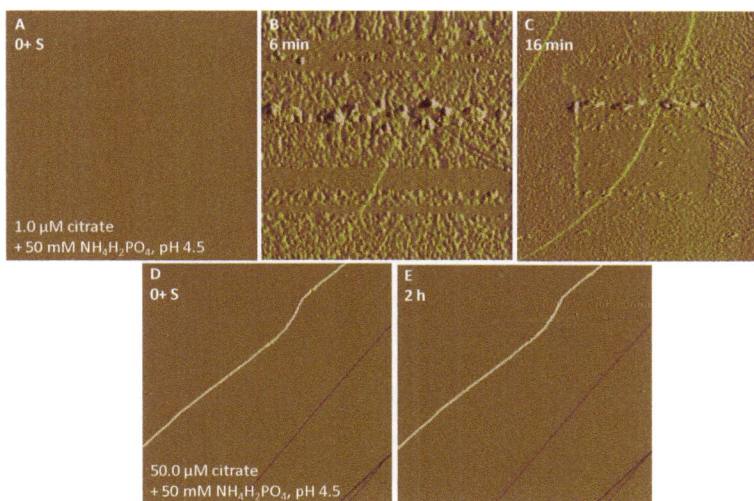

Figure 7. Nucleation kinetics of Fe-P phases on goethite in the presence of (**A–C**) 1.0 μM citrate and (**D,E**) 50 μM citrate with 50 mM $NH_4H_2PO_4$ at pH 4.5. AFM images, (**A,B**) 5 μm × 5 μm; (**C–E**) 10 μm × 10 μm.

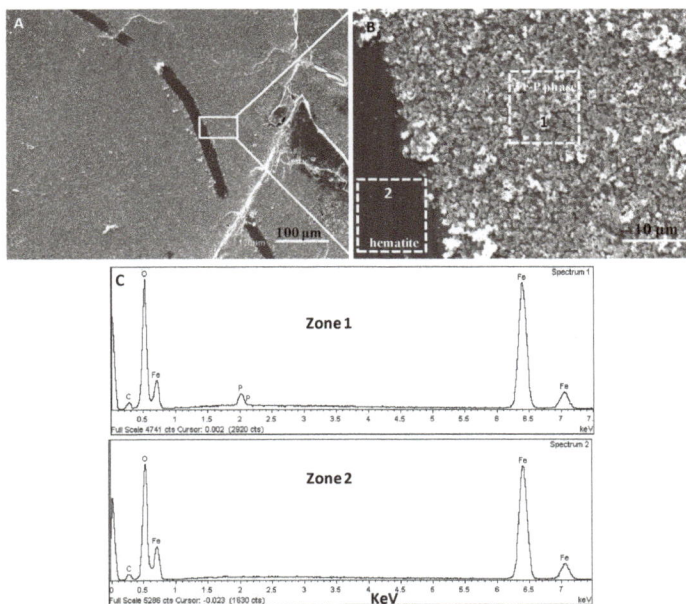

Figure 8. (**A,B**) SEM images showing the ex situ formation of Fe-P precipitates on hematite after 3 days of interfacial reaction in a 50 mM $NH_4H_2PO_4$ (pH 4.5) solution. The hematite substrate was almost fully covered with a thick layer of precipitates. The dotted rectangles in (**B**) indicate areas chosen for EDX analyses. (**C**) EDX spectra taken from zone 1 and 2 shown by dotted rectangles in (**B**), indicating an iron phosphate phase and a hematite substrate, respectively.

The end effect is that Fe is bound as a ligand in solution with a strong chelation effect of citrate and therefore is not available to form Fe-phosphates. It is impossible to prove the likely citrate-binding mechanism. Within the limits of our analytical possibilities we can only reasonably suggest what may be the mechanism. Given the high chelate-binding nature of citrate, the above-mentioned mechanism is the best approach that satisfies thermodynamic considerations. Finally, since surface roughness modifications may be responsible for the local variability of surface charge, the potential impact of small-scale surface roughness and the resulting landscape of surface charge may be of critical importance to the efficiency of particle adsorption [28].

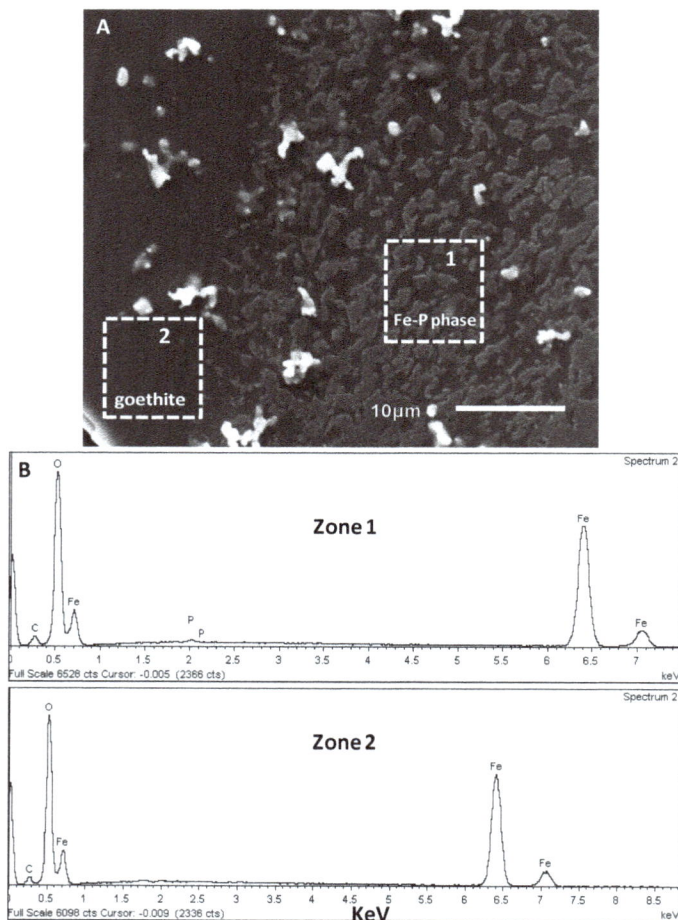

Figure 9. (**A**) SEM image showing the ex situ formation of Fe-P precipitates on goethite after 3 days of interfacial reaction in a 50 mM $NH_4H_2PO_4$ (pH 4.5) solution. The goethite substrate was covered with a thin layer of precipitates. The dotted rectangles in (**A**) indicate areas chosen for EDX analyses. (**B**) EDX spectra taken from zone 1 and 2 shown by dotted rectangles in (**A**), indicating an iron phosphate phase and a goethite substrate, respectively.

4. Conclusions

To summarize, Fe-P precipitation at the goethite-water or hematite-water interface is dependent on solution composition, including the activity of phosphate species, pH, ionic strength, and organic additives, thereby influencing the dissolution of hematite or goethite. An obvious decrease in the dissolved phosphate concentration in the soil solution is due to increased phosphate adsorption and subsequent precipitation on goethite or hematite. The bioavailability of phosphate will be rapidly reduced at high salt ($AlCl_3$) concentrations due to phosphate immobilization. If there is complete coverage by surface precipitates, the hematite and goethite surfaces will be passivated, resulting in the inhibition of phosphate immobilization. In effect, the presence of Fe mineral phases such as goethite and hematite in acid soils will sequester phosphate and therefore reduce P availability for plant uptake, but this can be modified by the presence of other ionic species and additives, such as other background ions or organic additives like citrate. These in situ observations may improve our understanding of goethite and hematite (as typical sources of Fe in soils) mineral surface-induced Fe-P precipitation, with possible implications for P fertilizer application management.

Supplementary Materials: The following are available online at http://www.mdpi.com/2075-163X/8/5/207/s1, Figure S1: AFM time sequence (deflection images) showing in situ nucleation kinetics of Fe-P phases on a dissolving hematite surface in 50.0 mM $NH_4H_2PO_4$ (pH 2.0) at (**A**) t = 10 min, (**B**) 48 min, (**C**) 127 min, and (**D**) 2 d, respectively. After 127 min reaction, the scan area was almost fully covered with the nucleating fine particles, which passivated the reacting surfaces. AFM images, 5 μm × 5 μm, Table S1: Main solution speciation of goethite and hematite in equilibrium with $NH_4H_2PO_4$ reaction solutions at pH 2.0 and 4.5 and saturation indices (*SI*) with respect to different iron phosphate phases of various $NH_4H_2PO_4$ reaction solutions after equilibration with goethite and hematite, Table S2: Main solution speciation of hematite and goethite in equilibrium with $NH_4H_2PO_4$ reaction solutions in the presence of $AlCl_3$ and saturation indices (*SI*) with respect to different iron phosphate phases of $NH_4H_2PO_4$ reaction solutions after equilibration with hematite and goethite, Table S3: Main solution speciation of hematite in equilibrium with $NH_4H_2PO_4$ reaction solutions in the presence of citrate and saturation indices (*SI*) with respect to different iron phosphate phases of $NH_4H_2PO_4$ reaction solutions after equilibration with hematite, Table S4: Main solution speciation of goethite in equilibrium with $NH_4H_2PO_4$ reaction solutions in the presence of citrate and saturation indices (*SI*) with respect to different iron phosphate phases of $NH_4H_2PO_4$ reaction solutions after equilibration with goethite.

Author Contributions: L.W., and C.V.P. conceived and designed the experiments; L.W., C.V.P., and J.H. performed the experiments; L.W., J.H. and A.P. analyzed the data; L.W., and C.V.P. wrote the paper.

Acknowledgments: This work was supported by the National Natural Science Foundation of China (41471245 and 41071208), the Fundamental Research Funds for the Central Universities (2662015PY206 and 2662017PY061). C.V.P. and A.P. acknowledge funding through the EU seventh Framework Marie S. Curie ITNs: Minsc; CO_2 react; and Flowtrans. J.H. acknowledges the financial support of the Helmholtz Recruiting Initiative to Liane G. Benning.

Conflicts of Interest: The authors declare no conflict of interest.

References

1. Filippelli, G.M. The global phosphorous cycle: Past, present, and future. *Elements* **2008**, *4*, 89–95. [CrossRef]
2. Shen, J.B.; Yuan, L.; Zhang, J.L.; Li, H.; Bai, Z.; Chen, X.; Zhang, W.; Zhang, F.S. Phosphorus dynamics: From soil to plant. *Plant Physiol.* **2011**, *156*, 997–1005. [CrossRef] [PubMed]
3. Gilbert, N. The disappearing nutrient. *Nature* **2009**, *461*, 716–718. [CrossRef] [PubMed]
4. Wang, L.J.; Ruiz-Agudo, E.; Putnis, C.V.; Menneken, M.; Putnis, A. Kinetics of calcium phosphate nucleation and growth on calcite: Implications for predicting the fate of dissolved phosphate species in alkaline soils. *Environ. Sci. Technol.* **2012**, *46*, 834–842. [CrossRef] [PubMed]
5. Wang, L.J.; Putnis, C.V.; Ruiz-Agudo, E.; Hovelmann, J.; Putnis, A. In situ imaging of interfacial precipitation of phosphate on goethite. *Environ. Sci. Technol.* **2015**, *49*, 4184–4192. [CrossRef] [PubMed]
6. Kwon, K.D.; Kubicki, J.D. Molecular orbital theory study on surface complex structures of phosphates to iron hydroxides: Calculation of vibrational frequencies and adsorption energies. *Langmuir* **2004**, *20*, 9249–9254. [CrossRef] [PubMed]
7. Khare, N.; Hesterberg, D.; Martin, J.D. XANES investigation of phosphate sorption in single and binary systems of iron and aluminum oxide minerals. *Environ. Sci. Technol.* **2005**, *39*, 2152–2160. [CrossRef] [PubMed]

8. Rahnemaie, R.; Hiemstra, T.; van Riemsdijk, W.H. Geometry, charge distribution, and surface speciation of phosphate on goethite. *Langmuir* **2007**, *23*, 3680–3689. [CrossRef] [PubMed]

9. Weng, L.P.; Vega, F.A.; Van Riemsdijk, W.H. Competitive and synergistic effects in pH dependent phosphate adsorption in soils: LCD modeling. *Environ. Sci. Technol.* **2011**, *45*, 8420–8428. [CrossRef] [PubMed]

10. Kim, J.; Li, W.; Philips, B.L.; Grey, C.P. Phosphate adsorption on the iron oxyhydroxides goethite, akaganeite, and lepidocrocite: A ^{31}P NMR study. *Energy Environ. Sci.* **2011**, *4*, 4298–4305. [CrossRef]

11. Liu, H.; Chen, T.; Frost, R.L. An overview of the role of goethite surfaces in the environment. *Chemosphere* **2014**, *103*, 1–11. [CrossRef] [PubMed]

12. Jonasson, R.G.; Martin, R.R.; Giuliacci, M.E.; Tazaki, K. Surface reactions of goethite with phosphate. *J. Chem. Soc. Faraday Trans.* **1988**, *1*, 2311–2315. [CrossRef]

13. Li, L.; Stanforth, R. Distinguishing adsorption and surface precipitation of phosphate on goethite (a-FeOOH). *J. Colloid Interface Sci.* **2000**, *230*, 12–21. [CrossRef] [PubMed]

14. Ler, A.; Stanforth, R. Evidence for surface precipitation of phosphate on goethite. *Environ. Sci. Technol.* **2003**, *37*, 2694–2700. [CrossRef] [PubMed]

15. Cornell, R.M.; Schwertmann, U. *The Iron Oxides: Structure, Properties, Reactions, Occurrence and Uses*, 2nd ed.; Wiley-VCH: Weinheim, Germany, 2003.

16. Torrent, J. Interactions between phosphate and iron oxide. *Adv. Geoecol.* **1997**, *30*, 321–344.

17. Parkhurst, D.L.; Appelo, C.A.J. *Users Guide to PHREEQC (Version 2): A Computer Program for Speciation, Batch Reaction, One-Dimensional Transport, and Inverse Geochemical Calculations*; Water-Resources Investigations Report 99-4259; US Geological Survey: Reston, VA, USA, 1999; 312p.

18. Nriagu, J.O.; Dell, C.I. Diagenetic formation of iron phosphates in recent lake sediments. *Am. Mineralogist* **1974**, *59*, 934–946.

19. Vieillard, P.; Tardy, Y. Stability fields of clays and aluminum phosphates: Parageneses in lateritic weathering of argillaceous phosphatic sediments. *Am. Mineral.* **1979**, *64*, 626–634.

20. Putnis, A.; Putnis, C.V. The mechanism of reequilibration of solids in the presence of a fluid phase. *J. Solid State Chem.* **2007**, *180*, 1783–1786. [CrossRef]

21. Putnis, A. Why mineral interfaces matter. *Science* **2014**, *343*, 1441–1442. [CrossRef] [PubMed]

22. Giuffre, A.J.; Hamm, L.M.; Han, N.; De Yoreo, J.J.; Dove, P.M. Polysaccharide chemistry regulates kinetics of calcite nucleation through competition of interfacial energies. *Proc. Natl. Acad. Sci. USA* **2013**, *110*, 9261–9266. [CrossRef] [PubMed]

23. Ruiz-Agudo, E.; Kowacz, M.; Putnis, C.V.; Putnis, A. The role of background electrolytes on the kinetics and mechanism of calcite dissolution. *Geochim. Cosmochim. Acta* **2010**, *74*, 1256–1267. [CrossRef]

24. Stack, A.G. Molecular dynamics simulations of solvation and kink site formation at the {001} barite-water interface. *J. Phys. Chem. C* **2009**, *113*, 2104–2110. [CrossRef]

25. Qin, L.H.; Zhang, W.J.; Lu, J.W.; Stack, A.G.; Wang, L.J. Direct imaging of nanoscale dissolution of dicalcium phosphate dihydrate by an organic ligand: Concentration matters. *Environ. Sci. Technol.* **2013**, *47*, 13365–13374. [CrossRef] [PubMed]

26. Qin, L.H.; Wang, L.J.; Wang, B.S. Role of alcoholic hydroxyls of dicarboxylic acids in regulating nanoscale dissolution kinetics of dicalcium phosphate dihydrate. *ACS Sustain. Chem. Eng.* **2017**, *5*, 3920–3928. [CrossRef]

27. Matzapetakis, M.; Raptopoulou, C.P.; Tsohos, A.; Papaefthymiou, V.; Moon, S.N.; Salifoglou, A. Synthesis, spectroscopic and structural characterization of the first mononuclear, water soluble iron-citrate complex, $(NH_4)_5Fe(C_6H_4O_7)_2 \cdot 2H_2O$. *J. Am. Chem. Soc.* **1998**, *120*, 13266–13267. [CrossRef]

28. Fischer, C.; Karius, V.; Weidler, P.G.; Lüttge, A. Relationship between micrometer to submicrometer surface roughness and topography variations of natural iron oxides and trance element concentration. *Langmuir* **2008**, *24*, 3250–3266. [CrossRef] [PubMed]

© 2018 by the authors. Licensee MDPI, Basel, Switzerland. This article is an open access article distributed under the terms and conditions of the Creative Commons Attribution (CC BY) license (http://creativecommons.org/licenses/by/4.0/).

![minerals logo]

MDPI

Article

Aqueous Fe(II)-Induced Phase Transformation of Ferrihydrite Coupled Adsorption/Immobilization of Rare Earth Elements

Yingheng Fei [1,2], Jian Hua [1], Chengshuai Liu [3,*], Fangbai Li [1], Zhenke Zhu [4], Tangfu Xiao [2], Manjia Chen [1], Ting Gao [1,3], Zhiqi Wei [1] and Likai Hao [3]

1   Guangdong Key Laboratory of Integrated Agro-environmental Pollution Control and Management, Guangdong Institute of Eco-Environmental Sciences & Technology, Guangzhou 510650, China; yhfei@gzhu.edu.cn (Y.F.); huajianyh1211@163.com (J.H.); cefbli@soil.gd.cn (F.L.); mjchen@soil.gd.cn (M.C.); gaoting_gt0@163.com (T.G.); weizhiq123@126.com (Z.W.)
2   Key Laboratory for Water Quality and Conservation of the Pearl River Delta, Ministry of Education, School of Environmental Science and Engineering, Guangzhou University, Guangzhou 510006, China; xiaotangfu@vip.gyig.ac.cn
3   State Key Laboratory of Environmental Geochemistry, Institute of Geochemistry, Chinese Academy of Sciences, Guiyang 550081, China; haolikai@mail.gyig.ac.cn
4   Key Laboratory of Agro-Ecological Processes in Subtropical Region, Institute of Subtropical Agriculture, Chinese Academy of Sciences, Changsha 410125, China; zhuzhenke@isa.ac.cn
*   Correspondence: liuchengshuai@vip.gyig.ac.cn; Tel.: +86-851-8589-1334

Received: 30 June 2018; Accepted: 15 August 2018; Published: 18 August 2018

check for updates

Abstract: The phase transformation of iron minerals induced by aqueous Fe(II) ($Fe(II)_{aq}$) is a critical geochemical reaction which greatly affects the geochemical behavior of soil elements. How the geochemical behavior of rare earth elements (REEs) is affected by the $Fe(II)_{aq}$-induced phase transformation of iron minerals, however, is still unknown. The present study investigated the adsorption and immobilization of REEs during the $Fe(II)_{aq}$-induced phase transformation of ferrihydrite. The results show that the heavy REEs of Ho(III) were more efficiently adsorbed and stabilized compared with the light REEs of La(III) by ferrihydrite and its transformation products, which was due to the higher adsorptive affinity and smaller atomic radius of Ho(III). Both La(III) and Ho(III) inhibited the Fe atom exchange between $Fe(II)_{aq}$ and ferrihydrite, and sequentially, the $Fe(II)_{aq}$-induced phase transformation rates of ferrihydrite, because of the competitive adsorption with $Fe(II)_{aq}$ on the surface of iron (hydr)oxides. Owing to the larger amounts of adsorbed and stabilized Ho(III), the inhibition of the $Fe(II)_{aq}$-induced phase transformation of ferrihydrite affected by Ho(III) was higher than that by La(III). Our findings suggest an important role for the $Fe(II)_{aq}$-induced phase transformation of iron (hydr)oxides in assessing the mobility and transfer behavior of REEs, as well as for their occurrence in earth surface environments.

Keywords: ferrihydrite; recrystallization; REEs; stabilization; Fe atom exchange

1. Introduction

The aqueous Fe(II) ($Fe(II)_{aq}$)-induced phase transformation of iron (hydr)oxides is one of the most important reactions in the soil iron cycle and imposes great effects on the environmental behaviors of metals [1]. It was proposed in the 1980s that $Fe(II)_{aq}$ could catalyze the phase transformation of iron (hydr)oxides through electron transfer and atom exchange with structural Fe(III) ($Fe(III)_{oxide}$) [2]. However, only in the last two decades has this process been confirmed by ^{57}Fe isotopic tracer and spectroscopic investigations [3,4]. In the presence of $Fe(II)_{aq}$, atom exchange occurs between $Fe(II)_{aq}$

and $Fe(III)_{oxide}$, when $Fe(II)_{aq}$ was sorbed on or into the oxide crystal and then oxidized and became a secondary $Fe(III)_{oxide}$, while a portion of primary $Fe(III)_{oxide}$ would be reduced and released as $Fe(II)_{aq}$. Such reactions would accelerate the phase transformation of iron (hydr)oxides to more stable phases, e.g., goethite, magnetite and hematite [4,5]. Coexisting metallic ions in the system might also be involved via adsorption, enwrapping and substitution, and hence influence the phase transformation process; once the metallic ions were fixed in the iron (hydr)oxides, their environmental mobility and bioavailability were eventually altered [5,6].

Ferrihydrite is a form of iron (hydr)oxide which is usually first formed in the course of iron mineralization and with relatively poor crystallinity [7]. Featuring a small particle size, large specific surface area, as well as high surface activity, ferrihydrite is often favored for metal adsorption and plays critical roles in soil element cycling [8,9]. Compared with other crystalline iron (hydr)oxides, ferrihydrite is thermodynamically unstable and can be readily recrystallized to lepidocrocite, goethite and magnetite [7]. Under anaerobic conditions in sub soils, $Fe(II)_{aq}$ generated by dissimilatory iron reduction is an important driving force for ferrihydrite transformation. Rather than the simple redox cycling of iron, the $Fe(II)_{aq}$-induced phase transformation of ferrihydrite plays an important role in controlling the co-existing metal ions' mobilization/immobilization in the environment, since its adsorption, immobilization, transport and transformation are heavily interfered with by the interplay between $Fe(II)_{aq}$ and $Fe(III)_{oxide}$.

Rare earth elements (REEs) include the lanthanide family, with atomic numbers from 57 to 71, as well as scandium and yttrium. They are predominantly trivalent (i.e., Ln^{3+}, Eu^{3+}), except for some which have other oxidation states which are less stable, such as cerium (Ce^{4+}) and europium (Eu^{2+}) [10]. As a result of their similar atomic radii and oxidation state, REEs can substitute each other in various mineral forms, such as in halides, carbonates, oxides, phosphates, and silicates, which results in their wide distribution within the Earth's crust [11]. Usually, REEs are also utilized as tracers of geochemical processes [12]. China, as the biggest supplier in the world, features high contents of REEs in soils, as well as using them long-term as fertilizer additives, and with uncontrolled and unplanned release by mining activities [11,13]. Iron (hydr)oxide is a promising potential carrier for REEs in the environment [14]. During the iron oxide formation and transformation, the uptake, release, and oxidation states of REEs also change accordingly [15]. Colloidal and sedimentary iron (hydr)oxides were reported to be a major factor influencing the fractionation, transport, and mobilization/immobilization of REEs in the aquatic environment [12,16]. Coexisted or structurally incorporated metal ions are universal within iron (hydr)oxides in soils. The affinity of REEs ions to iron (hydr)oxides might cause interference with the phase transformation of iron (hydr)oxides. However, related work has rarely been reported previously. Both the effects caused by REEs on $Fe(II)_{aq}$ and $Fe(III)_{oxide}$ interplay and the effect of iron (hydr)oxides' transformation on the environmental behavior of REEs are still unknown.

In a previous study, we found that divalent metal ions, e.g., Ba(II), Ca(II), Mg(II), Mn(II), Co(II), Ni(II) and Zn(II), inhibited the $Fe(II)_{aq}$-induced phase transformation of ferrihydrite, while the metal ions were immobilized during this process [5]. The binding affinity of metal ions to ferrihydrite was the main factor that controlled the inhibition effects and immobilization efficiency. Besides the difference in valence, REEs are lithophile elements that are more readily combined with oxygen, while the most previously studied heavy metals are often chalcophile elements or siderophile elements, which are usually combined with sulfur or iron in the earth crust [17]. The unique characteristics of REEs imply that the interaction between REEs and iron (hydr)oxide transformation might be different, which requires specific investigation.

Therefore, in the present study, we focused on the cation effects of two typical REE ions, lanthanum (La) and holmium (Ho), as representatives of light REEs (LREEs) and heavy REEs (HREEs), respectively, on the phase transformation of ferrihydrite, as well as the stabilization of these two REEs metals during the formation of secondary iron (hydr)oxides. The mechanisms of the interaction between REE cations and the ferrihydrite phase transformation were explored under fixed initial Fe(II) and ion concentrations at buffered constant pH conditions. The fate of REE cations and the transformation rates and products of ferrihydrite were the main concerns of the study. This investigation is novel in finding out the coupling mechanisms of the phase transformation of ferrihydrite with REE immobilization.

2. Materials and Methods

2.1. Ferrihydrite Preparation

Ferrihydrite was prepared in the laboratory according to previous studies [18,19]. Firstly, 20 g of $Fe(NO_3)_3 \cdot 9H_2O$ was dissolved in 250 mL of double distilled de-ionized (DDI) water. Afterwards, the solution was titrated by 1 M KOH to pH 7–8. The resulting suspension was then centrifuged. After decanting the supernatant, the remained solids were washed with DDI water. This process was repeated until the pH of the slurry reached the pH_{pzc} (~pH 7.5) of ferrihydrite. The stock suspension of ferrihydrite was stored at 4 °C and used for the Fe(II)-induced phase transformation experiments within three days. A portion of the synthesized ferrihydrite was freeze-dried for the X-ray diffraction (XRD) analysis.

2.2. Experiments of Fe(II)$_{aq}$-Induced Transformation of Ferrihydrite

To maintain anaerobic conditions, all transformation experiments were carried out inside an anoxic chamber (Coy LAB, Grass Lake, MI, USA 7% H_2/93% N_2). All the solutions used in this study were bubbled with high purity N_2 for 2 h to remove oxygen before being transferred into the anoxic chamber, then exposed to the anoxic atmosphere in the chamber for 48 h. A stock solution of 100 mM Fe(II) was prepared by dissolving $FeCl_2 \cdot 9H_2O$ in deoxygenated water inside the anoxic chamber. Similarly, the 100 mM stock solutions of the studied REE ions (denoted by Ln(III) in this study), i.e., La(III) and Ho(III), were prepared by dissolving their corresponding chloride salts in deoxygenated water, respectively.

Transformation experiments of ferrihydrite were conducted in 15 mL plastic tubes as reactors in the anoxic chamber. Each reactor contained 30 mM ferrihydrite (~3.2 g/L), 0 or 2.0 mM Fe(II), and 0 or 1.0 mM Ln(III). The pH of the reaction suspensions was buffered to around 6.5 by 20 mM 1,4-piperazinediethanesulfonic acid (PIPES), which was the typical soil pH condition [20] and also the favored pH for a high-efficiency Fe(II)$_{aq}$-induced reaction of iron (hydr)oxides [21]. Background electrolyte was provided by 20 mM KBr in the buffer. The reagents were added in the order of ferrihydrite, buffer solution, and Ln(III) solution. After equilibrating for 10 min, the Fe(II) solution was finally added to initialize the reactions. The reactors were then immediately sealed with Teflon-coated butyl rubber stoppers and crimp seals. All the reactors were wrapped in Al-foil and placed on an end-to-end rotator. Triplicates out of dozens of incubating reactors were sacrificed to be sampled for analyses at reaction time intervals of 5, 10, 15, 30, and 60 days.

2.3. Analyses of Ln(III)

Two 5 mL aliquots of suspension were taken from each sample and then placed into centrifuge tubes. After being sealed with an O-ring Teflon tape and covered tightly, the tubes were centrifuged outside the anoxic chamber at 4500 rpm for 10 min. After that, the centrifuge tubes were immediately transferred back into the anoxic chamber, and the supernatant was filtered through 0.22 μm nylon filters (Millipore, Burlington, MA, USA) and acidified with 20 μL of concentrated HCl [22,23]. The acidified solution was used to determine the concentrations of dissolved Ln(III). The solids from the retrieved two aliquots in the centrifuged tubes were extracted by 5 mL of 0.4 M HCl for 10 min. After being centrifuged again, the filtrated supernatant was used to determine the adsorbed Ln(III). The solids were then completely dissolved by concentrated HCl for 1.5 h, and the solutions were used to analyze the stabilized Ln(III) in the solids [21]. The aqueous concentrations of Ln(III) were determined by inductively coupled plasma-optical emission spectroscopy (ICP-OES, Perkin-Elmer optima 2000, Waltham, MA, USA).

2.4. Characterization of Solid Phase

The solid phases separated from the centrifugation process of reaction suspensions were washed with DDI water three times and then freeze-dried for characterization. Phase identification and quantification were conducted by XRD analyses on a D8 Advanced Diffractometer (Bruker AXS, Karlsruhe, Germany) with a Lynxeye detector, operating at 40 kV and 40 mA with Cu-Kα radiation at room temperature. The standard powder diffraction database of the International Center for Diffraction Data (ICDD PDF-2, Release 2008 [24]) was employed to match the obtained XRD patterns in this study [25,26]. Ferrihydrite was considered as amorphous due to its poor crystalline structure in this analysis. In order to quantify both crystalline and amorphous phases, 20% (*w/w*) of CaF_2 (449717-25G, Merck, Darmstadt, Germany) was added to all samples as the standard reference in the Rietveld quantitative analysis by the total patter solution (TOPAS) program [27,28].

Transmission electron microscopy (TEM, Philips-CM12, FEI, Hillsboro, OR, USA) was applied to observe the morphologies and crystallite sizes of the solid samples before and after Fe(II)$_{aq}$-induced reactions. According to the methods reported previously [7], the solid samples were dispersed in absolute ethanol ($\geq$99.5%) followed by ultrasonic shaking, and then deposited on holey-carbon film Cu grids. Afterwards, the images were recorded at 200 kV with the dried grids.

2.5. Isotope Tracer Experiments for Fe Atom Exchange

To evaluate the iron atom exchange between Fe(II)$_{aq}$ and Fe(III)$_{oxide}$, an iron isotope tracer was employed. One hundred millimole per liter of ^{57}Fe(II) stock solution was prepared by dissolving ^{57}Fe(0) powder (Isoflex > 96.0%) in 5 M HCl. Except for using ^{57}Fe(II) solution instead of normal $FeCl_2 \bullet 9H_2O$, all other parameters were used as described for the above procedure in Fe(II)$_{aq}$-induced ferrihydrite transformation. Since ^{56}Fe accounts for 91.8% of the total abundances of stable iron isotopes in the natural environment, once the atom exchange between Fe(II)$_{aq}$ and Fe(III)$_{oxide}$ occurred, the replacement of ^{57}Fe(II)$_{aq}$ by ^{56}Fe(II)$_{oxide}$ would be detected accordingly.

Thus, the iron isotopic fraction was analyzed for the reacted solution using inductively coupled plasma mass spectrometry (ICP-MS, Perkin Elmer NexION 300D, Waltham, MA, USA), operating in a collision cell mode with a glass concentric nebulizer and a HEPA filtered auto-sampler. A collision cell gas containing 93% He and 7% H_2 (>99.999% pure) was used to eliminate the polyatomic argide molecules (ArO$^+$ and ArN$^+$) at a flow rate of 1 mL·min^{-1}. All solutions were diluted to around 1 μM Fe with 2% HNO$_3$ (trace metal grade). Iron isotope fractions (*f*) were calculated by dividing the counts in each isotope channel by the sum of the total counts over all four channels (i.e., ^{54}Fe, ^{56}Fe, ^{57}Fe and ^{58}Fe).

Based on the measurements, the iron atom exchange rates were then calculated according to the following equation [29]:

$$\text{Fe atom exchange (\%)} = \frac{N_{aq} \times \left(f_{aq}^i - f_{Fe(II)}^t \right)}{N_{Fh}^{Tot} \times \left(f_{Fe(II)}^t - f_{Fh}^i \right)} \times 100 \qquad (1)$$

where N_{aq} and N_{Fh}^{Tot} are the moles of Fe(II)$_{aq}$ in the solution and total moles of Fe in ferrihydrite, f_{aq}^i and f_{Fh}^i are the initial isotopic fractions of Fe(II)$_{aq}$ and ferrihydrite, respectively, and $f_{Fe(II)}^t$ is the isotopic fraction of Fe(II)$_{aq}$ at time t.

3. Results and Discussion

3.1. Phase Transformation of Ferrihydrite Induced by Fe(II)$_{aq}$

Under natural conditions, ferrihydrite is relatively stable except for aging [7]. As examined in our preliminary experiment, no phase transformation of ferrihydrite was observed until 60 days, when Fe(II) or other cations were absent (data not shown). While under the anaerobic condition with the presence of 2.0 mM Fe(II)$_{aq}$, a significant phase transformation was detected. As shown in Figure 1A, peaks for goethite and magnetite could be identified from the XRD patterns after 5 days. Peak intensities increased with time, suggesting a continuous transformation from ferrihydrite to goethite and magnetite. With 1.0 mM La(III) added, the formation of secondary minerals differed. Peaks indicate that lepidocrocite appeared, while goethite and magnetite were not found (Figure 1B). With 1.0 mM Ho(III) added, XRD patterns clearly shown that ferrihydrite first transformed to lepidocrocite and then finally aged to goethite (Figure 1C). Peak intensities of lepidocrocite increased first and then disappeared later, and goethite peaks kept growing during the whole process. At the end of 60 days of incubation, Fe(II)$_{aq}$-induced ferrihydrite transformation resulted in different compositions of secondary minerals with or without Ln(III) presence. A majority of goethite with a minority of magnetite was obtained without Ln(III). Ferrihydrite was transformed to lepidocrocite once La(III) was present. Goethite was the only end product of the transformation under the effect of Ho(III); however, the amount was rather small compared with the absence of Ho(III) and La(III) (CK).

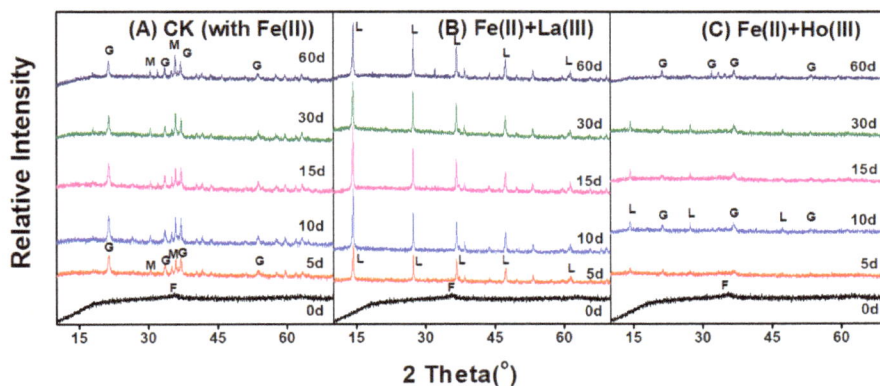

Figure 1. X-ray diffraction (XRD) patterns of the transformed ferrihydrite samples: (**A**) Fe(II)$_{aq}$ induced without any Ln(III); (**B**) Fe(II)$_{aq}$ induced with the presence of La(III); and (**C**) Fe(II)$_{aq}$ induced with the presence of Ho(III). The experiments were conducted at pH with 2.0 mM of Fe(II)$_{aq}$ and 1.0 mM of La(III) or Ho(III) under anoxic conditions. G: goethite, M: magnetite, L: lepidocrocite, F: ferrihydrite, respectively.

To elucidate the Fe(II)$_{aq}$-induced transformation processes and rates of ferrihydrite with or without REE cations, the TOPAS program was employed to quantitatively analyze the phase compositions of the incubated solids. Changes of three iron (hydr)oxide phase transformation products were obtained in a relative quantity to ferrihydrite (Figure 2). The quantity percentage of ferrihydrite decreased as a result of Fe(II)$_{aq}$-induced transformation. It was completely transformed to secondary minerals within 10 days in CK, while it took 50 days longer once La(III) was present, suggesting the transformation rate was retarded by La(III). Under the presence of Ho(III), about 20% of ferrihydrite was transformed in the first 10 days, while the transformation seemed stagnant afterwards, suggesting the transformation extent might be inhibited by Ho(III).

Lepidocrocite was detected in La(III) or Ho(III)-treated solids during the incubation (Figure 2). Under the treatment of La(III), consistent with the ferrihydrite decreasing process, the relative quantity of lepidocrocite increased quickly at the beginning, and then the rate slowed down until 100% of ferrihydrite was transformed to lepidocrocite at 60 days. In contrast, for Ho(III)-treated solids, the relative quantity of lepidocrocite first increased in the first 10 days to around 15%, and later decreased until it completely disappeared at 60 days, suggesting lepidocrocite was further transformed to more stable minerals.

Goethite was the major final product during the Fe(II)$_{aq}$-induced secondary mineral formation from ferrihydrite without REE cations. About 80% of ferrihydrite was transformed to goethite within the first 10 days, after which the relative quantity of goethite decreased to around 70%, and eventually transformed to magnetite. No goethite was observed with La(III) present, while it was found with additional Ho(III) present. About 20% of ferrihydrite was transformed to goethite at the end of 60 days with Ho(III) added. For magnetite, 30% of the transformed quantity was identified in CK, but a small fraction was found to be affected by additional REE ions.

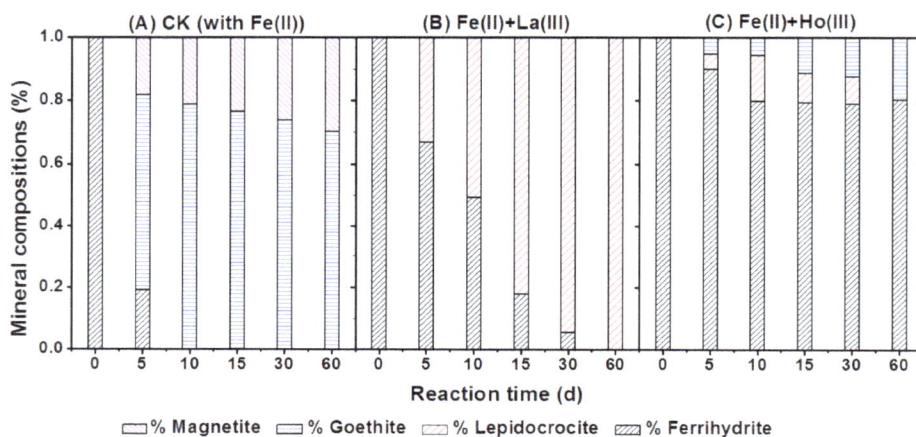

Figure 2. The compositions of iron (hydr)oxides of the transformed ferrihydrite: (**A**) Fe(II)$_{aq}$ induced without any Ln(III); (**B**) Fe(II)$_{aq}$ induced with the presence of La(III); and (**C**) Fe(II)$_{aq}$ induced with the presence of Ho(III).

The TEM images for the final solids obtained at 60 days confirmed the significant differences among the structural forms of the solid products from Fe(II)$_{aq}$-induced ferrihydrite under conditions without the effects of REEs (CK) and with the effects of La(III) and Ho(III) (Figure 3).The samples displayed remarkably different appearances under the microscope. Based on the comparison, the star-like particles, indicating the formation of goethite and magnetite [19,30], could be recognized from CK and Ho(III)-added samples, while lath-like lepidocrocite [30] was detected in the La(III)-added sample.

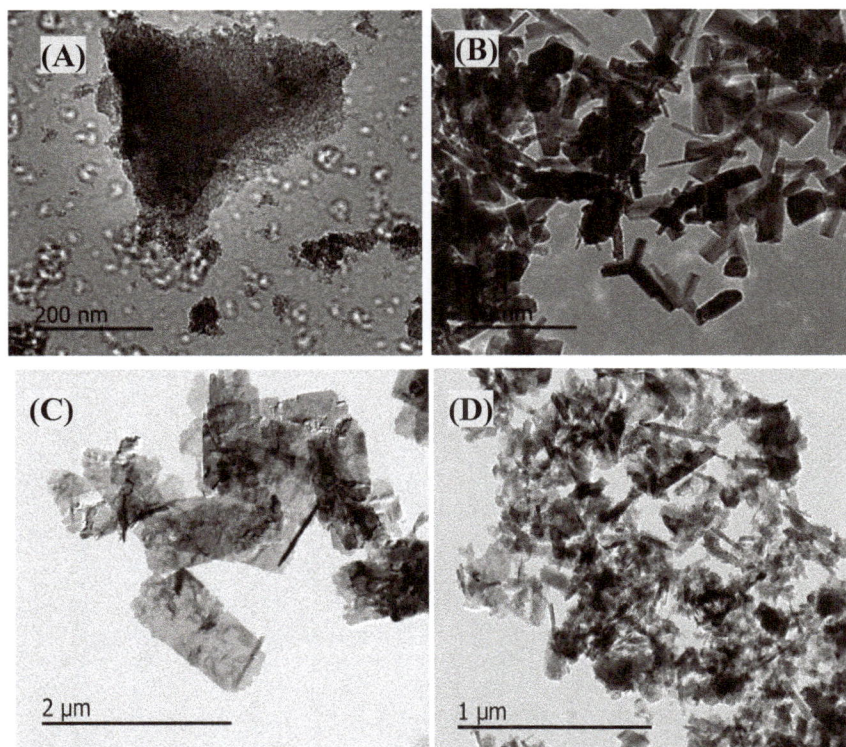

Figure 3. Transmission electron microscope images of ferrihydrite before and after reaction for 60 days in Fe(II)-induced phase transformation: (**A**) ferrihydrite before reaction; (**B**) after the Fe(II)-induced reaction without effects of Ln(III); (**C**) after the Fe(II)-induced reaction with La(III); and (**D**) after the Fe(II)-induced reaction with Ho(III).

3.2. Iron Atom Exchange between Fe(II)$_{aq}$ and Fe(III)$_{oxide}$

Atom exchange is one of the key steps in the process of the phase transformation of ferrihydrite. After Fe(II)$_{aq}$ was adsorbed on the surface of ferrihydrite, Fe(II)$_{aq}$ was oxidized to Fe(III)$_{oxide}$ due to the electron transfer and Fe atom exchange, which sequentially drive the transformation of ferrihydrite to more stable minerals [5]. In ^{57}Fe solution, the atom exchange would result in the exchange between ^{57}Fe$_{aq}$ and ^{56}Fe$_{oxide}$, which could be detected through isotopic composition changes. Therefore, according to the decrease of ^{57}Fe concentration and increase of ^{56}Fe concentration in the solution, as well as the increase of ^{57}Fe content and decrease of ^{56}Fe content in the mineral phases, the rate of atom exchange can be calculated.

As shown in Figure 4, the calculated atom exchange rates increased during the phase transformation process. The increase was fast at the beginning, and then gradually slowed down, revealing that the most dynamic change occurred within 3 days under all three treatments. After 30 days, the Fe(II)$_{aq}$-induced ferrihydrite transformation reached 75% of iron atom exchange when no Ln(III) co-existed (CK treatment). The co-existent La(III) and Ho(III) reduced the atom exchange rate to 60% and 41%, respectively. The co-existent La(III) and Ho(III) both inhibited the iron atom exchange rates, but Ho(III) inhibited it more. This result was consistent with that of previously studied heavy metals [5,8], which resulted from the competitive sorption of La(III) and Ho(III) with

Fe(II). The competition for surface access to ferrihydrite imposed by co-existing cations might be one of the main causes of the inhibited iron atom exchange and, therefore, the inhibited phase transformation.

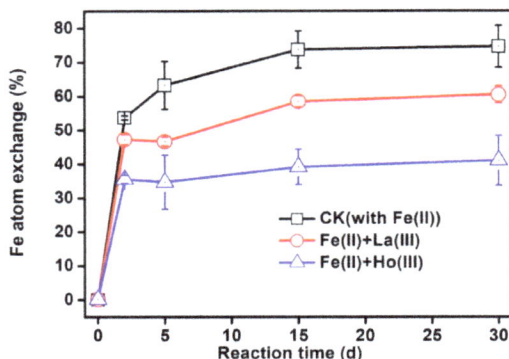

Figure 4. The atom exchange rates between Fe(II)$_{aq}$ and Fe(III)$_{oxide}$ of ferrihydrite during the Fe(II)$_{aq}$-induced phase transformation of ferrihydrite with or without the coexistence of Ln(III).

3.3. Distribution of Ln(III) Species During Fe(II)$_{aq}$-Induced Transformation of Ferrihydrite

The process of the Fe(II)$_{aq}$-induced phase transformation of ferrihydrite would also change the distribution of metallic ion species through adsorption, enwrapping, or substitution [5,6]. To examine the influence of phase transformation on Ln(III) adsorption and immobilization, a differential extraction protocol (described in Section 2.3) was conducted on Ln(III) ions. Beside the dramatic changes within the first 5 days, the distribution of Ln(III) species changed slightly afterwards (Figure 5), while the phase transformation of iron (hydr)oxides occurred dynamically during the whole period of 60 days (Figure 2). The results suggest a high variety between La(III) and Ho(III) (Figure 5). An overwhelming majority of supplemented La(III)$_{aq}$ remained dissolved in the solution after reacting with ferrihydrite. A smaller portion of La(III) was stabilized to become structurally incorporated into the secondary iron (hydr)oxide, and the smallest portion was adsorbed. For Ho(III), conversely, the greatest amount was the structural species, followed by adsorbed species, and a very small portion remain dissolved in the aqueous phase.

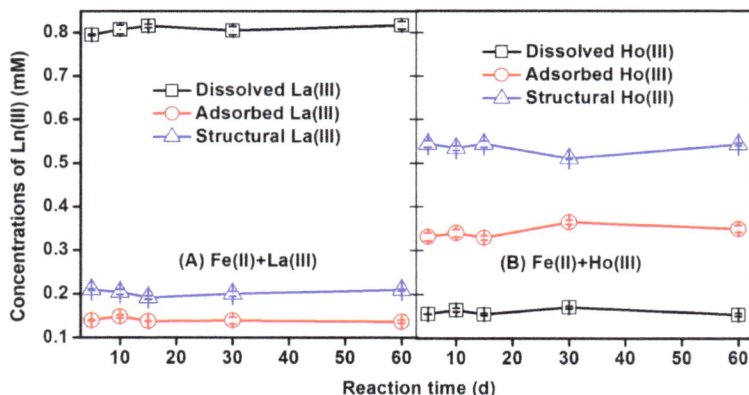

Figure 5. The concentrations of La(III) (**A**) and Ho(III) (**B**) species during the Fe(II)$_{aq}$-induced phase transformation of ferrihydrite in the presence of La(III) and Ho(III), respectively.

Remarkably, despite the variety in the amount, the distribution of structural Ln(III) had a good consistency with adsorbed Ln(III), as with the less adsorbed La(III) with less structural La(III) and the more adsorbed Ho(III) with more structural Ho(III). Similarly to Fe(II)$_{aq}$, the attachment to the mineral surfaces was supposed to be the first and critical step for ions to be able to be further incorporated or enwrapped into the lattice of secondary minerals and thus finally immobilized.

3.4. The Mechanism of Fe(II)$_{aq}$-Induced Ferrihydrite Transformation Coupled REE Immobilization

Environmental factors including pH, temperature, Fe(II)$_{aq}$ concentration, and ionic strength were reported to influence the Fe(II)$_{aq}$-induced phase transformation of ferrihydrite [19,25]. Previous studies stated that metal-substitution in iron (hydr)oxides slowed down the phase transformation [31,32]. The substituted metal could block the reductive dissolution of iron and reduce the bulk electron conduction, thus inhibiting the iron atom exchange or electron exchange between Fe(II)$_{aq}$ and Fe(III)$_{oxide}$ [6,33]. The surface adsorption of silica, organic matter, and metal ions would also influence the transformation rate and pathway of ferrihydrite; alternatively, this may change the surface accessibility and reactivity [5,8,34].

Adsorption to the surface of ferrihydrite might be the most important step during the phase transformation process. It was implied that competition with co-existing ions for surface adsorption would be crucial to the phase transformation. As our previous study found [5], the adsorption of Fe(II)$_{aq}$ had a positive relationship with the rate of the Fe(II)$_{aq}$-induced transformation of ferrihydrite, while the binding ability of different divalent metal ions (Me(II)), i.e., Mn(II), Mg(II), Ca(II), Ni(II), Ba(II), Co(II) and Zn(II), imposed negative effects on the ferrihydrite transformation rate. This might have similar effects by Ln(III) in the present study. It was also found by other researchers that REEs could be adsorbed on the fine grain surface of ferrihydrite, and the adsorption would then impede the crystallization of ferrihydrite [35].

The family of REEs are usually classified to LREEs and HREEs according to their varied behaviors in complexation with ligands, such as carbonate, phosphate, sulphate, oxides as well as organic ligands [36]. LREEs are usually less stably sorbed with organic ligands and minerals than HREEs [37]. In our study, La, as the 'lightest' REE, was indeed sorbed significantly less than the representative HREE, Ho (Figure 5). The difference in the adsorption behavior between La(III) and Ho(III) might be the first cause of the difference in their effect on ferrihydrite transformation. The large surface coverage caused by Ho(III) adsorption would significantly block the accessibility for Fe(II)$_{aq}$ and thus limit the ferrihydrite transformation extent, which provides a better explanation for the fact that Ho(III)-treated solids had a lower atom exchange rate than La(III)-treated ones, which were both less than CK (Figure 4).

However, the case of Ln(III) appeared to be more complicated than the previously studied Me(II). These results could not simply be explained by binding abilities or competitive adsorption. For the formerly studied Me(II), ferrihydrite transformed from lepidocrocite to goethite and eventually magnetite, while only the transformation rates were different [5]. However, in the present study, the presence of La(III) significantly slowed the transformation process, taking 60 days longer to achieve the 100% transformation of original ferrihydrite than that without the effects of Ln(III) (Figure 2A,B). At the end of incubation, however, this was still half-way through the transformation at the lepidocrocite phase (Figures 1 and 3), whilst for Ho(III), about 80% of ferrihydrite did not go through the transformation, though the other 20% reached the transformed phase of goethite (Figure 2C).

Evidence of the incorporation of adsorbed metals into the secondary iron minerals has been reported [38,39]. REEs such as neodymium and lutetium were also found to be incorporated into hematite when transformed from ferrihydrite [40,41]. Different to Fe(II) and Me(II), the trivalent Ln(III) used in this study could substitute for Fe(III)$_{oxide}$ directly without requiring oxidation to a higher valence. Compared with La (104 pm), Ho has a smaller atomic radius (90 pm) and may have advantages in atom exchange and substitution.

However, the incorporation or metal-substitution of Ho(III) in iron (hydr)oxides in the present study was found to be limited. As compared with La(III) in this study and Me(II) in our previous work [5], the adsorbed species took a small percentage in the species distribution, while a large portion of Ho(III) remained in the adsorbed species (Figure 5). If the incorporation, enwrapping or substitution occurred once the cations were adsorbed, a smaller amount of adsorbed Me(II) or Ln(III) would be detected. Hence, it could be implied that the alteration of adsorbed Ho(III) to structural Ho(III) was limited. It was reasonable that REEs behaved differently with Me(II) when interplayed with iron minerals due to the varied atomic radii, dominant valences and affinities to iron-bearing matters, as well as other physicochemical properties. The successful transformation of the 20% Ho-treated ferrihydrite to goethite did not bring most of the adsorbed Ho(III) into the crystalline structure, suggesting the adsorption ability of Ho(III) was larger than the saturated capacity of structural incorporation, resulting in a significant portion of Ho(III) remaining on the mineral surfaces and limiting the surface access to Fe(II)$_{aq}$. The inhibition of 80% of the ferrihydrite phase transformation might be strongly related with this.

Based on the discussion above, it is proposed here that the two-step process, i.e., surface adsorption followed by structural incorporation into minerals, imposed significant impacts on the iron (hydr)oxide phase transformation. Given the differences in adsorption affinities and incorporation behaviors, La(III) and Ho(III) showed distinguishable performances in influencing the ferrihydrite phase transformation. For La(III), the competition with Fe(II)$_{aq}$ in surface adsorption was the main mechanism that retarded the ferrihydrite phase transformation. Afterwards, the incorporation of adsorbed La(III) into the crystalline structure appeared to be smooth. The recovered surface access allowed for the completion of the 100% transformation of ferrihydrite, though at a much slower rate. But then, it was possible that the La-substitution made lepidocrocite refractory to be further transformed. Assuming ferrihydrite was transformed through the pathway of lepidocrocite–goethite/magnetite [5], the transformation of La-treated ferrihydrite was not completed. A longer incubation is still necessary to see whether the phase transformation was blocked half-way or not. For Ho(III), the incorporation was relatively small. Hence, Ho(III)-treated ferrihydrite could go through the whole transformation pathway in 60 days. However, the surface adsorption was rather large, and the limited surface access to Fe(II)$_{aq}$ resulted in the largely inhibited secondary transformation. It was inferred here that La(III) and Ho(III) blocked different steps during the normal phase transformation of ferrihydrite. More studies are required to examine this assumption. Detailed mechanisms involving the incorporation of Ln(III) into iron (hydr)oxides and its influence in the subsequent phase transformation still need further investigations.

Remarkably, it was also observed that the changes of Ln(III) species were most dynamic within the first 5 days (Figure 5), while changes in iron (hydr)oxide phases lasted longer (Figure 2). As formerly reported, when ferrihydrite was transformed to finer crystallized phases, i.e., goethite and magnetite, the sorption of REEs was decreased; that is, the formerly scavenged Ln(III)$_{aq}$ would be released to the solution, and LREEs, especially La, were the most sensitive elements to this change [14,42]. However, the changes of La(III) after 5 days seemed to be random fluctuations, suggesting that the species were not affected by the iron transformation thereafter. The direct inclusion to structural La(III) might protect the adsorbed La(III) from being desorbed during iron (hydr)oxide transformation, since results showed more structural species than adsorbed species (Figure 5).

4. Conclusions

Coexistent REEs play important roles in the Fe(II)$_{aq}$-induced phase transformation of ferrihydrite as well as resulting in the different adsorption and structural immobilization behaviors of different REEs. The two studied REEs, i.e., the LREE of La(III) and the HREE of Ho(III), both inhibited the phase transformation of ferrihydrite due to their competitive adsorption with Fe(II)$_{aq}$ on the surface of ferrihydrite and its transformed iron (hydr)oxide products. Because of the low adsorptive affinity of LREEs and the smaller atomic radius, more Ho(III) was adsorbed on the surface of iron minerals, which led to the lower phase transformation rates of ferrihydrite than that with La(III). Accordingly, more Ho(III) ions were structurally incorporated into the iron mineral products of transformed ferrihydrite induced by Fe(II)$_{aq}$. These findings highlight the important roles of iron minerals and the phase transformation processes in evaluating the geochemical behavior of different REEs, especially in iron-rich environments.

Author Contributions: Writing-Original Draft preparation, Y.F.; Data curation, J.H.; Investigation, C.L.; Conceptualization, C.L. and F.L.; Writing–Review and Editing, C.L., L.H. and Z.Z.; Validation, F.L. and T.X.; Methodology, M.C., T.G. and Z.W.

Funding: This work was funded by the National Natural Science Foundation of China (U1701241, 41673135, and U1612442), the Frontier Science Research Program of the Chinese Academy of Sciences (CAS) (QYZDB-SSW-DQC046), the Science and Technology Foundation of Guangdong Province, China (2016B020242006 and 2016TX03Z086), the Science and Technology Foundation of Guangzhou, China (201704020200 and 201710010128), the Innovation Team Program of Modern Agricultural Science and Technology of Guangdong Province (2016LM2149), and the One Hundred Talents Program of the CAS.

Conflicts of Interest: The authors declare no conflict of interest.

References

1. Latta, D.E.; Gorski, C.A.; Scherer, M.M. Influence of Fe^{2+}-catalysed iron oxide recrystallization on metal cycling. *Biochem. Soc. Trans.* **2012**, *40*, 1191–1197. [CrossRef] [PubMed]
2. Suter, D.; Siffert, C.; Sulzberger, B.; Stumm, W. Catalytic dissoluion of iron(III)(hydr)oxides by oxalic acid in the presence of Fe(II). *Naturwissenschaften* **1988**, *75*, 571–573. [CrossRef]
3. Williams, A.G.; Scherer, M.M. Spectroscopic evidence for Fe(II)-Fe(III) electron transfer at the iron oxide—Water interface. *Environ. Sci. Technol.* **2004**, *38*, 4782–4790. [CrossRef] [PubMed]
4. Pedersen, H.D.; Postma, D.; Jakobsen, R.; Larsen, O. Fast transformation of iron oxyhydroxides by the catalytic action of aqueous Fe(II). *Geochim. Cosmochim. Acta* **2005**, *69*, 3967–3977. [CrossRef]
5. Liu, C.; Zhu, Z.; Li, F.; Liu, T.; Liao, C.; Lee, J.J.; Shih, K.; Tao, L.; Wu, Y. Fe(II)-induced phase transformation of ferrihydrite: The inhibition effects and stabilization of divalent metal cations. *Chem. Geol.* **2016**, *444*, 110–119. [CrossRef]
6. Frierdich, A.J.; Scherer, M.M.; Bachman, J.E.; Engelhard, M.H.; Rapponotti, B.W.; Catalano, J.G. Inhibition of trace element release during Fe(II)-activated recrystallization of Al-, Cr-, and Sn-substituted goethite and hematite. *Environ. Sci. Technol.* **2012**, *46*, 10031–10039. [CrossRef] [PubMed]
7. Cornell, R.M.; Schwertmann, U. *The Iron Oxides: Structure, Properties, Reactions, Occurrences and Uses*, 2nd ed.; Wiley-VCH: Weinhein, Germany, 2003.
8. Liu, C.S.; Li, F.B.; Chen, M.J.; Liao, C.Z.; Tong, H.; Hua, J. Adsorption and stabilization of lead during Fe(II)-catalyzed phase transformation of ferrihydrite. *Acta Chim. Sin.* **2017**, *75*, 621–628. [CrossRef]
9. Du, H.; Peacock, C.L.; Chen, W.; Huang, Q. Binding of Cd by ferrihydrite organo-mineral composites: Implications for Cd mobility and fate in natural and contaminated environments. *Chemosphere* **2018**, *207*, 404–412. [CrossRef] [PubMed]
10. Freslon, N.; Bayon, G.; Toucanne, S.; Bermell, S.; Bollinger, C.; Cheron, S.; Etoubleau, J.; Germain, Y.; Khripounoff, A.; Ponzevera, E.; et al. Rare earth elements and neodymium isotopes in sedimentary organic matter. *Geochim. Cosmochim. Acta* **2014**, *140*, 177–198. [CrossRef]
11. Wang, L.Q.; Liang, T. Geochemical fractions of rare earth elements in soil around a mine tailing in Baotou, China. *Sci. Rep.* **2015**, *5*, 12483. [CrossRef] [PubMed]
12. Leybourne, M.I.; Johannesson, K.H. Rare earth elements (REE) and yttrium in stream waters, stream sediments, and Fe-Mn oxyhydroxides: Fractionation, speciation, and controls over REE plus Y patterns in the surface environment. *Geochim. Cosmochim. Acta* **2008**, *72*, 5962–5983. [CrossRef]
13. Li, X.; Chen, Z.; Chen, Z.; Zhang, Y. A human health risk assessment of rare earth elements in soil and vegetables from a mining area in Fujian Province, Southeast China. *Chemosphere* **2013**, *93*, 1240–1246. [CrossRef]
14. Bolanz, R.M.; Kiefer, S.; Göttlicher, J.; Steininger, R. Hematite (alpha-Fe$_2$O$_3$)—A potential Ce^{4+} carrier in red mud. *Sci. Total Environ.* **2018**, *622*, 849–860. [CrossRef] [PubMed]

15. Nedel, S.; Nedel, K.; Dideriksen, B.C.; Christiansen, N.; Bovet, S.L.S.; Stipp, S.L. Uptake and Release of Cerium During Fe-Oxide Formation and Transformation in Fe(II) Solutions. *Environ. Sci. Technol.* **2010**, *44*, 4493–4498. [CrossRef] [PubMed]

16. Steinmann, M.; Stille, P. Controls on transport and fractionation of the rare earth elements in stream water of a mixed basaltic-granitic catchment basin (Massif Central, France). *Chem. Geol.* **2008**, *254*, 1–18. [CrossRef]

17. Migaszewski, Z.M.; Galuszka, A. The Characteristics, Occurrence, and Geochemical Behavior of Rare Earth Elements in the Environment: A Review. *Crit. Rev. Environ. Sci. Technol.* **2015**, *45*, 429–471. [CrossRef]

18. Das, S.; Hendry, M.J.; Essilfie-Dughan, J. Effects of Adsorbed Arsenate on the Rate of Transformation of 2-Line Ferrihydrite at pH 10. *Environ. Sci. Technol.* **2011**, *45*, 5557–5563. [CrossRef] [PubMed]

19. Boland, D.D.; Collins, R.N.; Miller, C.J.; Glover, C.J.; Waite, T.D. Effect of solution and solid-phase conditions on the Fe(II)-accelerated transformation of ferrihydrite to lepidocrocite and goethite. *Environ. Sci. Technol.* **2014**, *48*, 5477–5485. [CrossRef] [PubMed]

20. Liu, C.S.; Wang, Y.; Li, F.B.; Chen, M.J.; Zhai, G.; Tao, L.; Liu, C. Influence of geochemical properties and land-use types on the microbial reduction of Fe(III) in subtropical soils. *Environ. Sci. Process. Impacts* **2014**, *16*, 1938–1947. [CrossRef] [PubMed]

21. Reddy, T.R.; Frierdich, A.J.; Beard, B.L.; Johnson, C.M. The effect of pH on stable iron isotope exchange and fractionation between aqueous Fe(II) and goethite. *Chem. Geol.* **2015**, *397*, 118–127. [CrossRef]

22. Guelke, M.; Blanckenburg, F.; Schoenberg, R.; Staubwasser, M.; Stuetzel, H. Determining the stable Fe isotope signature of plant-available iron in soils. *Chem. Geol.* **2010**, *277*, 269–280. [CrossRef]

23. Frierdich, A.J.; Beard, B.L.; Scherer, M.M.; Johnson, C.M. Determination of the Fe(II)aq–magnetite equilibrium iron isotope fractionation factor using the three-isotope method and a multi-direction approach to equilibrium. *Earth Planet. Sci. Lett.* **2014**, *391*, 77–86. [CrossRef]

24. ICDD. *Powder Diffraction File (PDF 2), Release 2008*; International Centre for Diffraction Data: Newtown Square, PA, USA, 2008.

25. Das, S.; Hendry, M.J.; Essilfie-Dughan, J. Transformation of Two-Line Ferrihydrite to Goethite and Hematite as a Function of pH and Temperature. *Environ. Sci. Technol.* **2011**, *45*, 268–275. [CrossRef] [PubMed]

26. Lu, X.W.; Shih, K.M.; Liu, C.S.; Wang, F. Extraction of Metallic Lead from Cathode Ray Tube (CRT) Funnel Glass by Thermal Reduction with Metallic Iron. *Environ. Sci. Technol.* **2013**, *47*, 9972–9978. [CrossRef] [PubMed]

27. De La Torre, A.G.; Bruque, S.; Aranda, M.A.G. Rietveld quantitative amorphous content analysis. *J. Appl. Crystallogr.* **2001**, *34*, 196–202. [CrossRef]

28. Bernasconi, A.; Dapiaggi, M.; Gualtieri, A.F. Accuracy in quantitative phase analysis of mixtures with large amorphous contents. The case of zircon-rich sanitary-ware glazes. *J. Appl. Crystallogr.* **2014**, *47*, 136–145. [CrossRef]

29. Frierdich, A.J.; Helgeson, M.; Liu, C.; Wang, C.; Rosso, K.M.; Scherer, M.M. Iron atom exchange between hematite and aqueous Fe(II). *Environ. Sci. Technol.* **2015**, *49*, 8479–8486. [CrossRef] [PubMed]

30. Liu, H.; Li, P.; Zhu, M.Y.; Wei, Y.; Sun, Y.H. Fe(II)-induced transformation from ferrihydrite to lepidocrocite and goethite. *J. Solid State Chem.* **2007**, *180*, 2121–2128. [CrossRef]

31. Massey, M.S.; Lezama-Pacheco, J.S.; Michel, F.M.; Fendorf, S. Uranium incorporation into aluminum-substituted ferrihydrite during iron(II)-induced transformation. *Environ. Sci. Process. Impacts* **2014**, *16*, 2137–2144. [CrossRef] [PubMed]

32. Masue-Slowey, Y.; Loeppert, R.H.; Fendorf, S. Alteration of ferrihydrite reductive dissolution and transformation by adsorbed As and structural Al: Implications for As retention. *Geochim. Cosmochim. Acta* **2011**, *75*, 870–886. [CrossRef]

33. Latta, D.E.; Bachman, J.E.; Scherer, M.M. Fe Electron Transfer and Atom Exchange in Goethite: Influence of Al-Substitution and Anion Sorption. *Environ. Sci. Technol.* **2012**, *46*, 10614–10623. [CrossRef] [PubMed]

34. Jones, A.M.; Collins, R.N.; Rose, J.; Waite, T.D. The effect of silica and natural organic matter on the Fe(II)-catalysed transformation and reactivity of Fe(III) minerals. *Geochim. Cosmochim. Acta* **2009**, *73*, 4409–4422. [CrossRef]

35. Riley, E.; Dutrizac, J.E. The behaviour of the rare earth elements during the precipitation of ferrihydrite from sulphate media. *Hydrometallurgy* **2017**, *172*, 69–78. [CrossRef]

36. Byrne, H.R.; Li, B. Comparative complexation behavior of the rare earths. *Geochim. Cosmochim. Acta* **1995**, *59*, 4575–4589. [CrossRef]

37. Coppin, F.; Berger, G.; Bauer, A.; Castet, S.; Loubet, M. Sorption of lanthanides on smectite and kaolinite. *Chem. Geol.* **2002**, *182*, 57–68. [CrossRef]

38. Karthikeyan, K.G.; Elliott, H.A.; Cannon, F.S. Adsorption and coprecipitation of copper with the hydrous oxides of iron and aluminum. *Environ. Sci. Technol.* **1997**, *31*, 2721–2725. [CrossRef]

39. Sun, T.; Paige, C.R.; Snodgrass, W.J. The effect of cadmium on the transformation of ferrihydrite into crystalline products at pH 8. *Water Air Soil Pollut.* **1996**, *91*, 307–325. [CrossRef]

40. Nagano, T.; Mitamura, H.; Nakayama, S.; Nakashima, S. Formation of goethite and hematite from neodymium-containing ferrihydrite suspensions. *Clays Clay Miner.* **1999**, *47*, 748–754. [CrossRef]

41. Dardenne, K.; Schäfer, T.; Lindqvist-Reis, P.; Denecke, M.A.; Plaschke, M.; Rothe, J.; Kim, J.I. Low temperature XAFS investigation on the lutetium binding changes during the 2-line ferrihydrite alteration process. *Environ. Sci. Technol.* **2002**, *36*, 5092–5099. [CrossRef] [PubMed]

42. Bozau, E.; Gottlicher, J.; Stark, H.J. Rare earth element fractionation during the precipitation and crystallisation of hydrous ferric oxides from anoxic lake water. *Appl. Geochem.* **2008**, *23*, 3473–3486. [CrossRef]

© 2018 by the authors. Licensee MDPI, Basel, Switzerland. This article is an open access article distributed under the terms and conditions of the Creative Commons Attribution (CC BY) license (http://creativecommons.org/licenses/by/4.0/).

Communication

Identification of Surface Processes in Individual Minerals of a Complex Ore through the Analysis of Polished Sections Using Polarization Microscopy and X-ray Photoelectron Spectroscopy (XPS)

Dhamelyz Silva-Quiñones [1,2], **Chuan He** [2], **Melissa Jacome-Collazos** [1], **Carsten Benndorf** [1,3], **Andrew V. Teplyakov** [2] **and Juan Carlos F. Rodriguez-Reyes** [1,*]

[1] Department of Bioengineering and Chemical Engineering, Universidad de Ingeniería y Tecnología—UTEC, Jr. Medrano Silva 165, Lima 04, Peru; dsilvaq@udel.edu (D.S.-Q.); melissajacome6@gmail.com (M.J.-C.); benndorf@chemie.uni-hamburg.de (C.B.)

[2] Department of Chemistry and Biochemistry, University of Delaware, Newark, DE 19716, USA; river@udel.edu (C.H.); andrewt@udel.edu (A.V.T.)

[3] Institute of Physical Chemistry, Department of Chemistry, University of Hamburg, Bundesstr. 45, D-20146 Hamburg, Germany

[*] Correspondence: jcrodriguez@utec.edu.pe; Tel.: +51-1-230-5000 (ext. 4220)

Received: 10 September 2018; Accepted: 26 September 2018; Published: 28 September 2018

check for
updates

Abstract: Understanding the changes of a mineral during ore processing is of capital importance for the development of strategies aimed at increasing the efficiency of metal extraction. This task is often difficult due to the variability of the ore in terms of composition, mineralogy and texture. In particular, surface processes such as metal re-adsorption (preg-robbing) on specific minerals are difficult to evaluate, even though they may be of importance as the re-adsorbed material can be blocking the valuable mineral and negatively affect the extraction process. Here, we show a simple yet powerful approach, through which surface processes in individual minerals are identified by combining polarization microscopy (MP) and X-ray photoelectron spectroscopy (XPS). Taking as an example a silver-containing polymetallic sulfide ore from the Peruvian central Andes (pyrite-based with small amounts of galena), we track the changes in the sample during the course of cyanidation. While polarization microscopy is instrumental for identifying mineralogical species, XPS provides evidence of the re-adsorption of lead on a pyrite surface, possibly as lead oxide/hydroxide. The surface of pyrite does not show significant changes after the leaching process according to the microscopic results, although forms of oxidized iron are detected together with the re-adsorption of lead by XPS. Galena, embedded in pyrite, dissolves during cyanide leaching, as evidenced by PM and by the decrease of XPS signals at the positions associated with sulfide and sulfate. At the same time, the rise of a lead peak at a different position confirms that the re-adsorbed lead species cannot be sulfides or sulfates. Interestingly, lead is not detected on covellite surfaces during leaching, which shows that lead re-adsorption is a process that depends on the nature of the mineral. The methodology shown here is a tool of significant importance for understanding complex surface processes affecting various minerals during metal extraction.

Keywords: leaching; cyanide; pyrite; polarization microscopy; XPS; surface; re-adsorption

---

1. Introduction

Understanding the metallurgical behavior of an ore is key in determining the viability of a project in mineral processing. Ore characterization is important not only for making a correct assessment of reserves and resources, but also for exploring novel processing strategies [1]. Polarization microscopy (PM) is a characterization technique that differentiates minerals based on their different optical properties under polarized light. Minerals with transition metals will have characteristic reflection indexes; other minerals (such as carbonates and silicates) do not reflect light and appear opaque. Thin sections of the latter minerals, however, are transparent to transmitted light, while the first group of minerals are non-transparent. Thus, by combining experiments with reflected and transmitted polarized light, this technique allows for the identification of textures, alterations, intergrowths, among other characteristics of the mineral [2]. This technique is often complemented by more detailed microscopies, such as scanning electron microscopy (including its variant for quantitative evaluation of minerals, QEM-SCAN), which together provide complete information regarding the mineralogical composition of ores. Applied to different stages of a metallurgical process, these techniques allow for a better understanding of the effect of such process on specific minerals within the ore [3].

Since techniques for mineralogical characterization often analyze information over a few micrometers within the mineral, surface-related processes are rarely detected. This is a significant drawback, considering that important metallurgical processes occur at the topmost layers of minerals. For example, preg-robbing is a detrimental process through which a metal is re-adsorbed after it has been successfully leached [4]. X-ray photoelectron spectroscopy (XPS) has a depth sensitivity on the order of 10 nm and provides information on the oxidation states and the chemical environment of different elements on a surface. There have been several investigations relying on XPS to obtain information regarding surface transformation of individual, pure minerals [5–8]. Even though this strategy provides important information regarding surface processes, many processes occur only when one mineral is in the presence of another. One of the best known examples of this is the formation of galvanic pairs, where a stable mineral (such as pyrite) can promote the dissolution of a more reactive mineral (such as sphalerite) [9,10]. Another example is the formation of passivating layers that can hinder the metal extraction by blocking the surface. For example, passivation may involve the formation of surface by-products (formation of sulfur during the oxidation of sulfide minerals) or the re-adsorption of dissolved metal ions (preg-robbing of $AuCN_{(ads)}$ hinders the cyanidation) [11,12]. Since passivating layers may be as thin as a few atomic layers, identification of these species is a difficult task with traditional mineralogical techniques, and it is in these cases that XPS can provide the most valuable information [13].

Here, a method to identify surface processes occurring on surfaces of specific minerals within a complex ore is presented. By using a polished section of minerals embedded into a resin, minerals within the section are identified with polarization microscopy and their surfaces are characterized with XPS. The same characterization tools are employed after the sample is subjected to a hydrometallurgical process (leaching with cyanide). The matrix responds well to the leaching medium and is compatible with ultra-high vacuum conditions, which allows us to use both techniques, namely XPS and PM, on the same samples [14].

2. Materials and Methods

The ore sample analyzed was generously donated by Volcan Cia. Minera, and is a polymetallic sulfide from the Central Andes of Peru which enters a leaching circuit for extracting silver. The ore composition is mainly pyrite (FeS_2), associated with galena (PbS), covellite (CuS) and sphalerite (ZnS), as well as with sulfosalts of copper and silver. The ore also contains non-metallic minerals, such as carbonates and silicates. The analysis via inductively-coupled plasma mass spectrometry (ICP-MS) shows a composition of 45.2 wt.% S, 37.1 wt.% Fe, Pb (4487 ppm), Cu (992 ppm), Zn (807 ppm) and Ag (136 ppm). The study focusing on the response of this silver-containing ore to cyanidation is a subject of a separate publication. The ore size used was mesh +35 ($\approx$500 μm) to make identification of the minerals via PM reliable. Polished sections were prepared by embedding ore samples in an epoxy resin, which after curing was polished to ensure flatness so that it can be characterized by polarization microscopy and XPS. In the ultra-high vacuum system, the resin was stable at the base pressure in the XPS chamber.

For the leaching tests the polished sections were suspended by a nylon thread and immersed in 150 mL of cyanide solution (2 g/L) at pH ≈ 11.5 using a magnetic stirring plate at 300 RPM for 6 h. The amount of cyanide was kept constant throughout the experiments, as confirmed by measuring the cyanide consumption via titration (with silver nitrate and potassium iodide as indicator, at times 15, 30, 60, 120, 180, 240 and 360 min) and replenishing the amount consumed. CAUTION: Experiments with cyanide must be conducted by trained personnel and special care should be exercised, as this product is extremely poisonous. Solutions must be kept at all times at pH above 10 to avoid the risk of formation of gaseous hydrogen cyanide (HCN). Ore samples and all materials exposed to cyanide should be treated with strong oxidants (bleach, hydrogen peroxide) afterwards to ensure decomposition of residual cyanide.

Polarization microscopy was performed with an OLYMPUS BX51 microscope (Olympus, Tokyo, Japan). The polished sections were marked on the back by two perpendicular axes to reproducibly locate the point chosen for analysis. The samples were analyzed in the magnification range from 5× to 100× and 4 different minerals were identified as pyrite (FeS_2), sphalerite (ZnS), covellite (CuS) and pyrite with inclusions of galena (PbS). The micrographs were taken before and after a 6-h leaching.

XPS measurements were performed on a Thermo Scientific K-Alpha+ instrument (Thermo Fisher, Waltham, MA, USA) equipped with an Al Kα source (hν = 1486.6 eV), and a 35.3° take-off angle normal to the surface was used. The pass energy used was 58.5 eV. Measurements were performed at a base pressure of 5×10^{-9} torr. Samples were transferred from solution to the XPS chamber as fast as possible to minimize changes due to exposure to ambient. No additional cleaning was employed following the transfer. Polished sections (with the points of interested marked) were introduced into the chamber and then the points chosen were located using the built-in microscope of the XPS instrument. Figure S2 shows the identification of the minerals using PM and the camera operating in UHV, together with the area selected for analysis (which was set to 300 μm in diameter). The analysis was done before and after 30 and 360 min of leaching. The C 1s peak at 284.6 eV was used to calibrate all spectra. Data was collected with the software Thermo Avantage v. 5.89 (Thermo Fisher, Waltham, MA, USA) and processed with the Casa XPS software version 2.3.16 (Casa Software Ltd, Teignmouth, UK) and Origin 2017 (OriginLab, Northampton, MA, USA).

The behavior of the resin employed to prepare the polished sections in the leaching solution was followed by XPS and PM. The XPS analysis of the resin showed C, O and Si as main components, which did not change after exposure to the leaching solution (data not shown). Micrographs of the resin showed small pits in some areas, likely from a reaction with the basic medium, although they were not apparent by simple visual inspection. Even though it could be argued that species from the epoxy resin can be interfering during the mechanical preparation of samples and/or during the leaching process (which cannot be tested adequately by ex-situ XPS as C and O from ambient interfere with the measurements of epoxy resins), it will be shown here that the main surface process identified, lead re-adsorption, is also observed during the leaching of epoxy-free powdered samples.

3. Results and Discussion

Polarization microscopy allowed for the identification of the most representative minerals in the ore. Figure 1 shows that individual minerals, identified within a polished section with PM, can be analyzed by XPS. The presence of less-abundant minerals such as covellite and sphalerite was also confirmed by PM and XPS (Supplementary Materials, Figure S1). For XPS measurements, the assignment of species from Fe, S and Pb is shown in Table 1, together with selected information available from the literature [10,15–18]. To show the reliability of the XPS signal assignments, the calibration standards are listed in the table as well for all the references cited. Despite small differences arising from different calibration standards, all the assignments of the features observed by XPS, especially the core level shifts and different oxidation states of the elements analyzed are not affected noticeably by these different calibration standards.

Table 1. Position of XPS signals for species relevant for this work, as obtained from the literature. All values are given in eV.

Species	This Work	Ref [15]	Ref [10]	Ref [16]	Ref [17]	Ref [18]
S 2p$_{3/2}$ (Disulfide; Pyrite)	162.0	162.1–162.6	162.6	-	-	-
S 2p$_{3/2}$ (Sulfide; Galena)	160.7	160.1–160.7				
Fe 2p$_{3/2}$ (Disulfide)	706.7		707.4	-	-	-
Fe 2p$_{3/2}$ (O/OH or sulfate)	710.8		711.4	710.8	-	-
Pb 4f$_{7/2}$ (sulfide)	137.4		-	-	138.7	-
Pb 4f$_{7/2}$ (sulfate)	138.5		-	-	139.4	-
Pb 4f$_{7/2}$ (oxide/thiosulfate)	138.1–138.2		-	-	138.0	138.0–138.1
B.E. calibration	C 1s 284.6	C1s 284.6	Au 4f$_{7/2}$ 83.9	C1s 284.6	C1s 284.8	C1s 285.1

Since in Figure 1 the measurements were made over a pyrite grain, the XPS spectra of this mineral show the presence of iron and sulfur, while other metals, such as lead (Pb), are not present. It is interesting to notice that iron features a Fe 2p$_{3/2}$ signal at a position corresponding to sulfides (Table 1), which is consistent with the fact that sulfur features a S 2p$_{3/2}$ signal with its maximum intensity around values corresponding to sulfide species (162.0 eV). The sulfur peak seems to be composed of two pairs of signals, from which one has been attributed to disulfide and the other may be associated with polysulfides or other sulfur species in higher oxidation states [15]. No sulfate is detected in this spot, which indicates that pyrite has not been oxidized after the sample was prepared for experiments (polished), as it would be expected since pyrite is relatively stable with respect to oxidation in air [19]. After 30 min of cyanidation, however, the pyrite surface shows evidence of a reaction represented by the increase of a Fe 2p$_{3/2}$ signal at 710.8 eV, which is attributed to oxides and/or hydroxides. The sulfur signal, however, features an increased sharpness, suggesting that species different to disulfide have been removed from the surface. The most striking finding is the appearance of lead on the pyrite surface during leaching (at surface concentrations of 5.7 and 17.4 At % after 30 min and six hours, respectively), probably in the form of oxide/hydroxide or thiosulfate. Since we are focusing the study on a pyrite grain, the only explanation for the presence of lead is a two-step process, where lead is leached from particles of galena (also embedded in the polished resin sample) and it later re-adsorbs on the surface of pyrite. Senanayake [12] has considered the following equations occurring during cyanidation in presence of galena:

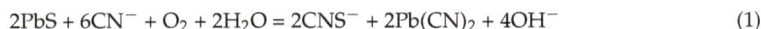

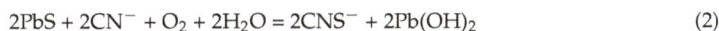

$$2PbS + 6CN^- + O_2 + 2H_2O = 2CNS^- + 2Pb(CN)_2 + 4OH^- \tag{1}$$

$$2PbS + 2CN^- + O_2 + 2H_2O = 2CNS^- + 2Pb(OH)_2 \tag{2}$$

$$2PbS + 2CN^- + O_2 + 2Ca(OH)_2 = 2CNS^- + 2CaPbO_2 + 2H_2O \tag{3}$$

Lead products from Equations (1)–(3) are only slightly soluble in water, so it is possible that forming lead-containing products are impregnated (re-adsorbed) on pyrite. Reactions (2) and (3) have also been proposed by Prestidge et al. [20], suggesting that plumbates (PbO$_2$$^{-2}$) can finally decompose to form lead oxide, PbO.

As the process continues for a total of six hours, the surface seems more oxidized, which is suggested by the intense Fe 2p$_{3/2}$ signal at 710.8 eV. No other changes are observed, as the S 2p$_{3/2}$ and Pb 4f signals remain essentially unchanged. Polarization microscopy after leaching indicates that on a mineralogical scale no change can be identified, so lead re-adsorption seems to be a surface-related process, probably taking place over a scale of a few nanometers.

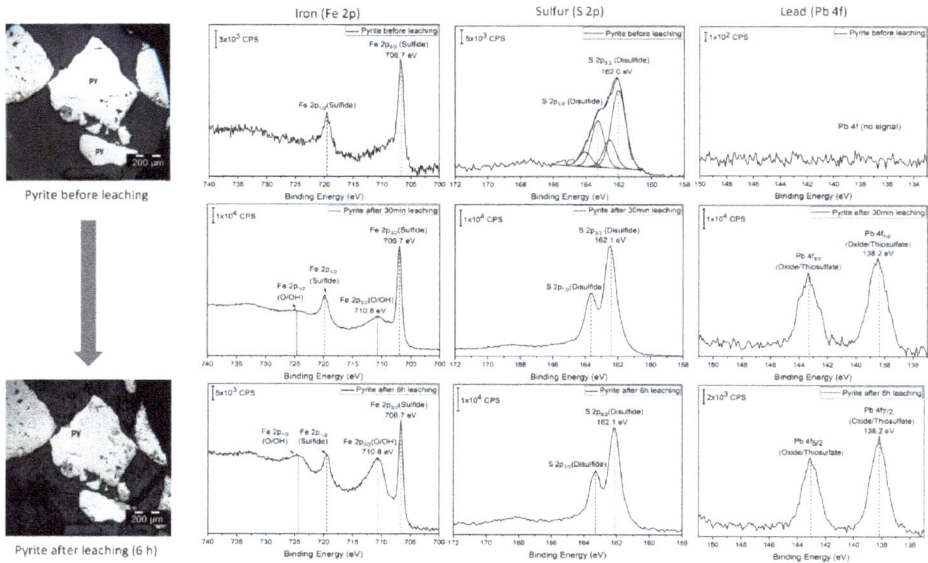

Figure 1. Changes observed during cyanidation of pyrite. The top panels represent the initial results of characterization with polarization microscopy (left) and XPS signals for iron, sulfur and lead (from left to right). The middle panels correspond to XPS signals after the first 30 min of leaching and the bottom panels show the state of the pyrite grain investigated with PM and XPS after six-hour leaching.

To better identify the nature of the lead layer grown on the pyrite surface, a different grain was also followed during cyanidation, namely the one corresponding to pyrite with inclusions of galena. Identification of this grain was possible with PM, and micrographs are shown in Figure 2. Before leaching, iron is mostly present as pyrite, as evidenced by the sharp peaks in the Fe $2p_{3/2}$ region at 706.8 eV corresponding to sulfide. Lead, on the other side, shows a broad signal which suggests the presence of sulfides and sulfates (137.4 and 138.5 eV, respectively) and, indeed, the S $2p_{3/2}$ signal shows the presence of small amounts of sulfates on the surface. After a six-hour leaching, the PM micrographs show evidence of a reaction, which promoted the loss of portions of the grain, particularly associated with galena. This suggests that leaching of galena, given by Equations (1)–(3), takes place at a fast rate, likely due to the formation of a galvanic pair pyrite–galena, which enhances the reactivity of galena while pyrite becomes more inert [21].

As observed in Figure 1 for a pure grain of pyrite, the XPS data in Figure 2 shows that iron oxide is produced as a result of leaching (Fe $2p_{3/2}$ signal at 710.8 eV). The surface concentration of lead varies from 20.90 At % in the original sample to 34.96 and 26.59 At % after 30 min and 6 h of leaching, respectively. In this case, however, the surface concentration of lead is only of relative importance, since lead is present before and after the reaction. The most important feature in XPS is the narrowness of the signal for lead (Pb $2p_{7/2}$) at 138.1 eV, a position which was not associated with sulfides nor sulfates. According to the literature, this position can be attributed to oxides and/or thiosulfates [18,22] and evidences that new lead species are present on the surface while sulfides and sulfates are leached out. This fact is noticeable when inspecting the micrographs in Figure 2. It is important to highlight that in spite of the similar surface concentration of lead throughout the leaching process, the intensity of the Pb $4f_{7/2}$ signal shows approximately a ten-fold increase with respect to the original mineral, further supporting the fact that lead re-adsorbs on the pyrite surface. This experiment confirms that the position of the peak corresponding to adsorbed lead (138.2 eV) in Figure 1 cannot be assigned to sulfide or sulfate.

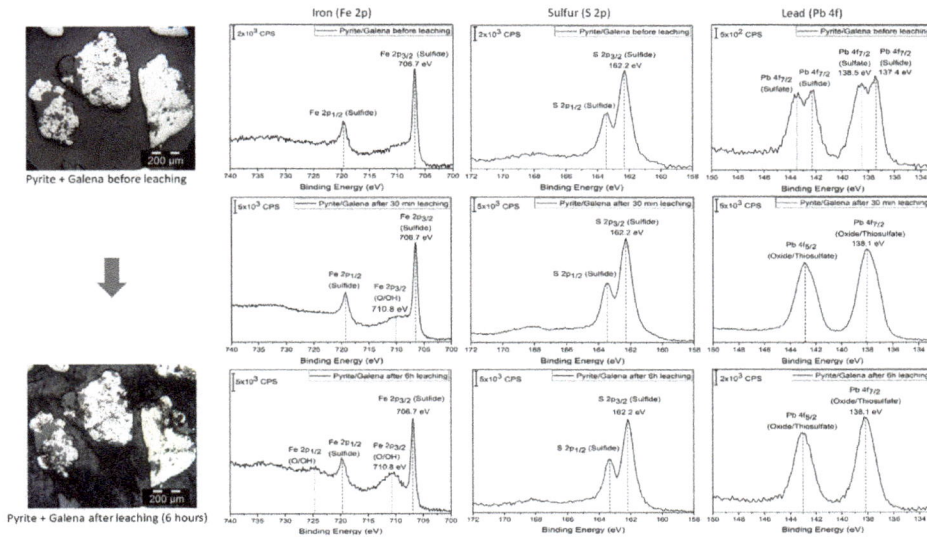

Figure 2. Changes observed during cyanidation of pyrite with inclusions of galena. The top panels represent the initial results of characterization with polarization microscopy (left) and XPS signals for iron, sulfur and lead (from left to right). The bottom panels show the state of the grain after a six-hour leaching with PM and XPS. The high reactivity of galena inclusions leads to the loss of portions of the grain after leaching.

Re-adsorption is a spontaneous process which depends on the affinity of the adsorbate to interact with a surface. Therefore, it can be expected that it is not a universal process which can happen on all surfaces. Indeed, Figure 3 shows that covellite does not feature significant evidence of lead re-adsorption. From our XPS measurements, the signals of lead after 30 min and 5 h represent only ~5 at %. The intensity of the areas of lead on covellite with respect to the case of pyrite and pyrite–galena is significantly smaller, which also supports the case that on this mineral the re-adsorption is significantly smaller. According to Breuer [23], covellite is leached using cyanide following the reaction below:

$$2CuS + 5CN^- + H_2O = 2Cu(CN)_2^- + SCN^- + OH^- + SH^- \tag{4}$$

This is a reductive dissolution mechanism, where copper passes from Cu^{+2} in covellite to Cu^{+1} in the cyanide complex. The leaching of covellite is possible but the process is slow, hence the clear change in the PM micrograph as the reaction proceeds (from an original white color to a tarnished surface) without seeing a damaged surface as in the case of galena. It may be likely that in the case of covellite, which presents an intermediate reactivity between that of pyrite (essentially inert) and galena (highly reactive in presence of pyrite), the leaching reaction is sufficiently fast to impede the adsorption of lead species.

Since lead re-adsorption was observed to be a selective process, taking place preferably over pyrite surfaces, it was expected to be also observed on the fine-milled, pyrite-based mineral entering the industrial leaching circuit. Indeed, leaching experiments done with non-embedded samples also showed evidence of lead re-adsorption (Figure S3). In accordance with the data in Figure 2, the two original Pb $4f_{7/2}$ peaks, associated with lead sulfide and sulfate (137.4 and 138.5 eV, respectively), become less intense as a new component, located at 138.1 eV, which rises during leaching. This data will be further discussed in the light of silver extraction and cyanide consumption in a separate work.

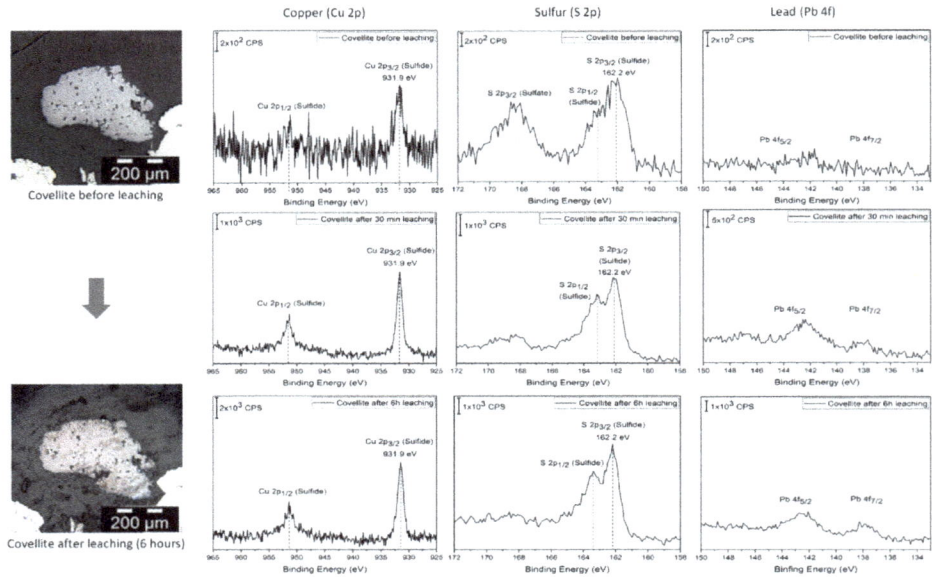

Figure 3. Changes observed during cyanidation of covellite. The top panels represent the initial results of characterization with polarization microscopy (left) and XPS signals for copper, sulfur and lead (from left to right). Identification of copper and sulfur species goes beyond the goal of this work, and they are reported collectively as "sulfide". The middle panels correspond to XPS signals after the first 30 min of leaching and the bottom panels show the state of the pyrite grain investigated with PM and XPS after six-hour leaching. In this case there is no evidence of lead re-adsorption.

4. Summary and Outlook

Through three relatively simple examples, it has been demonstrated that re-adsorption of lead takes place during cyanide leaching of a polymetallic sulfide. This process cannot be identified easily with traditional techniques for mineral characterization (such as polarization microscopy and scanning electron microscopy) since they are not sufficiently surface-sensitive. Polarization microscopy allows for a rapid identification of mineralogical species, while X-ray photoelectron spectroscopy can identify chemical changes on a surface. By working with minerals from polished sections, it is possible to monitor in time the response of a specific mineral towards a hydrometallurgical agent. In some cases, as in Figure 2, the response of minerals is easily observed simply using polarization microscopy. However, in other cases, as in Figure 1, the mineral seems unreactive towards a hydrometallurgical agent by polarization microscopy. In this case, XPS proves to be an important tool for identifying surface processes such as the re-adsorption of leached metals. The impact of lead re-adsorption on the metallurgical behavior of the ore (silver extraction with cyanidation and consumption of cyanide) will be the subject of a separate publication.

Supplementary Materials: The following are available online at http://www.mdpi.com/2075-163X/8/10/427/s1, Figure S1: Polarization micrographs of minerals within the samples ore, Figure S2: Comparison of micrographs from polarization microscopy and from the camera coupled to the XPS system, Figure S3: XPS data for free mineral particles before and after leaching, showing that re-adsorption of lead also takes place in the absence of resin.

Author Contributions: J.C.F.R.-R. and A.V.T. conceived and designed the experiments; D.S.-Q., C.H. and M.J.-C. performed the experiments; D.S.-Q., C.H., M.J.-C., C.B., A.V.T. and J.C.F.R.-R. analyzed the data and wrote the paper.

Funding: This research was supported by Peru's National Council for Science, Technology and Technological Innovation (FONDECYT-CONCYTEC) and the British Embassy in Lima (contracts 154-2015 and 002-2016), as well as by the Phosagro/UNESCO/IUPAC Partnership in Green Chemistry for Life (Contract 4500245048). The authors acknowledge the NSF (9724307; 1428149) and the NIH NIGMS COBRE program (P30-GM110758) for partial support of activities in the University of Delaware Surface Analysis Facility.

Acknowledgments: We acknowledge the support of Volcan Cía. Minera for facilitating the ore samples employed in this research. Luis Loaiza (Volcan) is especially thanked for his decisive support for this academia-industry collaboration. D.S.-Q. acknowledges the support provided by the Department of Chemistry and Biochemistry at the University of Delaware during her stay in the USA. Humberto Chirif, Renzo Portilla and Karinna Visurraga (UTEC) are acknowledged for useful discussions and suggestions. Karinna Visurraga is also acknowledged for technical and administrative support.

Conflicts of Interest: The authors declare no conflict of interest. The funding sponsors had no role in the design of the study; in the collection, analyses, or interpretation of data; in the writing of the manuscript, and in the decision to publish the results.

References

1. Baum, W. Ore characterization, process mineralogy and lab automation a roadmap for future mining. *Miner. Eng.* **2014**, *60*, 69–73. [CrossRef]

2. Klein, C.; Philpotts, A.R. *Earth Materials—Introduction to Mineralogy and Petrology*, 2nd ed.; Cambridge University Press: New York, NY, USA, 2012; ISBN 9780521761154.

3. Celep, O.; Devec, H.; Vicil, M. Characterization of refractory behaviour of complex gold/silver ore by diagnostic leaching. *Trans. Nonferr. Met. Soc. China* **2008**, *19*, 707–713. [CrossRef]

4. Rees, K.L.; Van Deventer, J.S.J. Preg-robbing phenomena in the cyanidation of sulphide gold ores. *Hydrometallurgy* **2000**, *58*, 61–80. [CrossRef]

5. Mwase, J.M.; Petersen, J. Characterizing the leaching of sperrylite (PtAs 2) in cyanide-based solutions. *Hydrometallurgy* 2017. [CrossRef]

6. Parker, A.; Klauber, C.; Kougianos, A.; Watling, H.R.; Van Bronswijk, W.; Parker, A.J. An X-ray photoelectron spectroscopy study of the mechanism of oxidative dissolution of chalcopyrite. *Hydrometallurgy* **2003**, *71*, 265–276. [CrossRef]

7. Boulton, A.; Fornasiero, D.; Ralston, J. Characterisation of sphalerite and pyrite flotation samples by XPS and ToF-SIMS. *Int. J. Miner. Process.* **2003**, *70*, 205–219. [CrossRef]

8. Jiang, H.; Xie, F.; Dreisinger, D.B. Comparative study of auxiliary oxidants in cyanidation of silver sulfide. *Hydrometallurgy* **2015**, *158*, 149–156. [CrossRef]

9. Bas, A.D.; Ghali, E.; Choi, Y. A review on electrochemical dissolution and passivation of gold during cyanidation in presence of sulphides and oxides. *Hydrometallurgy* **2017**, *172*, 30–44. [CrossRef]

10. Derycke, V.; Kongolo, M.; Benzaazoua, M.; Mallet, M.; Barrès, O.; De Donato, P.; Bussière, B.; Mermillod-Blondin, R. Surface chemical characterization of different pyrite size fractions for flotation purposes. *Int. J. Miner. Process.* **2013**, *118*. [CrossRef]

11. Jeffrey, M.I.; Breuer, P.L. Cyanide leaching of gold in solutions containing sulfide. *Miner. Eng.* **2000**, *13*, 1097–1106. [CrossRef]

12. Senanayake, G. A review of effects of silver, lead, sulfide and carbonaceous matter on gold cyanidation and mechanistic interpretation. *Hydrometallurgy* **2008**, *90*, 46–73. [CrossRef]

13. Harmer, S.L.; Thomas, J.E.; Fornasiero, D.; Gerson, A.R. The evolution of surface layers formed during chalcopyrite leaching. *Geochim. Cosmochim. Acta* **2006**, *70*, 4392–4402. [CrossRef]

14. Lane, G.R.; Martin, C.; Pirard, E. Techniques and applications for predictive metallurgy and ore characterization using optical image analysis. *Miner. Eng.* **2008**, *21*, 568–577. [CrossRef]

15. Smart, R.S.C.; Skinner, W.M.; Gerson, A.R. XPS of sulphide mineral surfaces: Metal-deficient, polysulphides, defects and elemental sulphur. *Surf. Interface Anal.* **1999**. [CrossRef]

16. Mills, P.; Sullivan, J.L. A study of the core level electrons in iron and its three oxides by means of X-ray photoelectron spectroscopy. *J. Phys. D Appl. Phys.* **1983**, *16*, 723–732. [CrossRef]

17. Bastl, Z.; Republic, C. X-ray photoelectron spectroscopy study of galena dissolution in ferric chloride media. *J. Mater. Sci. Lett.* **1993**, *12*, 789–790. [CrossRef]

18. Abdel-Samad, H.; Watson, P.R. An XPS study of the adsorption of lead on goethite (alpha-FeOOH). *Appl. Surf. Sci.* **1998**, *136*, 46–54. [CrossRef]

19. La Brooy, S.R.; Linge, H.G.; Walker, G.S. Review of gold extraction from ores. *Miner. Eng.* **1994**, *7*, 1213–1241. [CrossRef]

20. Prestidge, C.A.; Ralston, J.; Smart, R.S.C. The role of cyanide in the interaction of ethyl xanthate with galena. *Colloids Surf. Physicochem. Eng. Asp.* **1993**, *81*, 103–119. [CrossRef]

21. Aghamirian, M.M.; Yen, W.T. Mechanisms of galvanic interactions between gold and sulfide minerals in cyanide solution. *Miner. Eng.* **2005**, *18*, 393–407. [CrossRef]

22. Manocha, A. Flotation related ESCA studies on PbS surfaces. *Appl. Surf. Sci.* **1977**, *1*, 129–141. [CrossRef]

23. Breuer, P.L.; Dai, X.; Jeffrey, M.I. Leaching of gold and copper minerals in cyanide deficient copper solutions. *Hydrometallurgy* **2005**, *78*, 156–165. [CrossRef]

© 2018 by the authors. Licensee MDPI, Basel, Switzerland. This article is an open access article distributed under the terms and conditions of the Creative Commons Attribution (CC BY) license (http://creativecommons.org/licenses/by/4.0/).

minerals

MDPI

Review

Tracing Mineral Reactions Using Confocal Raman Spectroscopy

Helen E. King [1],* and Thorsten Geisler [2]

1   Department of Earth Sciences, Universiteit Utrecht, 3584 CB Utrecht, The Netherlands
2   Steinmann Institute, Rheinische Friedrich-Wilhelms Universität Bonn, 53115 Bonn, Germany;
    tgeisler@uni-bonn.de
*   Correspondence: h.e.king@uu.nl; Tel.: +31-30-253-5070

Received: 7 March 2018; Accepted: 9 April 2018; Published: 13 April 2018

check for
updates

Abstract: Raman spectroscopy is a powerful tool used to identify mineral phases, study aqueous solutions and gas inclusions as well as providing crystallinity, crystallographic orientation and chemistry of mineral phases. When united with isotopic tracers, the information gained from Raman spectroscopy can be expanded and includes kinetic information on isotope substitution and replacement mechanisms. This review will examine the research to date that utilizes Raman spectroscopy and isotopic tracers. Beginning with the Raman effect and its use in mineralogy, the review will show how the kinetics of isotope exchange between an oxyanion and isotopically enriched water can be determined in situ. Moreover, we show how isotope tracers can help to unravel the mechanisms of mineral replacement that occur at the nanoscale and how they lead to the formation of pseudomorphs. Finally, the use of isotopic tracers as an in situ clock for mineral replacement processes will be discussed as well as where this area of research can potentially be applied in the future.

Keywords: Raman spectroscopy; mineralogy; replacement reaction; isotopes

---

1. Introduction

Some of the first materials in which the Raman effect was documented were minerals [1]. Now, modern data compilations, such as The RRUFF™ Project database [2], contain Raman spectra for over 3000 different minerals making Raman spectroscopy a key mineral identification tool for geoscientists. Raman spectroscopy has the additional benefits that, unlike X-ray diffraction, it is well-suited to study amorphous solids such as glasses [3]. One of the main advantages, however, is that minerals as small as a few nanometers can be analyzed non-destructively by modern confocal Raman spectrometer systems. In addition, Raman spectroscopic analysis can be conducted on liquid and gas samples making it an interesting tool for studying inclusions in minerals [4,5]. This means that Raman spectroscopy can provide information about the factors controlling mineral nucleation [6,7] as well as be used to find cryptic clues for mineral formation. Critically, Raman spectroscopy is also sensitive to isotopic substitution, particularly of light isotopes such as oxygen or hydrogen isotopes. This sensitivity has been used previously in the chemical sciences to assign spectral bands to specific vibrational modes [8,9] or to follow reactions [10]. However, it is only in the past couple of decades that geoscientists have begun to use the combination of light, stable isotopic tracers and Raman spectroscopy to probe mineral reactions. This article will review the different information that can be gained from coupling stable isotopic tracers with Raman spectroscopy and where we could go in the future with this kind of Raman mass spectroscopy in mineralogy.

2. Background to Raman Spectroscopy

In descriptive terms, Raman bands arise due to energy gain or loss from a monochromatic incoming light source as it is scattered by the electron field of vibrating molecular groups. Therefore, in Raman spectroscopy we examine the wavelength shift of the scattered light compared to that of the incident light (set at 0 in a Raman spectrum). Typically, Raman spectroscopy focuses on the energy lost by the incoming light due to inelastic scattering, called Stokes scattering, because these bands are more intense at room temperature than anti-Stokes bands that reflect a gain in energy from the molecule. As the spectral bands directly reflect vibration within the material, the energy shift of the scattered light is dependent on the energy required for the atoms to oscillate around their equilibrium position as they vibrate. Therefore, the amount of energy loss, and hence, the Raman spectrum, depend on bond strength, mass of the vibrating atoms, site symmetry within a structure (point groups), as well as short- and long-range symmetry. Specific vibrations will only be present in the Raman spectra if the polarization ellipsoid is modified during a vibration; in other words the electron cloud around a group of atoms changes its shape. Given that Raman spectroscopy is dependent on atomic structure, it is probably no surprise that the selection rules for Raman activity are based on quantum mechanics. We will not cover this level of detail in this review; for an in-depth description of the quantum mechanics behind Raman spectroscopy the reader is referred to [11].

An example of the sensitivity of Raman spectroscopy to different mineral phases is shown in Figure 1. The spectra can be split into two different regions: below ~400–500 cm^{-1} where bands called external lattice modes occur and internal lattice mode bands at wavenumbers >~500 cm^{-1}. The external modes arise due to long-range translational symmetry within the example mineral structure, whereas the internal modes reflect the coupling of vibrations within molecular groups. There are multiple internal modes for a molecular group: stretching, twisting and bending, all of which can occur either symmetrically or antisymmetrically around the central axis of the vibration. Each of these motions requires a different amount of energy and, therefore, multiple bands within a spectrum are related to a specific molecular group. For example, the spectra shown in Figure 1 are all from carbonate minerals. In these examples, the high energy bands between 1075 and 1150 cm^{-1} reflect the symmetrical stretching of the carbonate (CO_3^{2-}) molecular group [12]. Similarly, the band close to 700 cm^{-1} is also related to the carbonate molecular group, but in this case, it reflects the antisymmetric bending mode. Both calcite ($CaCO_3$) and magnesite ($MgCO_3$) show the same number of bands in their spectra due to their identical structure (space group $R\bar{3}c$), thus, exhibiting same site symmetries. However, the bands in magnesite are shifted to higher wavenumbers, reflecting the ability of Mg to more strongly bind to oxygen while at the same time Mg is lighter than Ca, both of which increase the energy of the molecular vibration. In contrast, changing the structure of the calcite to its polymorph aragonite ($CaCO_3$, space group Pmcn) produces additional bands in comparison to calcite. Therefore, the overall Raman spectrum is characteristic of a mineral phase, i.e., can be considered as a fingerprint.

The height difference that can be observed between the bands within a single Raman spectrum is related to the modulation of the polarizability associated with the vibrating atoms. This varies between different vibrational modes of the same molecular group producing different band heights. In addition, in crystalline solids, such as those in Figure 1, the position of the atoms is fixed with respect to the incoming, polarized laser light. As the polarizability related to a specific vibration is directly related to the symmetry of the molecular group, the polarized laser light interacts differently along different crystal axes. This causes the relative intensities of different vibrational modes to vary with crystal orientation to the laser light and, if the polarization of the laser is known, can be used to evaluate crystallographic orientation. We already use the phenomenon of changes in the interaction of light with crystal structures when we examine minerals in thin section under crossed polarizers in the light microscope. The Raman effect was documented for the highly birefringent mineral calcite in 1940 [13] and has since been shown for a variety of mineral phases [14,15]. However, unlike light microscopy, this effect can also be observed in structurally isotropic minerals such as magnetite [16]. In contrast to solid phases, for freely moving molecules, e.g., molecules in a gas, the Raman spectrometer probes the molecular group in a range of different orientations with respect to the incoming laser light. Therefore, the intensity of a specific band is proportional to the concentration of the species of interest in the medium examined. This effect has been used to estimate concentrations of different ions within a solid structure with the same orientation [17]. Systematic variations of spectral bands are also particularly useful in the analysis of minerals, as band shifts can be linked to solid solution composition [18].

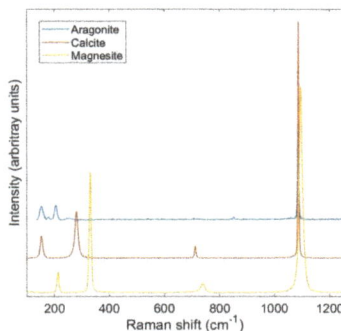

Figure 1. Example Raman spectra from The RRUFF™ Project database of the minerals aragonite (R080142), calcite (R040010), and magnesite (R040114). These spectra illustrate the change of the number, position, and width of the bands that occur in the Raman spectra due to changes in structure (aragonite-calcite) and chemistry (calcite-magnesite).

3. Isotopes in Raman Spectroscopy

As for changes in the elemental composition shown in Figure 1, Raman spectroscopy is also sensitive to the isotopic composition of the material examined. This arises due to the dependence of the vibrational frequency on the mass of the atoms involved [19]. Therefore, isotopic substitution is reflected in the Raman spectra as a shift in the spectral bands upon incorporation of different isotopes into a vibrating group. For example, 100 at.% enrichment of ^{30}Si or ^{18}O into the mineral phase alpha-quartz produces shifts in the Raman bands of up to 20 cm^{-1} and 50 cm^{-1}, respectively [9]. Raman spectroscopy is not as sensitive to isotopic composition as more traditional methods for examining isotope incorporation such as mass spectrometry. Typically, band positions can be determined with an error down to ± 0.05 cm^{-1}, when the system is well-calibrated [20]. Therefore, per mill changes in isotope composition associated with small fractionation effects cannot be determined by Raman spectroscopy at present. Nevertheless, even if this precision is reached, the uncertainties associated with compositional or structural effects on the vibrational frequency are too large in solid and liquid phases. However, isotopic shifts can be used to determine which bands are affiliated to specific motions within a material by isotopically doping the material of interest. Critically, unlike mass spectrometry, the bands in Raman spectroscopy are site-specific. This means that isotope substitution into the crystal structure and the presence of isotopically enriched water that may become entrapped in pores or adsorbed during experiments can be clearly discriminated in the Raman spectra. In addition, the incorporation of different isotopes can be measured in situ and ex situ at μm spatial resolution in a non-destructive manner providing an unparalleled picture into the processes occurring during mineral reactions.

The response of spectral bands to isotope incorporation can be separated into two different forms. O-isotope incorporation into silicate phases that have Si–O–Si linkages, including the quartz example above as well as zeolites [21] and amorphous silica [22], has been shown to produce a gradual mass-related shift in the Si–O related bands. Assuming, as an approximation, a linear frequency shift with isotope composition, the amount of the isotope incorporated into a material can be determined [23]. In contrast, isolated molecular groups within a mineral structure, e.g., carbonates, produce new bands upon isotope substitution. These can be assigned to molecular groups with specific isotopic compositions, called isotopologues. For example, the carbonate symmetrical stretching band in calcite at 1086 cm^{-1} is split into four bands upon ^{18}O substitution into the carbonate group [24]. Each of these bands corresponds to the different isotopologue of the carbonate group, namely $C^{16}O_3$, $C^{16}O_2^{18}O$, $C^{16}O^{18}O_2$, and $C^{18}O_3$ as shown in Figure 2. We assume that the Raman scattering cross sections of the isotopologues are similar; hence, the incorporation of an isotope into an isolated molecular group can be calculated from each spectrum by measuring the intensity of the different isotopologue bands. However, the observed Raman intensities are not expected to correspond exactly to relative concentrations, but should obey the isotope intensity sum rule. This rule states for Raman intensities that A_i/v_i is a constant, where A_i is the integrated intensity of each isotopologue band (i) and v_i its frequency [25]. A similar band splitting upon O-isotope substitution has been found for other oxyanions including phosphate [25–27], sulfate [28], and vanadate [29]. Moreover, the $^{13}C^{16}O_2$ isotopologue of gaseous or liquid CO_2 generates its own Raman band at the lower frequency side of both $^{12}C^{16}O_2$ Fermi diad bands. This allows the $^{12}C/^{13}C$ isotope composition of CO_2 gas inclusions in minerals to be determined by Raman spectroscopy as first shown by Arakawa and coworkers [30]. However, ratios determined from band area for CO_2 gas samples were found to be associated with an error of 20‰, which is by far not precise enough for Raman spectroscopy to become a valuable tool in isotope geosciences. On the other hand, reports have been made that with modern Raman spectroscopic systems it should be possible to reach a precision in the order of ±2‰ if a standard gas is measured with identical instrumental settings and, importantly, at the same density as the inclusion [31,32]. Therefore, by analyzing isotopic tracers with Raman spectroscopy we have an in situ probe that can be tracked throughout the mineral material to evaluate mineral reactions or formation conditions. In addition, we can probe isotopic exchange as it occurs in solution, which is important for both mechanistic studies as well as probing rate-limiting steps in lower concentration solutions applicable for fractionation effects.

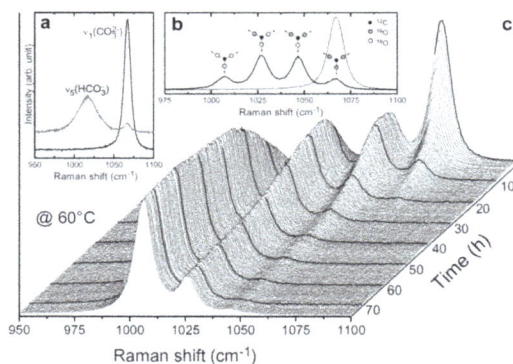

Figure 2. Isotopic exchange of aqueous carbonate ion with ^{18}O-enriched water. (**a**) Raman spectra from a 1 M sodium carbonate (lower spectrum) and bicarbonate (upper spectrum) solution, showing the $v_1(CO_3^{2-})$ and the $v_5(HCO_3^-)$ band near 1065 and 1016 cm^{-1}, respectively. (**b**) Isotopologue bands of CO_3^{2-} ion in solution. The dark line shows the enriched carbonate population present in a 1 M NaCO$_3$ solution after 10 h at 60 °C and the grey spectrum represents the initial spectrum of the solution shortly after being heated to 60 °C, showing only the $v_1(C^{16}O_3^{2-})$ band. (**c**) Time-resolved Raman spectra of the $v_1(CO_3^{2-})$ band, illustrating the O isotope exchange between aqueous CO_3^{2-} and water at 60 °C. Reprinted from Geochimica et Cosmochimica Acta, vol. 90, Geisler et al., "Real-time monitoring of the overall exchange of oxygen isotopes between aqueous CO_3^{2-} and H_2O by Raman spectroscopy", pages 1–11, (2012), with permission from Elsevier [20].

4. Examining the Properties of Ions in Solution Using Raman Spectroscopy

Probing reactions in aqueous solution is critical for many different aspects of mineralogy from ore geology to biomineralisation. For these reactions, the properties of the mineral constituent ions in solution will be controlling factors in how mineral reactions progress. Raman spectroscopy is a particularly powerful tool in this respect, as it can provide a unique image of ions in solution. Liquid water shows distinct bands in Raman spectra, a medium intensity band near 1630 cm^{-1} related to $\nu_2(H_2O)$ bending motions and an intense band that shows a maximum near 3600 cm^{-1}. This Raman band profile is composed of contributions from the symmetric $\nu_1(H_2O)$, the antisymmetric $\nu_3(H_2O)$ stretching vibration, as well as from a $2\nu_2$ combination mode [33]. There are many examples of Raman spectroscopy being used to identify hydrated ions [34,35], oxyanions [36,37], and ion complex formation [38–40], including oligomers of the same ions [41,42]. Formation of ion complexes is particularly interesting as this can aid precipitation if the different components of a mineral phase interact in solution. Alternatively, if ion pairs form between one component of a mineral and another solution component precipitation may be inhibited. An example of such behavior can be seen with the growth of magnesite. Addition of sulfate to a solution that would otherwise be highly oversaturated with respect to magnesite slows down the growth of the mineral phase dramatically. Thermodynamic calculations indicate that this is related to a lowering of the solution saturation with respect to magnesite through the formation of Mg-sulfate ion pairs [43]. This is supported by Raman spectroscopy, which shows the development of a new spectral band associated with Mg-sulfate ion pair formation in solution [38]. Similarly, the formation of immiscible, dense liquid phases at high temperatures has been identified in situ using the ratios of sulfate and OH Raman spectral band areas and the formation of new bands that correspond to polymeric mixtures [44,45]. These dense fluid phases play an important role in mineral nucleation [7] when it occurs via a non-classical route [46] as well as element transport [45].

We can also trace isotope incorporation in the dissolved components, for example ^{18}O incorporation into oxyanions. Experiments show that substitution of ^{18}O into aqueous phosphate ions produce distinct new bands at lower wavenumbers that reflect the presence of new isotopologue species [47], as observed for phosphate-bearing crystals [26]. By using heated fluid cells, we are now able to study the isotopic exchange in situ. For example, Geisler et al. [20] has examined the substitution of ^{18}O from enriched water into carbonate oxyanions at temperatures from 45 to 100 °C, shown in Figure 3. This has provided key information on the exchange rate, allowing the authors to extract the activation energy of the process. Therefore, the kinetics of isotopic exchange can be elucidated using a combination of Raman spectroscopy and isotopic enrichment. Raman spectroscopy has the additional benefit for these types of reactions, because it can clearly distinguish between different ion speciation in solution. The symmetrical stretching mode of aqueous CO_3^{2-}, for example, occurs at ~1060 cm^{-1} whereas the equivalent vibrational mode for HCO_3^- is found at ~1015 cm^{-1} [48], as can be seen in Figure 2a. The lower frequency of the stretching mode in solution compared to that of calcite, shown in Figure 1, arises due to the difference in the site symmetry of the molecular group in solution (D_{3h}) and the solid phase (D_3) as well as due to a different chemical environment [49]. In initial isotope substitution experiments the presence of different ion speciations in solution led to the erroneous prediction of shifts of ^{18}O bands to higher wavenumbers, contradictory to what is expected from theory [50,51]. Therefore, care must be taken when conducting such experiments to ensure minimal amount of the ion with different speciation is present in the solution. However, when conducted carefully, these experiments can provide key information to unravel isotopic signatures in more complicated systems, such as those produced during mineral replacement.

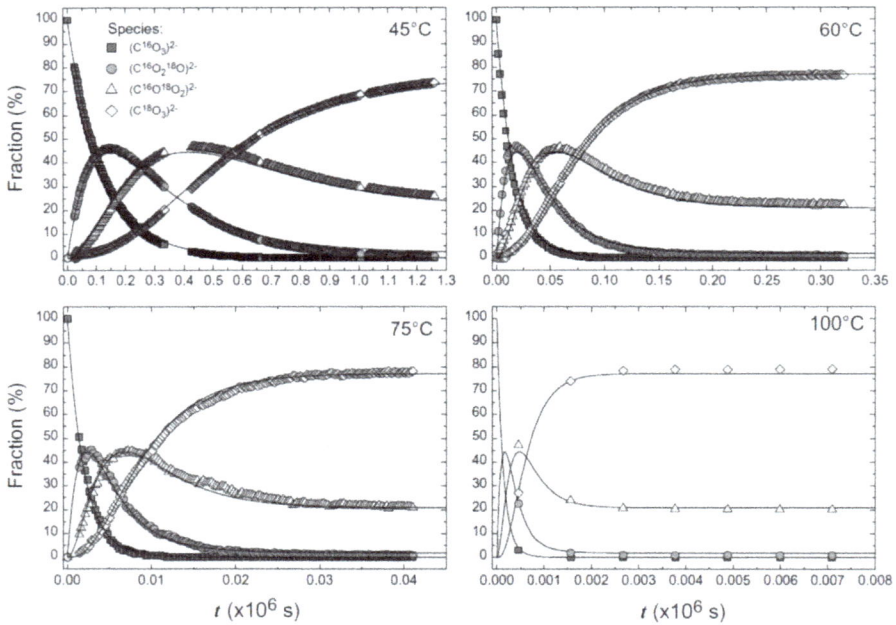

Figure 3. Fraction (%) of the different isotopologues of aqueous CO_3^{2-} determined by in situ Raman spectroscopy as a function of reaction time between $H_2^{18}O$ and $C^{16}O_3^{2-}$, shown in Figure 2, at different temperatures. Reprinted from Geochimica et Cosmochimica Acta, vol. 90, Geisler et al., "Real-time monitoring of the overall exchange of oxygen isotopes between aqueous CO_3^{2-} and H_2O by Raman spectroscopy", pages 1–11, Copyright (2012), with permission from Elsevier [20].

5. Isotopes and Raman Spectroscopy as a Probe for Mineral Replacement Reaction Mechanisms

Replacement of one mineral by another whilst retaining the habit of the original mineral, pseudomorphic replacement, is common in geological systems. Examples of natural pseudomorphic mineral replacement span from hydrothermal systems [52] to surface weathering [53] as well as biologically controlled processes [54]. Deciphering the nanoscale mechanism for these reactions helps geoscientists to predict reaction rates as well as the concomitant changes in system chemistry. Thus far, two different mechanisms have been proposed. In the first mechanism, it is assumed that specific ions are extracted through an exchange mechanism whilst the surrounding mineral structure remains intact. These ions then diffuse away from the reacting area and the remaining crystal structure can undergo rearrangement in the solid state to form a new material. This mechanism has been observed to form "leached layers" in high temperature experiments of feldspar ion exchange in a dry environment [55]. However, contradictory to the solid-state mechanism above, replacement of feldspar in an aqueous solution shows sharp chemical and isotopic signatures that cannot be explained by a diffusional exchange process [56]. Experiments of albite replacement by K-feldspar at 600 °C in ^{18}O-doped solutions followed by analysis using nanoscale secondary ion mass spectrometry show ^{18}O enrichment within the K-feldspar and µm sharp interfaces in cation and isotope chemistry [57]. Later work using similar experiments coupled with Raman spectroscopic analysis demonstrated that ^{18}O was incorporated into the aluminosilicate network itself [58]. Furthermore, the same textural and isotopic signatures have also been observed with oligoclase and labradorite replacement by albite [59]. The sharp change in isotopic and elemental composition across the reaction interface along with the incorporation of ^{18}O into the network structure both indicate that the reacting mineral is actually dissolving to release its cations, silica, and alumina into solution before a new feldspar can form.

"Leached" layers have also been observed in low temperature weathering environments where fluids are clearly active [60]. In the low-temperature solid-state scenario the pseudomorph is proposed to form via the attachment of active solution species, such as H^+ in acidic solutions, to surface atoms. These interactions break the underlying bonds to the structure releasing specific ions. For example, for silicate minerals such as olivine and pyroxene diffusion of H^+ to the unreacted mineral results in hydrolysis of the cation-O bond releasing the cation [61,62]. Exchange of the cation for an H^+ is proposed to be followed by condensation of the silica groups exposed at the mineral surface to produce a "leached" amorphous silica pseudomorphic layer [63,64]. The reaction is propagated by diffusion of the released cation through the layer to the solution and the counter diffusion of H^+ to the unreacted crystal. However, experiments on amorphous silica production during wollastonite [65] and olivine replacement [66] also show the enrichment of ^{18}O in the amorphous silica using Raman spectroscopy. This information in conjunction with a sharp reaction interface between the amorphous silica rim and initial mineral as well as lack of cation diffusion profiles indicates a fluid-mediated reaction is also active in low temperature systems.

Coupled dissolution-precipitation was originally proposed as a mechanism for mineral replacement by Cardew and Davey [67] and has been applied to natural and analogue mineralogical systems extensively by Putnis [68,69], Merino [53,70], and Glikin [71]. In this mechanism, dissolution occurs due to chemical [69] or physiochemical [53] driving forces. Release of the mineral components into the solution at the interface leads to supersaturation of the interfacial solution with respect to a new, more stable mineral that precipitates. This has resulted in this process also being referred to as an interface-coupled dissolution reprecipitation (ICDP) mechanism as it is the close spatial coupling that allows the system to produce a pseudomorph [71,72]. Isotope incorporation thus occurs during precipitation of the new mineral from the interfacial solution. For example, isotopically labelled K in solution is incorporated into the newly precipitated phase during the replacement of KBr by KCl [73]. Similarly, ^{18}O has been used to label phosphate that was incorporated into apatite during calcite replacement [74]. Critically, oxyanions that have been released from a dissolving phase can exchange O-atoms with the surrounding interfacial solution in situ, as described in Section 4. This means that isotope incorporation into the crystal structure can be used to demonstrate that isochemical reactions, such as the transformation of aragonite to calcite, are fluid mediated [75]. The incorporation of isotopic tracers has been demonstrated for many different mineral and amorphous systems (Table 1) across a range of temperatures and this list is likely to get longer in the future.

Table 1. Replacement reactions that have shown ^{18}O isotope incorporation into structure during mineral replacement by Raman spectroscopy.

Reactant	Product	Temp (°C)	Reference
Calcite	Whewellite	60	[76]
Wollasonite	Amorphous silica	90	[65]
Olivine	Amorphous silica	90	[66]
Borosilicate glass	Amorphous silica	150	[77]
Ilmenite	Rutile	150	[78]
Aragonite	Apatite	150	[74]
Calcite	Apatite	200	[79]
Aragonite	Calcite	200	[75]
Leucite	Analcime	200	[23]
Pyrochlore	Pyrochlore	200	[80]
Pyrochlore	Rutile	250	[81]
Oligoclase	Albite	600	[59]
Labradorite	Albite	600	[59]
Albite	K-feldspar	600	[58]

6. An In Situ Stop Clock for Mineral Replacement Reactions

If in situ analyses can be performed, then we can examine the formation or removal of different system components, such as crystals, ion species or gases, using Raman spectroscopy to obtain rates and determine reaction kinetics [82–84]. First reports about successful space- and time-resolved in situ Raman spectroscopic experiments designed to investigate the aqueous fluoridation of tooth material at 21 °C, corrosion of borosilicate glass at 70 °C, and the replacement of celestine by strontianite at 21 °C [85,86] have yielded new insights into solid–fluid reactions. However, in situ analyses are not always possible requiring samples to be studied ex situ after the reaction has been stopped. Thus, whilst we can probe reaction mechanisms using experiments, examining the kinetic controls of replacement reactions is challenging. Yet, this is a critical parameter for understanding geological system reactivity. Recent work has demonstrated that the kinetic inhibition of isotopic exchange into oxyanions, such as carbonate [20] and phosphate [74], can fill this void and be used as a timer of the reaction progress [76,79]. For example, fluid infiltration along different reactive pathways into a mineral system can be traced using O-isotope enrichment and Raman spectroscopy. This has been demonstrated using marble blocks in which individual calcite crystals are replaced by apatite using a 2 M $(NH_4)_2HPO_4$ solution at 200 °C [79]. In these experiments, ^{18}O-enriched water was prepared with a phosphate source with a natural abundance $^{18}O/^{16}O$ ratio, i.e., the experiment begins with a dominantly ^{16}O isotopologue signature in the dissolved phosphate. However, over the course of the five-day experiment, isotopic exchange between the dissolved phosphate and water produces an increasingly mature phosphate O-isotope signature. Upon precipitation of apatite, this signature becomes locked into the precipitated mineral phase (Figure 4). This produces an isotopic tracer of the solution within the newly precipitated apatite phase. Enrichment can be laterally resolved down to ~1 μm with a typical confocal Raman spectrometer set up, where the lateral resolution d_l is limited by the diffraction limit given by $d_l = 1.22\lambda/N.A.$, where λ and N.A. represent the excitation wavelength and the numerical aperture of the objective lens, respectively. Comparison of line profiles across reaction rims at different points along a grain contact is shown in Figure 4. As can be seen from the graph, profile A to D become increasingly flattened, indicating that at profile D the solution had reached isotopic equilibrium with the enriched water. We can use this signature as a stop clock for a period of time in the experiment and derive that in the time taken for the fluid to penetrate ~100 μm along the grain contact (length to D profile) the ICDP reaction proceeded 10 μm into the calcite crystal (point at which A profile reaches same value as in profile D).

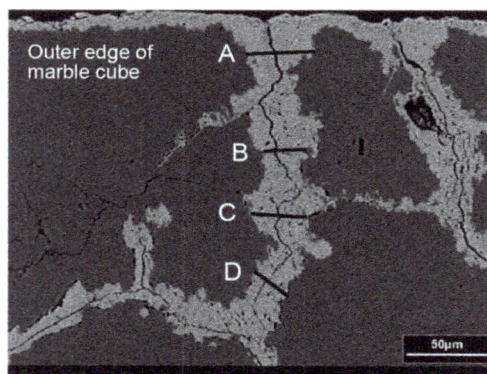

Figure 4. *Cont.*

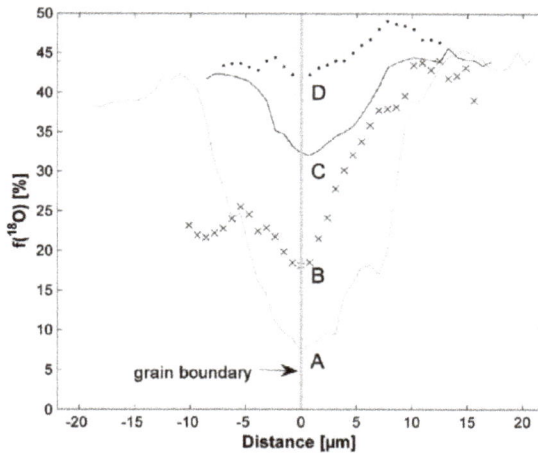

Figure 4. Backscattered electron image of calcite crystals within a cube of Carrara marble that have been partially replaced by apatite (bright phase). Top image shows the location of four line profiles measured using Raman spectroscopy that are depicted in the bottom graph. Isotopic enrichment increases along line A away from the grain contact (labelled grain boundary). Whereas, the apatite at the grain contact probed by line D shows the same level of isotopic enrichment as ~10 μm into the apatite rim along line A. Reprinted from Earth and Planetary Science Letters, vol. 386, Jonas et al., "The role of grain boundaries and transient porosity in rocks as fluid pathways for reaction front propagation", pages 64–74, (2014), with permission from Elsevier [79].

Information about relative timings can be united with the more traditional information gained from Raman spectroscopy (mineral identification, crystallinity, and crystal orientation) to produce an unparalleled view of the processes occurring during mineral replacement. This has been demonstrated by King et al. [76], who studied the formation of a new phase upon reaction of marble with an oxalate ($C_2O_4^{2-}$) bearing solution. In these experiments a complex replacement rim was produced showing three separate generations of rim material, labelled external crystals, outer and inner rim in Figure 5a,d,g. Despite the clear textural differences of the three replacement layer sections Raman spectroscopy identified them all as the monohydrate Ca-oxalate phase, whewellite. In addition, isotope incorporation into the oxalate oxyanion was retained in the rim. Figure 5b,e, demonstrates that the external crystals had the lowest amount of ^{18}O, followed by the inner rim with the highest amount of ^{18}O incorporation in the outer rim material. This allowed the authors to propose that the reaction was initially not coupled at the interface of the calcite, producing the external crystals. This was followed by dissolution and precipitation coupled at the reacting interface resulting in the formation of the inner rim. These crystals subsequently underwent a fluid-mediated transformation to produce the outer rim material, allowing the ^{18}O signature in the whewellite to be reset. Critically, crystallinity measurements [87] using the width of the main whewellite oxalate symmetrical stretching band at 1454 cm^{-1} showed that the inner rim had a much lower crystallinity than the outer rim material. This in turn allowed the authors to suggest that the rim underwent a textural re-equilibration driven by increasing crystallinity through the growth of larger crystals. These crystals were not templated on one another nor are they related to the original calcite structure as no preferred orientation could be deduced from the Raman spectra (Figure 5c,f,i), measured using the relative intensities of the 1454 and 1483 cm^{-1} bands of whewellite. Thus, Raman spectroscopy coupled to isotopic tracers is a powerful tool for probing mineral reactivity in different systems.

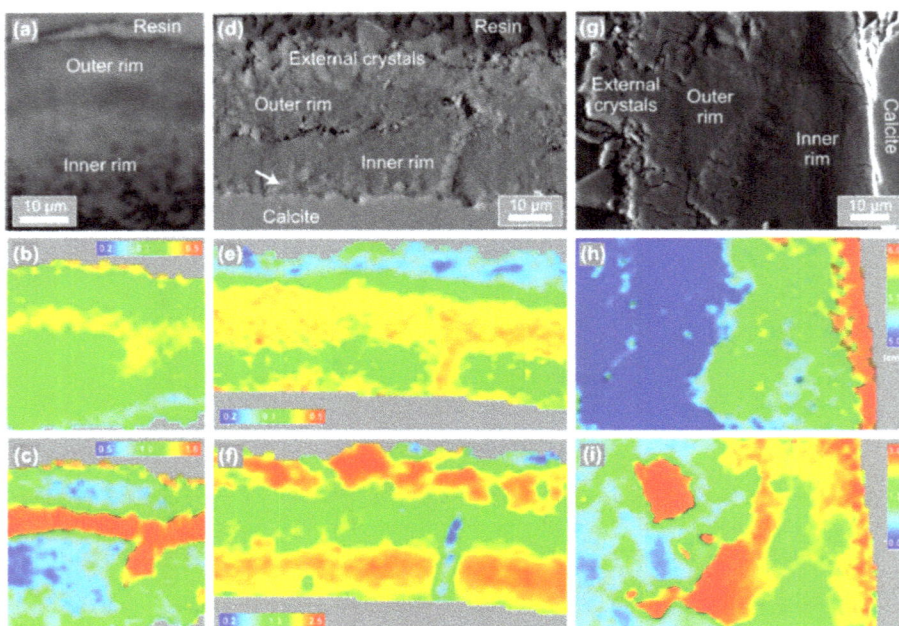

Figure 5. Reaction rims produced on marble cubes reacted with 0.1 M oxalic acid at 60 °C for 7 days. (**a**) shows a reflected light microscope image of a reaction rim generated in a 50 at.% ^{18}O-enriched solution. (**b**,**c**) show the ^{18}O fraction in the solid and the relative crystal orientation, respectively. (**d**) is an electron backscattered image of another section of the reaction rim, where (**e**) is a map of the ^{18}O fraction, and (**f**) is a map of the relative crystal orientation in the area in (**d**). (**g**) displays an equivalent area of the rim generated in an experiment with natural abundance water (^{16}O-rich). (**h**) shows the difference in crystallinity of the reaction rim area shown in (**g**), reflected by the half width of the most intense oxalate band, whereas (**i**) shows the relative crystal orientation for the same area. Reprinted with permission from Crystal Growth and Design, vol. 14, "Forming cohesive calcium oxalate layers on marble surfaces for stone conservation", King et al., pages 3910–3917. (2014) American Chemical Society [76].

7. The Future for Isotopes and Raman Spectroscopy

The work described in this review has laid the foundations for the use of isotopic tracers and Raman spectroscopy in many different mineralogical systems in the future. It has also provided the possibility to extend isotopic tracers and Raman spectroscopy to probe mechanisms of critical processes that control the reactivity of minerals and geological systems. For example, experiments with ^{18}O-enriched water examining amorphous silica formation after olivine [66] and wollastonite [65] show a different extent of ^{18}O-enrichment. Olivine and wollastonite have different silicate structures, where wollastonite contains chains of silica, whereas olivine is composed of individual silica tetrahedra that are joined together by cations. If we calculate the uptake of ^{18}O through hydrolysis of the wollastonite chain, we would expect a maximum of 17 at.% ^{18}O in the amorphous silica. Yet, amorphous silica after wollastonite had an enrichment of 33 at.%, indicating that the silica had been hydrolyzed and released into solution where it could undergo isotope exchange. This is still lower than the enrichment observed in amorphous silica that replaced olivine (54 at.%). Therefore, this signature could be evidence for longer chain sections that are released into solution, as suggested by Weissbart and Rimstidt [88].

Until now, many mineral replacement reactions have been identified that are governed by a coupled dissolution-reprecipitation process at mineral surfaces. This includes dissolution-reprecipitation due to physicochemical processes such as pressure solution creep, a key mechanism by which geological materials deform that is present from diagenetic to greenschist facies metamorphism [89]. These systems are often monomineralic, making differentiation of newly precipitated material challenging. Therefore, use of isotopic tracers in experimental systems that can incorporate information about the time in solution after dissolution and precipitation site have the potential to provide critical data for models of deforming systems. However, to truly probe mechanisms and rates of dissolution-reprecipitation reactions, we need a more fundamental understanding of the different processes that can lead to isotopic enrichment. This includes how condensation of silica potentially changes the isotopic signature. In addition, although computational simulations have demonstrated that oxygen exchange between water and the silica oxyanion is an energetically favorable and efficient process [90], we do not yet know how silica oligomers may affect isotope exchange in solution. Similarly, competition between different oxyanions in solution for O-isotopes may also affect the final signature observed. All these subjects can potentially be examined with Raman spectroscopy in situ, providing critical information that will not only enhance the use of isotopes as tracers at high concentrations, but also the controls of isotopic signatures that are important for paleoclimate and paleohabitat investigations.

Acknowledgments: The authors thank C.V. Putnis for the invitation to submit this paper to the special issue of Minerals on which she is a guest editor. The two reviewers are also thanked for their constructive comments that helped to improve the paper. Publication of this paper was made possible through internal funding at the Department of Earth Sciences, Utrecht University.

Author Contributions: H.E.K. and T.G. both wrote and edited the paper.

Conflicts of Interest: The authors declare no conflict of interest.

References

1. Landsberg, G. Eine neue Erscheinung bei der Lichtzerstreuung in Krystallen. *Naturwissenschaften* **1928**, *16*, 557–558.

2. Lafuente, B.; Downs, R.T.; Yang, H.; Stone, N. The power of databases: The RRUFF project. In *Highlights in Mineralogical Crystallography*; Walter de Gruyter GmbH: Berlin, Germany, 2016.

3. McMillan, P. Structural studies of silicate glasses and melts-applications and limitations of Raman spectroscopy. *Am. Miner.* **1984**, *69*, 622–644.

4. Frezzotti, M.L.; Tecce, F.; Casagli, A. Raman spectroscopy for fluid inclusion analysis. *J. Geochem. Explor.* **2012**, *112*, 1–20. [CrossRef]

5. Rosso, K.; Bodnar, R. Microthermometric and Raman spectroscopic detection limits of CO_2 in fluid inclusions and the Raman spectroscopic characterization of CO_2. *Geochim. Cosmochim. Acta* **1995**, *59*, 3961–3975. [CrossRef]

6. Prigiobbe, V. Estimation of nucleation and growth parameters from in situ Raman spectroscopy in carbonate systems. *J. Environ. Chem. Eng.* **2018**, *6*, 930–936. [CrossRef]

7. Wan, Y.; Wang, X.; Hu, W.; Chou, I.; Wang, X.; Chen, Y.; Xu, Z. In situ optical and Raman spectroscopic observations of the effects of pressure and fluid composition on liquid–liquid phase separation in aqueous cadmium sulfate solutions ($\leq$400 °C, 50 MPa) with geological and geochemical implications. *Geochim. Cosmochim. Acta* **2017**, *211*, 133–152. [CrossRef]

8. McMillan, P.F.; Poe, B.T.; Stanton, T.R.; Remmele, R.L. A Raman spectroscopic study of H/D isotopically substituted hydrous aluminosilicate glasses. *Phys. Chem. Miner.* **1993**, *19*, 454–459. [CrossRef]

9. Sato, R.K.; McMillan, P.F. An infrared and Raman study of the isotopic species of alpha-quartz. *J. Phys. Chem.* **1987**, *91*, 3494–3498. [CrossRef]

10. Weckhuysen, B.M.; Jehng, J.; Wachs, I.E. In situ Raman spectroscopy of supported transition metal oxide catalysts: $^{18}O_2 - ^{16}O_2$ Isotopic Labeling Studies. *J. Phys. Chem. B* **2000**, *104*, 7382–7387. [CrossRef]

11. Wilson, E.B.; Decius, J.C.; Cross, P.C. *Molecular Vibrations: The Theory of Infrared and Raman Vibrational Spectra*; Courier Corporation: New York, NY, USA, 1980.

12. Edwards, H.G.M.; Villar, S.E.J.; Jehlicka, J.; Munshi, T. FT–Raman spectroscopic study of calcium-rich and magnesium-rich carbonate minerals. *Spectrochim. Acta A* **2005**, *61*, 2273–2280. [CrossRef] [PubMed]

13. Bhagavantam, S. Effect of crystal orientation on the Raman spectrum of calcite. *Proc. Math. Sci.* **1940**, *11*, 62–71.

14. Farmer, V. Transverse and longitudinal crystal modes associated with OH stretching vibrations in single crystals of kaolinite and dickite. *Spectrochim. Acta. A* **2000**, *56*, 927–930. [CrossRef]

15. Ishibashi, H.; Arakawa, M.; Ohi, S.; Yamamoto, J.; Miyake, A.; Kagi, H. Relationship between Raman spectral pattern and crystallographic orientation of a rock-forming mineral: A case study of $Fo_{89}Fa_{11}$ olivine. *J. Raman Spectrosc.* **2008**, *39*, 1653–1659. [CrossRef]

16. Shebanova, O.N.; Lazor, P. Raman spectroscopic study of magnetite ($FeFe_2O_4$): A new assignment for the vibrational spectrum. *J. Solid State Chem.* **2003**, *174*, 424–430. [CrossRef]

17. Awonusi, A.; Morris, M.D.; Tecklenburg, M.M.J. Carbonate assignment and calibration in the Raman spectrum of apatite. *Calcif. Tissue Int.* **2007**, *81*, 46–52. [CrossRef] [PubMed]

18. Alia, J.; de Mera, Y.D.; Edwards, H.G.M.; Martín, P.G.; Andres, S.L. FT-Raman and infrared spectroscopic study of aragonite-strontianite ($Ca_xSr_{1-x}CO_3$) solid solution. *Spectrochim. Acta A* **1997**, *53*, 2347–2362. [CrossRef]

19. Herzberg, G. *Molecular Spectra and Molecular Structure*; D. Van Nostrand Company, Inc.: Princeton, NJ, USA, 1945.

20. Geisler, T.; Perdikouri, C.; Kasioptas, A.; Dietzel, M. Real-time monitoring of the overall exchange of oxygen isotopes between aqueous CO_3^{2-} and H_2O by Raman spectroscopy. *Geochim. Cosmochim. Acta* **2012**, *90*, 1–11. [CrossRef]

21. Geidel, E.; Krause, K.; Förster, H.; Bauer, F. Vibrational spectra and computer simulations of ^{18}O-labelled NaY zeolites. *J. Chem. Soc. Faraday Trans.* **1997**, *93*, 1439–1443. [CrossRef]

22. Galeener, F.L.; Mikkelsen, J., Jr. Vibrational dynamics in ^{18}O-substituted vitreous SiO_2. *Phys. Rev. B* **1981**, *23*, 5527. [CrossRef]

23. Putnis, C.V.; Geisler, T.; Schmid-Beurmann, P.; Stephan, T.; Giampaolo, C. An experimental study of the replacement of leucite by analcime. *Am. Miner.* **2007**, *92*, 19–26. [CrossRef]

24. Gillet, P.; McMillan, P.; Schott, J.; Badro, J.; Grzechnik, A. Thermodynamic properties and isotopic fractionation of calcite from vibrational spectroscopy of ^{18}O-substituted calcite. *Geochim. Cosmochim. Acta* **1996**, *60*, 3471–3485. [CrossRef]

25. Crawford, B., Jr. Vibrational intensities. II. The use of isotopes. *J. Chem. Phys.* **1952**, *20*, 977–981. [CrossRef]

26. Petreanu, E.; Pinchas, S.; Samuel, D. The infra-red absorption of 75% ^{18}O-barium phosphate. *J. Inorg. Nucl. Chem.* **1965**, *27*, 2519–2523. [CrossRef]

27. Corno, M.; Busco, C.; Civalleri, B.; Ugliengo, P. Periodic ab initio study of structural and vibrational features of hexagonal hydroxyapatite $Ca_{10}(PO_4)_6(OH)_2$. *Phys. Chem. Chem. Phys.* **2006**, *8*, 2464–2472. [CrossRef] [PubMed]

28. Montero, S.; Schmölz, R.; Haussühl, S. Raman spectra of orthorhombic sulfate single crystals I: K_2SO_4, Rb_2SO_4, Cs_2SO_4 and Tl_2SO_4. *J. Raman Spectrosc.* **1974**, *2*, 101–113. [CrossRef]

29. Pinchas, S.; Shamir, J. The anomalous behaviour of ^{18}O-labelled compounds. Part IV. Vibrational spectra of isotopic vanadate ions. *Isr. J. Chem.* **1969**, *7*, 805–811. [CrossRef]

30. Arakawa, M.; Yamamoto, J.; Kagi, H. Developing micro-Raman mass spectrometry for measuring carbon isotopic composition of carbon dioxide. *Appl. Spectrosc.* **2007**, *61*, 701–705. [CrossRef] [PubMed]

31. Menneken, M.; Geisler, T. An evaluation of the potential of using Raman spectroscopy to determine the carbon isotope composition of CO_2 inclusions. *J. Geochem. Explor.* **2009**, *101*, 70. [CrossRef]

32. Menneken, M.; Geisler, T.; Nemchin, A.A.; Strauss, H. Raman spectroscopic determination of the isotope composition of CO_2 inclusions. *Geochim. Cosmochim. Acta Suppl.* **2009**, *73*, A869.

33. Carey, D.M.; Korenowski, G.M. Measurement of the Raman spectrum of liquid water. *J. Chem. Phys.* **1998**, *108*, 2669–2675. [CrossRef]

34. Pye, C.C.; Rudolph, W.W. An ab initio and Raman investigation of magnesium (II) hydration. *J. Phys. Chem. A* **1998**, *102*, 9933–9943. [CrossRef]

35. Bulmer, J.T.; Irish, D.E.; Ödberg, L. The temperature dependence of Raman band parameters for aquated Mg(II) and Zn(II). *Can. J. Chem.* **1975**, *53*, 3806–3811. [CrossRef]

36. Martinez, I.; Sanchez-Valle, C.; Daniel, I.; Reynard, B. High-pressure and high-temperature Raman spectroscopy of carbonate ions in aqueous solution. *Chem. Geol.* **2004**, *207*, 47–58. [CrossRef]

37. Pye, C.C.; Rudolph, W.W. An ab initio, infrared, and Raman investigation of phosphate ion hydration. *J. Phys. Chem. A* **2003**, *107*, 8746–8755. [CrossRef]

38. Rudolph, W.W.; Irmer, G.; Hefter, G.T. Raman spectroscopic investigation of speciation in $MgSO_4$(aq). *Phys. Chem. Chem. Phys.* **2003**, *5*, 5253–5261. [CrossRef]

39. Wang, X.; Wang, X.; Chou, I.; Hu, W.; Wan, Y.; Li, Z. Properties of lithium under hydrothermal conditions revealed by in situ Raman spectroscopic characterization of Li_2O-SO_3-H_2O (D_2O) systems at temperatures up to 420 °C. *Chem. Geol.* **2017**, *451*, 104–115. [CrossRef]

40. Rudolph, W.W.; Brooker, M.H.; Tremaine, P.R. Raman spectroscopic investigation of aqueous $FeSO_4$ in neutral and acidic solutions from 25 °C to 303 °C: Inner-and outer-sphere complexes. *J. Solut. Chem.* **1997**, *26*, 757–777. [CrossRef]

41. Preston, C.M.; Adams, W.A. A laser Raman spectroscopic study of aqueous orthophosphate salts. *J. Phys. Chem.* **1979**, *83*, 814–821. [CrossRef]

42. Zotov, N.; Keppler, H. Silica speciation in aqueous fluids at high pressures and high temperatures. *Chem. Geol.* **2002**, *184*, 71–82. [CrossRef]

43. King, H.E.; Satoh, H.; Tsukamoto, K.; Putnis, A. Nanoscale observations of magnesite growth in chloride-and sulfate-rich solutions. *Environ. Sci. Technol.* **2013**, *47*, 8684–8691. [CrossRef] [PubMed]

44. Wang, X.; Chou, I.; Hu, W.; Burruss, R.C. In situ observations of liquid–liquid phase separation in aqueous $MgSO_4$ solutions: Geological and geochemical implications. *Geochim. Cosmochim. Acta* **2013**, *103*, 1–10. [CrossRef]

45. Wang, X.; Wan, Y.; Hu, W.; Chou, I.; Cao, J.; Wang, X.; Wang, M.; Li, Z. In situ observations of liquid–liquid phase separation in aqueous $ZnSO_4$ solutions at temperatures up to 400 °C: Implications for Zn^{2+}–SO_4^{2-} association and evolution of submarine hydrothermal fluids. *Geochim. Cosmochim. Acta* **2016**, *181*, 126–143. [CrossRef]

46. Gebauer, D.; Cölfen, H. Prenucleation clusters and non-classical nucleation. *Nano Today* **2011**, *6*, 564–584. [CrossRef]

47. Geisler, T.; Kasioptas, A.; Menneken, M.; Perdikouri, C.; Putnis, A. A preliminary in situ Raman spectroscopic study of the oxygen isotope exchange kinetics between H_2O and (PO_4) aq. *J. Geochem. Explor.* **2009**, *101*, 37. [CrossRef]

48. Rudolph, W.W.; Irmer, G.; Königsberger, E. Speciation studies in aqueous HCO_3^-–CO_3^{2-} solutions. A combined Raman spectroscopic and thermodynamic study. *Dalton Trans.* **2008**, 900–908. [CrossRef] [PubMed]

49. Gunasekaran, S.; Anbalagan, G.; Pandi, S. Raman and infrared spectra of carbonates of calcite structure. *J. Raman Spectrosc.* **2006**, *37*, 892–899. [CrossRef]

50. Pinchas, S.; Sadeh, D. Fundamental vibration frequencies of the main isotopic PO_4^{3-} ions in aqueous solutions. *J. Inorg. Nucl. Chem.* **1968**, *30*, 1785–1789. [CrossRef]

51. Campbell, N.J.; Flanagan, J.; Griffith, W.P. Vibrational spectra of $[B(^{18}OH)_4]^-$, $[V^{18}O_4]^{3-}$, and $[P^{18}O_4]^{3-}$: An anomaly resolved. *J. Chem. Phys.* **1985**, *83*, 3712–3713. [CrossRef]

52. Putnis, A.; John, T. Replacement Processes in the Earth's Crust. *Elements* **2010**, *6*, 159–164. [CrossRef]

53. Nahon, D.; Merino, E. Pseudomorphic replacement in tropical weathering: Evidence, geochemical consequences, and kinetic-rheological origin. *Am. J. Sci.* **1997**, *297*, 393–417. [CrossRef]

54. Henao, D.M.O.; Godoy, M.A.M. Jarosite pseudomorph formation from arsenopyrite oxidation using Acidithiobacillus ferrooxidans. *Hydrometallurgy* **2010**, *104*, 162–168. [CrossRef]

55. Laves, F. Artificial preparation of microcline. *J. Geol.* **1951**, *59*, 511–512. [CrossRef]

56. Fiebig, J.; Hoefs, J. Hydrothermal alteration of biotite and plagioclase as inferred from intragranular oxygen isotope-and cation-distribution patterns. *Eur. J. Miner.* **2002**, *14*, 49–60. [CrossRef]

57. Labotka, T.C.; Cole, D.R.; Fayek, M.; Riciputi, L.R.; Stadermann, F.J. Coupled cation and oxygen-isotope exchange between alkali feldspar and aqueous chloride solution. *Am. Miner.* **2004**, *89*, 1822–1825. [CrossRef]

58. Niedermeier, D.R.D.; Putnis, A.; Geisler, T.; Golla-Schindler, U.; Putnis, C.V. The mechanism of cation and oxygen isotope exchange in alkali feldspars under hydrothermal conditions. *Contrib. Miner. Petrol.* **2009**, *157*, 65–76. [CrossRef]

59. Hövelmann, J.; Putnis, A.; Geisler, T.; Schmidt, B.C.; Golla-Schindler, U. The replacement of plagioclase feldspars by albite: Observations from hydrothermal experiments. *Contrib. Miner. Petrol.* **2010**, *159*, 43–59. [CrossRef]

60. Velbel, M.A. Formation of protective surface layers during silicate-mineral weathering under well-leached, oxidizing conditions. *Am. Miner.* **1993**, *78*, 405–414.

61. Liu, Y.; Olsen, A.A.; Rimstidt, D. Mechanism for the dissolution of olivine series minerals in acidic solutions. *Am. Miner.* **2006**, *91*, 455–458. [CrossRef]

62. Zakaznova-Herzog, V.P.; Nesbitt, H.W.; Bancroft, G.M.; Tse, J.S. Characterization of leached layers on olivine and pyroxenes using high-resolution XPS and density functional calculations. *Geochim. Cosmochim. Acta* **2008**, *72*, 69–86. [CrossRef]

63. Schott, J.; Pokrovsky, O.S.; Spalla, O.; Devreux, F.; Gloter, A.; Mielczarski, J.A. Formation, growth and transformation of leached layers during silicate minerals dissolution: The example of wollastonite. *Geochim. Cosmochim. Acta* **2012**, *98*, 259–281. [CrossRef]

64. Pokrovsky, O.S.; Schott, J. Forsterite surface composition in aqueous solutions: A combined potentiometric, electrokinetic, and spectroscopic approach. *Geochim. Cosmochim. Acta* **2000**, *64*, 3299–3312. [CrossRef]

65. Ruiz-Agudo, E.; King, H.E.; Patiño-López, L.D.; Putnis, C.V.; Geisler, T.; Rodriguez-Navarro, C.; Putnis, A. Control of silicate weathering by interface-coupled dissolution-precipitation processes at the mineral-solution interface. *Geology* **2016**, *44*, 567–570. [CrossRef]

66. King, H.E.; Plümper, O.; Geisler, T.; Putnis, A. Experimental investigations into the silicification of olivine: Implications for the reaction mechanism and acid neutralization. *Am. Miner.* **2011**, *96*, 1503–1511. [CrossRef]

67. Cardew, P.T.; Davey, R.J. The kinetics of solvent-mediated phase transformations. *Proc. R. Soc. A Math. Phys.* **1985**, *398*, 415–428. [CrossRef]

68. Putnis, A. Mineral replacement reactions: From macroscopic observations to microscopic mechanisms. *Miner. Mag.* **2002**, *66*, 689–708. [CrossRef]

69. Putnis, A.; Putnis, C.V. The mechanism of reequilibration of solids in the presence of a fluid phase. *J. Solid State Chem.* **2007**, *180*, 1783–1786. [CrossRef]

70. Wang, Y.; Wang, Y.; Merino, E. Dynamic weathering model: Constraints required by coupled dissolution and pseudomorphic replacement. *Geochim. Cosmochim. Acta* **1995**, *59*, 1559–1570. [CrossRef]

71. Glikin, A.E. *Polymineral-Metasomatic Crystallogenesis*; Springer: Dordrecht, The Netherlands, 2009.

72. Xia, F.; Brugger, J.; Chen, G.; Ngothai, Y.; O'Neill, B.; Putnis, A.; Pring, A. Mechanism and kinetics of pseudomorphic mineral replacement reactions: A case study of the replacement of pentlandite by violarite. *Geochim. Cosmochim. Acta* **2009**, *73*, 1945–1969. [CrossRef]

73. Putnis, C.V.; Mezger, K. A mechanism of mineral replacement: Isotope tracing in the model system KCl-KBr-H_2O. *Geochim. Cosmochim. Acta* **2004**, *68*, 2839–2848. [CrossRef]

74. Kasioptas, A.; Geisler, T.; Perdikouri, C.; Trepmann, C.; Gussone, N.; Putnis, A. Polycrystalline apatite synthesized by hydrothermal replacement of calcium carbonates. *Geochim. Cosmochim. Acta* **2011**, *75*, 3486–3500. [CrossRef]

75. Perdikouri, C.; Kasioptas, A.; Geisler, T.; Schmidt, B.C.; Putnis, A. Experimental study of the aragonite to calcite transition in aqueous solution. *Geochim. Cosmochim. Acta* **2011**, *75*, 6211–6224. [CrossRef]

76. King, H.E.; Mattner, D.C.; Plümper, O.; Geisler, T.; Putnis, A. Forming cohesive calcium oxalate layers on marble surfaces for stone conservation. *Cryst. Growth Des.* **2014**, *14*, 3910–3917. [CrossRef]

77. Geisler, T.; Janssen, A.; Scheiter, D.; Stephan, T.; Berndt, J.; Putnis, A. Aqueous corrosion of borosilicate glass under acidic conditions: A new corrosion mechanism. *J. Non-Cryst. Solids* **2010**, *356*, 1458–1465. [CrossRef]

78. Janssen, A.; Putnis, A.; Geisler, T. The experimental replacement of ilmenite by rutile in HCl solutions. *Miner. Mag.* **2010**, *74*, 633–644. [CrossRef]

79. Jonas, L.; John, T.; King, H.E.; Geisler, T.; Putnis, A. The role of grain boundaries and transient porosity in rocks as fluid pathways for reaction front propagation. *Earth Planet. Sci. Lett.* **2014**, *386*, 64–74. [CrossRef]

80. Geisler, T.; Pöml, P.; Stephan, T.; Janssen, A.; Putnis, A. Experimental observation of an interface-controlled pseudomorphic replacement reaction in a natural crystalline pyrochlore. *Am. Miner.* **2005**, *90*, 1683–1687. [CrossRef]

81. Pöml, P.; Menneken, M.; Stephan, T.; Niedermeier, D.; Geisler, T.; Putnis, A. Mechanism of hydrothermal alteration of natural self-irradiated and synthetic crystalline titanate-based pyrochlore. *Geochim. Cosmochim. Acta* **2007**, *71*, 3311–3322. [CrossRef]

82. Li, J.; Chou, I.; Yuan, S.; Burruss, R.C. Observations on the crystallization of spodumene from aqueous solutions in a hydrothermal diamond-anvil cell. *Geofluids* **2013**, *13*, 467–474. [CrossRef]

83. Wan, Y.; Wang, X.; Chou, I.; Hu, W.; Zhang, Y.; Wang, X. An experimental study of the formation of talc through CaMg(CO$_3$)$_2$–SiO$_2$–H$_2$O interaction at 100–200 °C and vapor-saturation pressures. *Geofluids* **2017**, *2017*, 3942826. [CrossRef]

84. Hänchen, M.; Prigiobbe, V.; Baciocchi, R.; Mazzotti, M. Precipitation in the Mg-carbonate system—Effects of temperature and CO$_2$ pressure. *Chem. Eng. Sci.* **2008**, *63*, 1012–1028. [CrossRef]

85. Geisler, T.; Lenting, C.; Stamm, F.M.; Sulzbach, M. Real-time, in situ hyperspectral Raman imaging of mineral-fluid reactions. *Goldschm. Abstr.* **2015**, *2015*, 1016.

86. Sulzbach, M.; Geisler, T. The replacement of Celestine (SrSO$_4$) by Strontianite (SrCO$_3$) studied in situ, spatially resolved, and real-time by Raman spectroscopy. In Proceedings of the EGU General Assembly Conference, Vienna, Austria, 12–17 April 2015.

87. Nasdala, L.; Irmer, G.; Wolf, D. The degree of metamictization in zircons: A Raman spectroscopic study. *Eur. J. Miner.* **1995**, *7*, 471–478. [CrossRef]

88. Weissbart, E.J.; Rimstidt, J.D. Wollastonite: Incongruent dissolution and leached layer formation. *Geochim. Cosmochim. Acta* **2000**, *64*, 4007–4016. [CrossRef]

89. De Meer, S.; Spiers, C.J. On mechanisms and kinetics of creep by intergranular pressure solution. In *Growth, Dissolution and Pattern Formation in Geosystems*; Springer: Dordrecht, The Netherlands, 1999; pp. 345–366.

90. Felipe, M.A.; Kubicki, J.D.; Rye, D.M. Oxygen isotope exchange kinetics between H$_2$O and H$_4$SiO$_4$ from ab initio calculations. *Geochim. Cosmochim. Acta* **2004**, *68*, 949–958. [CrossRef]

© 2018 by the authors. Licensee MDPI, Basel, Switzerland. This article is an open access article distributed under the terms and conditions of the Creative Commons Attribution (CC BY) license (http://creativecommons.org/licenses/by/4.0/).

![minerals logo] *minerals*

MDPI

Article

Biomineral Reactivity: The Kinetics of the Replacement Reaction of Biological Aragonite to Apatite

Martina Greiner [1,*], Lurdes Férnandez-Díaz [2,3], Erika Griesshaber [1], Moritz N. Zenkert [1], Xiaofei Yin [1], Andreas Ziegler [4], Sabino Veintemillas-Verdaguer [5] and Wolfgang W. Schmahl [1]

[1] Department für Geo-und Umweltwissenschaften, Ludwig-Maximilians-Universität, 80333 Munich, Germany; e.griesshaber@lrz.uni-muenchen.de (E.G.); moritz.zenkert@lrz.uni-muenchen.de (M.N.Z.); xiaofei.yin@campus.lmu.de (X.Y.); wolfgang.w.schmahl@lrz.uni-muenchen.de (W.W.S.)

[2] Departamento de Mineralogía y Petrología, Universidad Complutense de Madrid, 28040 Madrid, Spain; lfdiaz@geo.ucm.es

[3] Instituto de Geociencias (IGEO), (UCM, CSIC), Ciudad Universitaria, 28040 Madrid, Spain

[4] Central Facility for Electron Microscopy, University of Ulm, Albert-Einstein-Allee 11, 89069 Ulm, Germany; andreas.ziegler@uni-ulm.de

[5] Instituto de Ciencia de Materiales de Madrid (ICMM, CSIC), Cantoblanco, 28049 Madrid, Spain; sabino@icmm.csic.es

* Correspondence: martina.greiner@lrz.uni-muenchen.de; Tel.: +49-089-2180-4337

Received: 18 June 2018; Accepted: 20 July 2018; Published: 26 July 2018

check for updates

Abstract: We present results of bioaragonite to apatite conversion in bivalve, coral and cuttlebone skeletons, biological hard materials distinguished by specific microstructures, skeletal densities, original porosities and biopolymer contents. The most profound conversion occurs in the cuttlebone of the cephalopod *Sepia officinalis*, the least effect is observed for the nacreous shell portion of the bivalve *Hyriopsis cumingii*. The shell of the bivalve *Arctica islandica* consists of cross-lamellar aragonite, is dense at its innermost and porous at the seaward pointing shell layers. Increased porosity facilitates infiltration of the reaction fluid and renders large surface areas for the dissolution of aragonite and conversion to apatite. Skeletal microstructures of the coral *Porites* sp. and prismatic *H. cumingii* allow considerable conversion to apatite. Even though the surface area in *Porites* sp. is significantly larger in comparison to that of prismatic *H. cumingii*, the coral skeleton consists of clusters of dense, acicular aragonite. Conversion in the latter is sluggish at first as most apatite precipitates only onto its surface area. However, the process is accelerated when, in addition, fluids enter the hard tissue at centers of calcification. The prismatic shell portion of *H. cumingii* is readily transformed to apatite as we find here an increased porosity between prisms as well as within the membranes encasing the prisms. In conclusion, we observe distinct differences in bioaragonite to apatite conversion rates and kinetics depending on the feasibility of the reaction fluid to access aragonite crystallites. The latter is dependent on the content of biopolymers within the hard tissue, their feasibility to be decomposed, the extent of newly formed mineral surface area and the specific biogenic ultra- and microstructures.

Keywords: bioaragonite; apatite; microstructure; dissolution-reprecipitation; mineral replacement

---

1. Introduction

Research of the last decades has shown that carbonate biological hard tissues of marine and terrestrial organisms are highly valuable for their use in medical applications (e.g., [1–6]) e.g., as bone implant and bone graft materials. This is called forth not only by their biocompatibility, biodegradability and osteoconductive capabilities but also by their unique architectures: their specific shapes, varying porosities and thus compactness as well as pore size distributions. Many carbonate biological hard tissues were tested for their applicability for medical tasks (e.g., sponges, corals, coralline algae, cuttlefish bone, echinoderms, marine and terrestrial bivalves (e.g., [5]). Two of these emerged as being highly valuable: bivalve shell nacreous and coral skeleton aragonite [5,7–15].

The ideal material for bone tissue engineering has to provide initial mechanical strength for support as well as the ability of gradual resorption for the replacement with newly synthesized tissue. The first materials that were used for bone graft substitutes were dense materials in granule or particulate form (e.g., [5]). However, as these were only partially resorbable, replacement by new bone was often incomplete. The desire to mimic the natural structure of bone more closely led to the use of porous biological and biomimetic hard tissues, as these allow an improved interdigitation with the host bone and promote fluid exchange. This led to the search of naturally occurring scaffolds with bone like structures, e.g., coral skeletons. For this purpose, several colonial coral species were studied (e.g., *Porites* sp., *Goniopora* sp., *Acropora* sp., *Lophelia* sp., *Madropora* sp. [5], some having skeletal architectures (e.g., that of *Porites* sp.) resembling spongy cancellous bone and, others with a dense skeleton (e.g., *Acropora* sp., *Madropora* sp.), mimicking cortical bone. However, clinical tests showed that when too much stimulus is applied, pure coral scaffolds resorb too rapidly during the process of new bone formation. Hence, for use in medical applications, coral aragonite had to be functionalized into a, for the medical task, more appropriate material. Prior to use as a bone graft material, coral aragonite has to be transformed hydrothermally to apatite (AP). The conversion is performed only partially, such that an inner bioaragonite core is covered with an outer AP-coating. As aragonite is more soluble than apatite, the controlled conversion of bioaragonite to AP ensures a guided biodegradation rate that, if necessary, can be modified and tuned to local medical requirements [4,13]. The partial conversion of coral aragonite into AP ensures that the obtained bone graft material retains its porous structure and remains biocompatible, but has also improved biodegradation properties that suit better bone remodeling and turnover.

Conditioning of sepia cuttlebone and coral aragonite, echinoderm carapace and sea urchin spine calcite via hydrothermal conversion to AP has shown that in all cases porosity characteristics (pore abundance, size and distribution) of the apatite product were ideal for their use in medical applications. However, conversion times to AP were highly variable and depended on both, the chosen experimental conditions and the specific biological hard tissue. In the present manuscript we investigate the impact of different biocarbonate hard materials on the rate and kinetics of bioaragonite to apatite (AP) conversion. We use an open system set up, thus we allow under boiling conditions the continuous re-equilibration with atmosphere (see also [16]). We discuss major factors that lead to the differential reactivity of the biocarbonate hard tissue when transformed to AP, with the main focus being centered around the impact of distinct bioaragonite microstructures: that being present in the shell of the fresh water bivalve *Hyriopsis cumingii*, in the shell of the deep water, marine bivalve *Arctica islandica*, the microstructure that forms the skeleton of the warm water scleractinian coral *Porites* sp. and that, that comprises the lightweight structure of the cuttlebone of the cephalopod *Sepia officinalis*. We discuss results of conversion experiments that lasted up to 14 days and explore the role and effect of mineral organization, skeleton density, skeletal primary porosity and biopolymer content, fabric and distribution in the hard tissue for the kinetics of bioaragonite to AP conversion.

2. Materials and Methods

2.1. Test Materials

The warm-water coral *Porites* sp. was collected in Moorea, French Polynesia. *Arctica islandica* shell samples were collected in Loch Etive waters in Scotland. *Hyriopsis cumingii* samples were collected from Gaobao lake in China. *Sepia officinalis* cuttlebone was collected close to Elba in the Mediterranean Sea. The geological aragonite comes from Molina de Aragon in Spain.

2.2. Experimental Setup

Mineral conversion reactions were performed in a 2000 mL round flask filled with 1500 mL 1 M $(NH_4)_2HPO_4$ solution (Figure A1a). The diammonium hydrogen phosphate (Sigma-Aldrich, St. Louis, MO, USA) solution was prepared with high purity deionized water (18.2 MΩ). The system was heated with a heating mantle and kept at a constant temperature of about 99 °C that guaranteed that the aqueous solution was permanently boiling. A reflux condenser with constant flow of cold water was used to avoid water evaporation during the experiments. However, this reflux condenser did not prevent the escape of gases (e.g., CO_2) to the atmosphere.

Similarly sized fragments of the hard tissues were immersed in the boiling solution. In the case of *Sepia officinalis* cuttlebone the investigated samples were 15 mm sided cubes, while in the case of *Porites* sp. we used cylindrically shaped sample pieces (12 × 8 mm). 20 × 10 mm thick fragments were used for *Arctica islandica* and *Hyriopsis cumingii*. Initial sample weights ranged between 0.3 and 1.2 g. As the aqueous solution volume to biomineral weight was very high, the small initial differences in sample weights are irrelevant. The samples were recovered from the solution after specific conversion reaction times (Figure A1b), were washed with distilled water and were stored in isopropanol. $CaCO_3$ biominerals interacted with the phosphate-bearing boiling solution in an open system. Hence, our experimental setup, shown in Figure A1a, allowed for a continuous re-equilibration with atmosphere.

2.3. Organic Matrix Preparation by Selective Etching

In order to image the organic matrix, sample pieces were mounted on 3 mm thick cylindrical aluminium rods using super glue. The samples were first cut using a Leica Ultracut ultramicrotome with glass knifes to obtain plane surfaces. The cut pieces were then polished with a diamond knife by stepwise removal of material in a series of 20 sections with successively decreasing thicknesses (90 nm, 70 nm, 50 nm). The polished samples were etched for 180 s using 0.1 M HEPES (pH = 6.5) containing 2.5% glutaraldehyde as a fixation solution. The etching procedure was followed by dehydration in 100% isopropanol three times for 10 min each, before specimens were critical point dried. The dried samples were rotary coated with 4 nm platinum and imaged using a Hitachi S5200 Field Emission-Scanning Electron Microscope (FE-SEM) at 4 kV.

2.4. Characterization Methods

X-ray patterns of all samples were collected on a Bragg-Brentano type X-ray diffractometer (XRD 3003 TT, GE Sensing & Inspection Technologies, Hürth, Germany) with Cu-$K_{\alpha 1}$-radiation and a 2θ angle from 10°–110° with a step size of 0.013°. Rietveld refinement was performed with the program FullProf [17] on all samples containing more than 5 wt % AP. Refinements were conducted using the structural model of aragonite published by Jarosch and Heger [18] and the model of AP published by Kay et al. [19]. For samples with an AP content lower than 5 wt % a semi-quantitative percentage estimation was performed using the Reference Intensity (RIR) method [20] using the structural phase models mentioned before.

For the calculation of crystallite sizes, the pristine aragonitic sample was mixed with an internal standard (LaB$_6$, 10 wt %) to determine the instrumental resolution function. Thereafter the Thompson-Cox-Hastings method [21] for convolution of instrumental resolution and isotropic microstrain broadening was applied on the XRD pattern of the pristine sample.

Infrared spectra were measured on a Perkin–Elmer (Waltham, MA, USA) ATR-FTIR Spectrum two instrument with a resolution of 4 cm^{-1} with 128 scans.

The samples were rotary-coated with 5 nm of platinum-palladium and imaged using a Hitachi SU5000 field emission-scanning electron microscope (Tokyo, Japan).

3. Results

3.1. X-Ray Diffraction Analysis

XRD measurements confirm that all investigated hard tissues consist of aragonite and evidence that their conversion into AP does not involve any intermediate calcium phosphate. All refinements give a good agreement between observed and calculated profile (Chi^2 < 2.8, an exemplary Rietveld fit is shown in Figure A2). All obtained XRD patterns are shown in the Supplementary section of this manuscript. Table 1 and Figure 1 provide the percentage of aragonite to AP conversion as a function of conversion time, calculated from Rietveld analyses. We find a significant difference in newly formed AP between the studied biological hard tissues, not only between themselves but also between the biogenic samples and geological aragonite. The latter appears to be highly unreactive compared to most biogenic aragonites (Table 1).

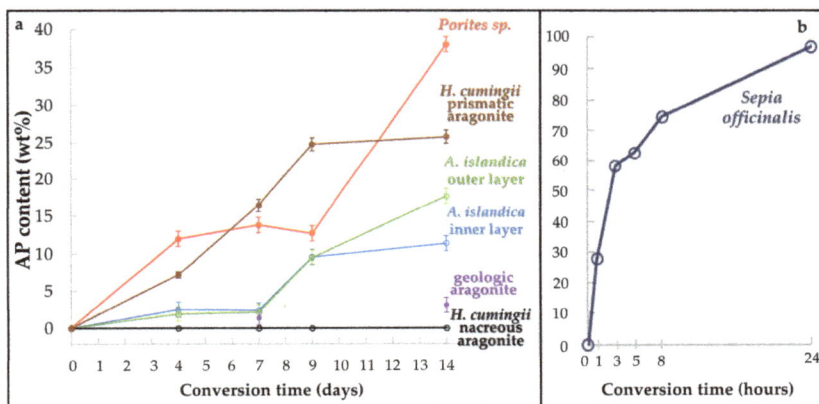

Figure 1. Newly formed apatite content of all samples relative to conversion time according to Rietveld analyses of the XRD measurements (**a**) 14 days *H. cumingii* nacre (brown line) and prismatic aragonite layer (black line), *A. islandica* outer (green line) and inner layer (blue line), geologic aragonite (violet data points) and *Porites* sp. (red line); (**b**) *Sepia officinalis*, conversion time 1–24 h. Error for *H. cumingii* nacre and *Sepia officinalis* is within the size of the corresponding data points.

Results of Rietveld refinements of XRD patterns of the aragonitic biogenic precursors versus time are shown in Figure 1. These as well highlight the striking differences in reactivity between the investigated samples. The highest reactivity is observed in *Sepia officinalis* cuttlebone, which consists of more than 60 wt % AP after only 5 h of conversion with the phosphate-bearing boiling solution. After 24 h, the conversion of the bioaragonite into AP is almost complete (Figure 1, Table 1). These results are in good agreement with those reported by Reinares-Fisac et al., 2017 [16]. All other investigated biological hard tissues transform at a much slower rate. Out of these we find the fastest conversion kinetics for the skeleton of *Porites* sp. The conversion trend is characterized by an initial rapid conversion speed, reaching 12 wt % AP after 4 days of reaction. With proceeding time, conversion does not progress further for the following 5 days. However, subsequently conversion kinetics accelerates again such that at 14 days of conversion 38 wt % AP is formed in the skeleton of *Porites* sp. (Figure 1). The shells of the bivalves, *Arctica islandica* and *Hyriopsis cumingii* react differently to conversion. Rietveld refinements of the XRD patterns of the shell of *Arctica islandica* show a similar conversion trend for both, the inner (shell portion next to the soft tissue of the animal) and the outer (seaward pointing) shell layer. Conversion starts first at a slow rate (~2 wt % of newly formed AP after 7 days of reaction), while after 9 days of reaction (~9.5 wt % of newly formed AP) it speeds up.

Alteration for the last five days is differently realized in the different shell portions of *Arctica islandica*. A fairly high conversion rate is observed for the seaward pointing shell part, while aragonite forming inner shell portions next to the soft tissue of the animal appears to be almost resistant to dissolution of aragonite and, hence, conversion to AP (Figure 1, Table 1).

Table 1. Samples used for conversion experiments, experiment time periods and apatite contents obtained for specific conversion times.

	AP Content (wt %) after 4 Days	AP Content (wt %) after 7 Days	AP Content (wt %) after 9 Days	AP Content (wt %) after 14 Days
Porites sp.	12.0(4)	13.8(3)	12.7(3)	38(1)
Hyriopsis cumingii (nacre)	< 1	< 1	< 1	< 1
Hyriopsis cumingii (prismatic aragonite)	7.2(4)	16.4(8)	24.8(9)	25.8(9)
Arctica islandica (shell layer next to seawater)	1.9	2.1	9.5(4)	17.6(5)
Arctica islandica (shell layer next to soft tissue)	2.5	2.4	9.6(7)	11.4(4)
geologic aragonite (single crystal)	not determined	1.4	not determined	3.1

	AP Content (wt %) after 1 Hour	AP Content (wt %) after 3 Hours	AP Content (wt %) after 5 Hours	AP Content (wt %) after 8 Hours	AP Content (wt %) after 24 Hours
Sepia officinalis	29.9(7)	58.2(8)	64.4(7)	74.5(8)	99(1)

Most striking is the very different conversion behavior of nacreous aragonite (next to the soft tissue of the animal) and the prismatic shell portion (seawater pointing shell layer) of *Hyriopsis cumingii* (Figure 1, Table 1). All diffraction peaks in XRD patterns of the nacre sample can be assigned to aragonite, regardless the time of interaction with the phosphate-bearing aqueous solution. This evidences that the nacre did not undergo any conversion to AP during the time span of our experiments (Figure A3). In contrast, the prismatic shell portion of *Hyriopsis cumingii* is readily transformed to apatite. We see a steady increase in conversion rate up to 9 days of alteration (Figure 1); about 25 wt % AP was found in the prismatic hard tissue of *Hyriopsis cumingii* after this alteration time. However, for the subsequent 5 days only minute amounts of bioaragonite transformed to AP in the prismatic shell portion of *Hyriopsis cumingii* (Figure 1, Table 1). In order to investigate the influence of the aragonite crystallite sizes on the reactivity, we calculated the crystallite size of *Sepia officinalis* cuttlebone and *Hyriopsis cumingii* nacreous aragonite by Rietveld refinement. The average size of nacreous aragonite crystals in *H. cumingii* is 504 Å; for *Sepia officinalis* the average granular aragonite crystallite size is 486 Å.

3.2. FTIR-Spectroscopy

IR-spectra of the four biological hard tissues being in contact with the phosphate-bearing solution for 14 days are shown in Figure 2.

Absorption bands that can be assigned to the carbonate group vibrations in the structure of aragonite are present in all samples, with the exception of the completely reacted *Sepia officinalis* cuttlebone. The bands appear at 700 and 713 cm^{-1} ($v_4CO_3^{2-}$), 855–858 cm^{-1} ($v_2CO_3^{2-}$), 1083 ($v_1CO_3^{2-}$) and 1450–1475 cm^{-1} ($v_3CO_3^{2-}$) [22–24] and show variable intensities depending on the specific hard tissue. They are sharp in the spectrum of *Hyriopsis cumingii* nacre, are intense and clearly visible in the spectra of both, the outer and inner shell layers of *Arctica islandica*, are weaker and broader in the spectra gained from the skeleton of *Porites* sp. and for prismatic aragonite in the shell of the bivalve *H. cumingii*.

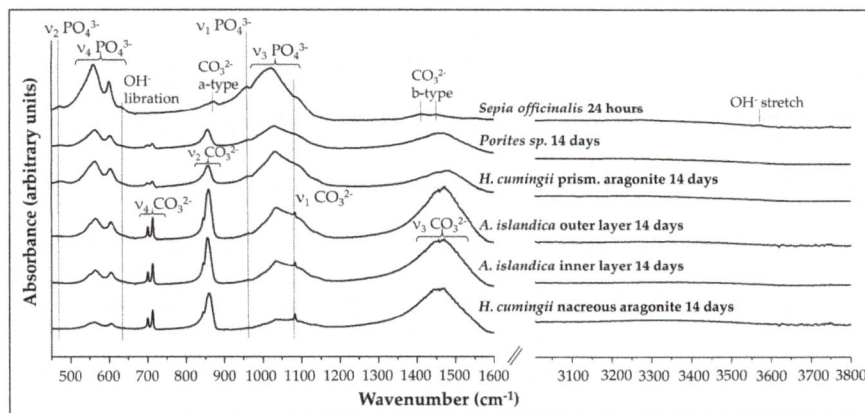

Figure 2. Infrared-spectra (450–1600 cm^{-1} and 3000–3800 cm^{-1}) for *Sepia officinalis* cuttlebone transformed for 24 h together with IR-spectra for *Porites* sp. coral aragonite, *H. cumingii* prismatic aragonite, *A. islandica* outer and inner shell layer aragonite and *H. cumingii* nacreous aragonite. Spectra are arranged according to AP content (wt %) determined with XRD and Rietveld analyses. AP content increases from bottom to top spectra.

Absorption bands that can be assigned to vibrations of the phosphate group in the structure of AP are clearly distinguishable at ~470 cm^{-1} ($v_2PO_4^{3-}$), 560–602 cm^{-1} ($v_4PO_4^{3-}$), ~960 cm^{-1} ($v_1PO_4^{3-}$) and at 1020–1100 cm^{-1} ($v_3PO_4^{3-}$) in the spectra of *Sepia officinalis* cuttlebone, *Porites* sp. skeleton, outer and inner shell layers of *Arctica islandica* and in prismatic aragonite in the shell of *H. cumingii* [25–27]. The intensity and definition of these bands is highly variable depending on the hard tissue. Thus, the v_2 phosphate vibration is visible as a broad, weak band in the spectra of *H. cumingii* prismatic aragonite and the skeleton of *Porites* sp. This band is sharp and appears well-defined in the spectrum of *Sepia officinalis* cuttlebone. The *Sepia officinalis* spectrum also shows well-defined v_4 and v_1 phosphate bands. In contrast, the bands are broad and weak in the spectra of the *Porites* sp. skeleton, outer and inner shell layers of *Arctica islandica* and these prismatic aragonite of *H. cumingii*. Finally, a broad v_3 phosphate band is detectable in all spectra shown in Figure 2, including *H. cumingii* nacre.

The observed variation in intensity and definition of the absorption bands assigned to the carbonate and phosphate groups is in good agreement with the different degrees of conversion of the studied hard tissues as deduced from Rietveld refinements of XRD patterns. *H. cumingii* nacre constitutes an exception. According to XRD analysis *H. cumingii* nacre does not undergo any conversion to AP, even after 14 days of interaction with the phosphate-bearing solution. However, the presence of very weak and broad phosphate-bands (v_3 and v_4) in the IR spectrum of *H. cumingii* nacre that was altered for 14 days supports that a minor amount of conversion to AP has taken place.

It is worthwhile to note that the IR spectrum of *Sepia officinalis* aragonite altered for 24 h shows bands that are not found in the spectra of the other altered samples. One of these bands appears at 873 cm^{-1} and can be assigned to vibrational frequencies of carbonate ions that substitute into OH^{-} sites in the apatite structure (a-type substitution). Further two bands, those that appear at 1405 and 1450 cm^{-1}, can be interpreted as corresponding to carbonate ions substituted into the phosphate site in apatite structure (b-type substitution) [24,28]. Moreover, the spectrum of *Sepia officinalis* hard tissue transformed for 24 h also shows a broad band at ~633 cm^{-1} which can be interpreted as the OH^{-} libration band [26], and a poorly developed shoulder appears at 3570 cm^{-1}. This can be attributed to the OH^{-} stretching mode [29–31].

3.3. Scanning Electron Microscopy (SEM)

Figures 3–10 present FE-SEM images showing crystallite and mineral unit arrangement preservation together with the distribution pattern of newly formed apatite (AP) for the investigated altered biogenic skeletons. As reference materials we used pristine equivalents of the geologic aragonite (Figure A4) and the altered skeletons (Figures A5–A16). Figure A4 shows structural features of a geologic aragonite single crystal that was altered for 14 days. We find that a thin layer of newly formed apatite has replaced the aragonite and covers the external surface of the crystal (Figure A4b). The apatite layer shows a multitude of cracks (Figure A4a) that most likely developed due to contraction when the sample was taken out of the boiling solution.

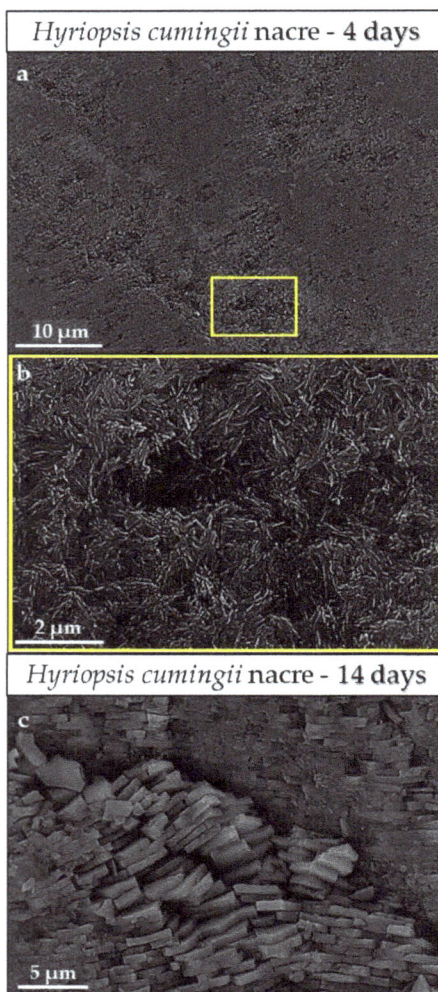

Figure 3. FE-SEM images of *Hyriopsis cumingii* nacre after (**a**,**b**) 4 days of conversion and (**c**) 14 days of conversion. Very little conversion to AP has taken place, the original arrangement of nacre tablets is well preserved (**c**). Even though, slight dissolution of aragonite induced that the surface of the nacreous shell layer became encrusted with a thin layer of newly formed AP crystals (**a**,**b**).

The shell of the bivalve *Hyriopsis cumingii* consists of aragonite that occurs in two distinct microstructures, nacreous aragonite next to the soft tissue of the animal and prismatic aragonite along the shell rim pointing to the seawater. The basic mineral units, prisms and tablets are encased by biopolymer membranes. Nacre tablets in the shell of the bivalve *H. cumingii* are arranged in a brick wall arrangement. As our XRD results (Figure A3) and the comparison between the pristine (Figure A5) and the 14 days altered (Figure 3c) shell portions show, the microstructure that was least transformed in the course of our experiments is nacreous aragonite. Nacre tablets as well as their microstructural arrangement remained well preserved even in the most altered sample (Figure 3). Even though, slight dissolution and reprecipitation has occurred as the surface of the 4 days altered and the nacre sample is covered with a layer of newly formed apatite crystals (Figure 3a,b). The weak and broad bands in the IR spectrum can be assigned to phosphate (ν_3 and ν_4). We identify them as AP, even though peaks that could be assigned to apatite were not found in the XRD pattern (Figure A3).

The microstructure of the shell of the bivalve *Arctica islandica* is also highly resistant to the formation of new apatite (compare images in Figure 4a,b to Figure A6a). The shell of pristine *Arctica islandica* (Figures A6 and A7) consists of densely packed and irregularly shaped mineral units (yellow stars in Figures A6b,c and A7) that are embedded into a network of biopolymer fibrils (Figure A6b,c and Casella et al., 2017 [32]). Mineral unit size, porosity and density of aragonite crystal packing are unevenly distributed within the shell (Figures A7 and A8), such that, relative to inner shell portions, mineral unit and pore sizes along the shell rim pointing to seawater, are increased. Mineral unit organization in the shell of *Arctica islandica* is little structured, especially in shell portions along the seawater pointing shell rim (Figure A8b). However, aragonite that constitutes inner shell layers is present in a crossed-lamellar microstructural arrangement (Figure A8a), especially that next to the soft tissue of the animal. Growth lines are frequent and are easily observable (white stars in Figure A7a,c), as, at these, biopolymer contents and mineral unit sizes are increased. The shell of *Arctica islandica* can be addressed as consisting of densely packed aragonite. However, it shows primary porosity (Figure A8), with the porosity being unevenly distributed. Along the seaward pointing shell portion, pores are abundant and large, while at shell parts that are inward and closer to the soft tissue of the animal, pores are small and significantly less frequent.

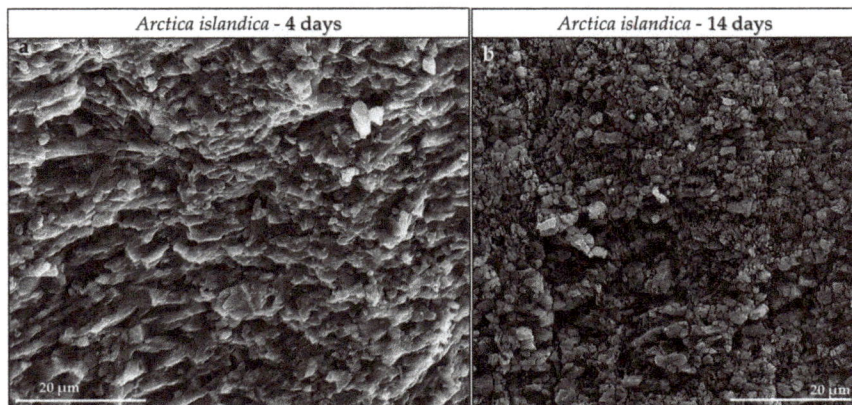

Figure 4. FE-SEM images of aragonite crystallite morphology and assembly after (**a**) 4 and (**b**) 14 days of conversion. We do not observe any major changes in crystallite morphology and size relative to that present in the pristine shell (compare Figure 4 to Figure A6a).

As Figure 1 highlights, up to 7 days of alteration the *Arctica islandica* microstructure is highly resistant to conversion of bioaragonite to apatite, the morphology of crystallites and mineral units remains well preserved and only a negligible amount of apatite formation takes place. Alteration becomes more marked from nine days onwards, when we find occasional occurrences of AP in pores and cavities that develop when the organic material within the shell becomes decomposed.

As the macroarchitecture of the skeleton of the scleractinian coral *Porites* sp. consists of a multitude of vertical and transverse elements (e.g., septa, columns, pali, theca, trabercular units), the bulk 2D appearance of the coral sample is characterized by many voids, gaps and recesses (Figure A9a) and effects that the skeleton has a high surface area.

However, the basic mineral units of scleractinian coral skeletons are differently sized. Oriented clusters (white stars in Figure A9b) of densely packed aragonitic needles and acicles nucleate at and grow outward from cavities within the coral skeleton, the centers of calcification (white arrows in Figure A9c). The shape of basic mineral units in scleractinian coral skeletons can be addressed as partial or even as full spherulites. Already at 4 days of conversion a noticeable amount of AP forms and covers the surface of the coral skeleton (Figure 5a,b). At 14 days of conversion the AP cover increases in denseness (Figure 5c,d). In addition, we also observe some conversion to AP within the skeleton.

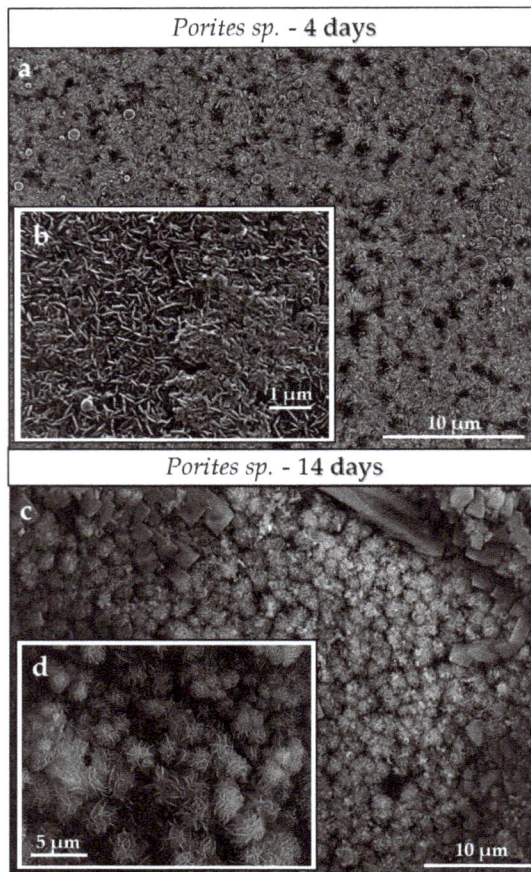

Figure 5. FE-SEM images depicting the coral skeleton *Porites* sp. after (**a**) 4 and (**b**) 14 days of conversion. Newly formed AP crystals cover the surface of the hard tissue already at 4 days of conversion. The compactness of the AP cover increases significantly with conversion time. SEM images shown in (**b**,**d**) depict AP crystal morphologies that form with different conversion times.

In contrast to the nacreous shell portion, prismatic aragonite in the shell of *Hyriopsis cumingii* is readily attacked. Aragonite prisms in modern *H. cumingii* (Figure A10a) are large units encased by thick biopolymer membranes (Figure A10d,e) consisting of aragonite crystallites (Figure A10b,c) placed within a network of organic fibrils (Figure A10f). Conversion from biogenic aragonite to apatite starts instantly and increases steadily with alteration time. At 14 days of alteration more than 25 wt % of prismatic aragonite of *H. cumingii* is transformed to AP, while the nacreous aragonite within the shell remains almost totally unaffected (Figure 1). At 4 of days of conversion we find that a layer of AP covers the prisms. When altered for 14 days, conversion affects both, the surface of the prisms (the cover with AP becomes thicker) and the aragonite crystallites within the prisms (Figure 6b–d).

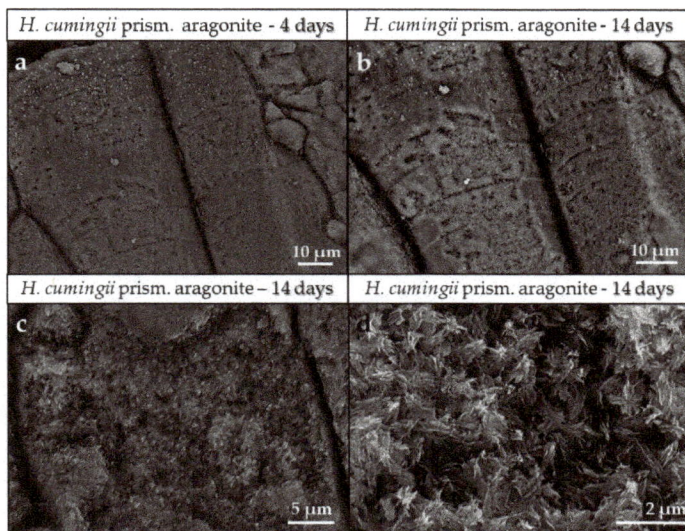

Figure 6. FE-SEM images showing *Hyriopsis cumingii* aragonite prisms altered for (**a**) 4 and (**b–d**) 14 days. Overall prism morphology does not change with alteration time. A layer of newly formed AP crystals covers the surface of the prisms already at 4 days of conversion. It increases in denseness with conversion time. Crystal morphologies in (**d**) clearly show that the aragonite within the prisms becomes converted to apatite at 14 days of alteration.

The most rapid and profound conversion takes place in the cuttlebone of the cephalopod *Sepia officinalis* (this study and [16]). The macrostructure of the cuttlebone resembles a carpark structure and consists of regularly spaced platforms that are interconnected by curved walls forming differently sized compartments within the cuttlebone (Figure A11). Nanoparticulate aragonite (Figure A11a,b) with sizes between 400 to 600 nm constitute all skeletal elements, the platforms and the walls. For the walls, we do not find a specific arrangement of the nanoparticulate aragonite, contrasting to the platforms that comprise different layers (white, yellow and blues stars in Figure A1d), consisting of differently oriented stacks of aragonite rods (Figure A12, white stars in Figure A12a), with rods being composed of nanoparticulate aragonite embedded into a biopolymer matrix (Figure A12b, yellow arrows in Figure A12c). Etching of the cuttlebone for 180 s exhibits the biopolymer fraction within the skeleton (Figures A15 and A16). The high amount of biopolymers present in the cuttlebone is well observable as well as the large variety of organic fabrics. For all biological hard tissues that we investigated in this study, the highest organic content is present in *Sepia officinalis* cuttlebone. This yields a huge interface area between the mineral and the biopolymer.

The aragonite within the cuttlebone is attacked instantly with the onset of the conversion process. Conversion for an hour already yields the formation of AP crystals within skeletal elements (white arrows in Figures 1b and 7a). At conversion for 5 h the original carpark structure of the cuttlebone is still preserved (Figure 7b), even though about 60 wt % of the aragonite is converted to AP. Up to this stage AP formation takes mainly place within skeletal elements (Figure 7d), while AP growth onto outer surfaces, e.g., the walls of compartments (white arrows in Figure 7c) occurs to a lesser degree. We find that AP growth onto surfaces is highly increased during late stages of the reaction process (Figures 8 and 9): AP crystals merge (Figure 9) and form larger units (white arrows in Figure 8b,c). This creates a dense and solid layer of AP crystals that reproduces the external shape of cuttlebone elements (Figure 8a). Figure 10 shows the macrostructure of *Sepia officinalis* cuttlebone that was exposed to the reaction with the phosphate-bearing solution for 3 days. Some ultrastructural features of the skeleton are still observable even though broken and deformed. The biopolymer membrane lining of the chambers is still present (yellow arrows in Figure 10d), however, now detached from the mineral. Even though the carpark structure of the skeleton is still preserved to some degree, the entire structure loses its cohesion as the walls detach from the platforms and the compartments become overgrown with newly formed precipitates.

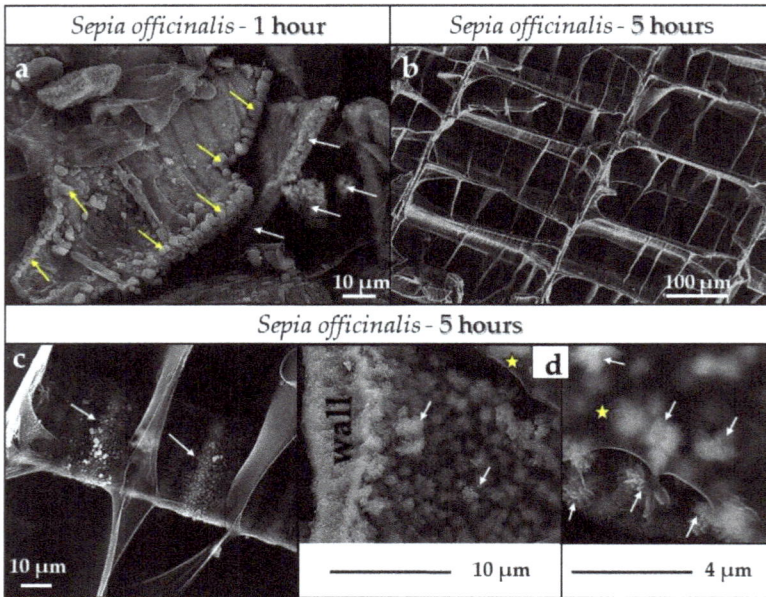

Figure 7. FE-SEM images showing *Sepia officinalis* cuttlebone altered for (**a**) 1 h and (**b–d**) 5 h. Even as early as 1 h of alteration of the cuttlebone, aragonite within skeletal elements starts to be converted to AP (white arrows in (**a**)), in addition to some newly formed crystals (black arrows in (**a**)) that precipitate onto skeletal element surfaces. At 5 h of conversion of aragonite to AP, AP formation within skeletal elements proceeds rapidly further (white arrows in (**d**)). In addition, newly formed AP starts to cover the surface of skeletal elements (white arrows in (**c**)). Yellow stars in (**d**) point to the biopolymer lining that covers the surface of a skeletal element, in this case, a wall.

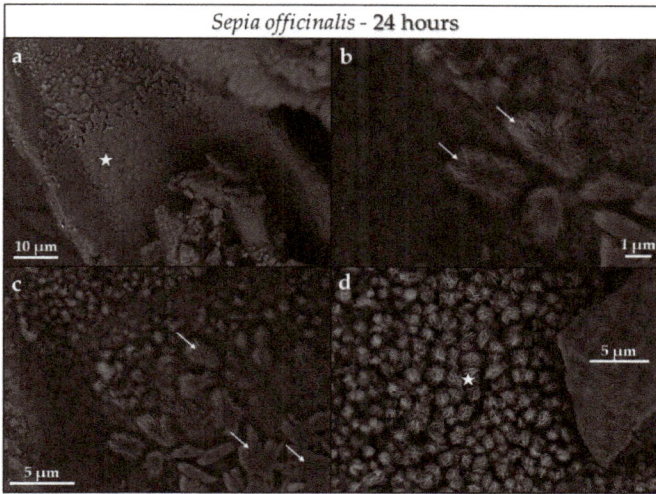

Figure 8. (a–d) FE-SEM images of *Sepia officinalis* cuttlebone after 24 h of conversion. Outer surfaces of skeletal elements (e.g., a chamber (**a**)) within the skeleton are covered with AP crystals. These become dense (white star in (**d**)), often fuse (white arrows in **b,c**) and finally encase the outer surface of the skeletal element (white star in (**a**)), conserving its outer morphology.

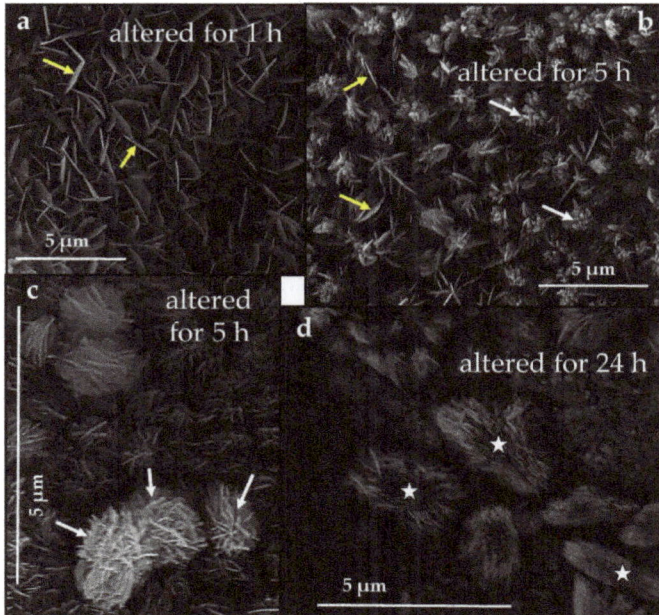

Figure 9. FE-SEM images showing the change in AP crystal morphology with progressive conversion of *Sepia officinalis* cuttlebone aragonite. (**a**) Conversion for 1 h; (**b,c**) conversion for 5 h and fusion of individual crystals; (**d**) fusion of some AP crystals and densification of the AP cover.

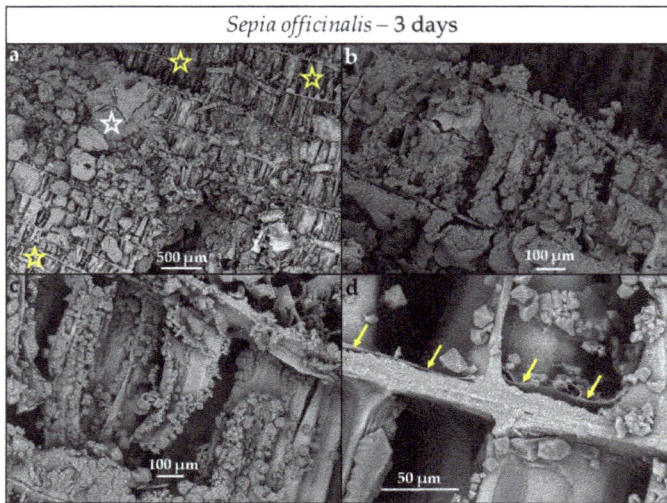

Figure 10. (**a**–**d**) FE-SEM images showing the *Sepia officinalis* cuttlebone ultrastucture when altered for 3 days. The original ultrastructure (yellow stars in (**a**)) is highly ruptured up to complete distortion (white star in (**a**)). Biopolymer membranes lining the chambers are not entirely decomposed yet (yellow arrows in (**d**)): However, they become detached from the surface of the skeletal element.

4. Discussion

4.1. The Phase Conversion Mechanism

The conversion of the studied aragonitic biominerals and that of geologic aragonite into apatite occurs with the preservation of morphologies and some microstructural features. Pseudomorphic conversion of biocarbonate in contact with phosphate-bearing solutions takes place via interface coupled dissolution-recrystallization [16,22,33,34]. The conversion mechanism involves the dissolution of primary aragonite and the precipitation of secondary apatite [35–40]. The preservation of external morphologies and microstructural features requires that the dissolution and the crystallization reactions are temporarily and spatially coupled [22,35–40] and that the dissolution of aragonite is the conversion rate-controlling step [22,35–44].

4.2. The Conversion of Geologic Aragonite

At 14 days of reaction we observe only a thin cover and a thin rim (~100 microns) of AP that replaces the outer surface of the aragonite crystal (Figure A4b). The transformed rim accurately reproduces the external shape of the aragonite crystal. The network of cracks within the newly formed AP layer covering the surface of the aragonite crystal (Figure A4a) is due to rapid cooling-related shrinkage and is most likely generated when the crystal is extracted from the boiling solution. The very limited conversion of geologic aragonite to AP is most probably not due to early surface passivation. Surface passivation during pseudomorphic mineral replacement occurs when (a) the structures of the parent and product phases show similarities that facilitate epitactic growth onto each other, and/or (b) the molar volume change associated with the conversion is positive [41]. Previous studies of the conversion of aragonite to AP under hydrothermal conditions demonstrated that AP grows oriented onto the aragonite surface [45]. Some authors interpreted this feature as evidence that the conversion is topotactic. However, strong arguments [33] support the notion that AP oriented growth develops with the advancement of the replacement front due to competitive growth. Hence, in those cases where an epitactic layer of the product phase forms, it often consists of an aggregate of crystals where several epitactic orientations coexist [46,47]. Consequently, the newly formed product layer commonly contains a certain amount of intrinsic intergranular porosity and does not seal completely the parent phase from further interaction with the solution [46,47]. In addition, the molar volume change associated with the aragonite—AP conversion is negative (−6%) [16,22,34]. As the external shape of the sample is preserved during conversion, porosity has to be generated to compensate the molar volume loss [35–40]. In addition, further porosity is generated due to the difference in solubility between the original and the product phases, e.g., AP is many orders of magnitude more soluble than geologic aragonite [37,40]. Hence, in the case of aragonite to apatite conversion, much porosity is generated, through which the fluid can communicate with the parent phase. In the case of geological aragonite, we must conclude that the negligible degree of conversion is not a consequence of lack of communication between the interface and the reaction fluid, where the dissolution-recrystallization reaction takes place. It is the consequence of a very slow dissolution rate of geologic aragonite under the experimental conditions used in this study. The slow dissolution of the geologic aragonite sample can be attributed to its low reactivity, to which its single crystallinity and its low surface area extent only adds up.

4.3. The Conversion of Biologic Aragonites

With the exception of *H. cumingii* nacre, which undergoes almost no phase change up to 14 days of alteration, all other biological hard tissues that were investigated in this study transform to AP at a much faster rate than geologic aragonite (Figure 1, Table 1). One difference that explains the higher rates of conversion is given by the fact that the biologic hard materials consist of nanometer-to micrometer sized biocrystals, while the geologic aragonite is a macroscopic single crystal. Hence, the surface area where the phosphate-bearing boiling solution can interact with the biogenic mineral is significantly higher in comparison to that of geologic aragonite. The direct consequence is a much faster dissolution of biogenic aragonite, and also a faster conversion to AP, as aragonite dissolution is the limiting step. In addition, the biogenic aragonite is more soluble than geological aragonite [48,49] since solubility is size-related and increases as the crystal size decreases—this also favors the faster conversion of biologic aragonite into AP.

Another major factor that plays a role in accelerating the conversion of biogenic aragonite to apatite relates to the composite nature of the biologic hard material. All extracellularly formed biological hard tissues contain occluded biopolymers [50–53], these being developed as organic membranes, fibers or networks. The steadily advancing degradation of the organic component effects that the biologic aragonite becomes progressively more porous, thus channels are formed, through which the solution can reach inner sample regions. This creates an additional increase in mineral surface area that is exposed to the fluid, and herewith contributes to a faster conversion kinetics. However, it should be kept in mind that this porosity is inherent to biological hard tissues and is independent of that, that develops due to the aragonite to AP conversion process itself.

4.3.1. *Sepia officinalis* Cuttlebone Aragonite

Most interesting is the very different conversion kinetics of the investigated biogenic aragonite samples (a) from geologic aragonite and (b) between the biological materials themselves (Figure 1, Table 1). This is called forth by an interplay between specific microstructure, inherent (original) porosity, biopolymer content, fabric, pattern of organics distribution and the extent of mineral surface that can interact with the reaction fluid.

Sepia officinalis cuttlebone is the biomaterial that shows the highest reactivity. This highest reactivity compared to other biominerals cannot be attributed to differences in average crystallite size, since differences in this parameter are negligible (504 Å for *H. cumingii* and 486 Å for *Sepia officinalis*, the slowest and the fastest transforming biomaterial, respectively). Of the investigated hard tissues the cuttlebone has the most lightweight skeleton with the aragonite being well protected by various biopolymers: (a) a thick outer organic membrane shielding the outermost surface of the entire cuttlebone, (b) resistant biopolymer linings occluded within the cuttlebone covering the outer surfaces of all building elements (walls, platforms) (Figures A13a, A14a and A16), and (c) biopolymer films and networks occluded within the building elements and in between the aragonite crystallites (Figures A12b,c, A13b and A14, especially Figures A15 and A16). Reaction, especially at the very beginning of the conversion process (up to three hours of reaction), takes place at an extremely fast rate and is facilitated by the very large surface area of the mineral that becomes exposed to interaction with the aqueous solution. The large surface area exposure is given by (a) the highly permeable carpark structure of the cuttlebone (Figure A11c), (b) the nanoparticulate nature of the biologic aragonite (Figure A11a), (c) the very delicate fabric (Figures A13b, A15 and A16) and (d) the rapid decomposition of organic material occluded within the building elements. Especially the latter contributes to an increase in porosity and escalates the conversion process.

The shape of the conversion curve for *Sepia officinalis* in our experiment very closely resemble that obtained by Kasioptas et al. [22] at similar temperature under hydrothermal conditions. These authors found that their experimental data fitted well the Avrami function. The profile of a curve that is given by the Avrami equation for phase transformation at constant temperature indicates that, subsequent to nucleation and initial rapid reaction, the transformation to a new phase steadily slows down as little unreacted material is left for the production of a new phase [54–56]. This is what we see for *Sepia officinalis* (Figure 1b). We find that two sections of the conversion curve can be distinguished according to different conversion rates (Figure 1b). Up to three hours of reaction the conversion rate of bioaragonite to AP is incredibly high. This is the time period where the minute aragonite crystallites within the structural elements of the hard tissue are transformed to AP (white arrows in Figure 7d). The reaction is driven by the fast decomposition of occluded organics (characterized by their delicate fabric) within the major structural elements. However, as it is well visible in Figure 7c,d, the membrane linings that protect the outer surfaces of structural elements remain unaffected up to that point (yellow stars in Figure 7d). Conversion slows significantly down after three hours of reaction (Figure 1b). This is the time period where the protecting biopolymer membrane linings start to become attacked as well. This facilitates AP nucleation onto the outer surfaces of the walls and the platforms (Figure 7c); progressive conversion induces their complete coverage with AP (Figure 8). When altered for a few days the compartments of *Sepia officinalis* are not just filled up with AP, as the conversion of bioaragonite to AP is complete in 24 h, thus there is no parent aragonite left to be transformed to AP. With ongoing conversion for a few days, we find that structural elements become highly deformed and the carpark structure becomes ruptured (Figure 10).

4.3.2. *Porites* sp. Acicular and *Hyriopsis cumingii* Prismatic Aragonite

Porites sp. and *Hyriopsis cumingii* prismatic aragonite are the biomaterials that reach the next highest degree of conversion to AP after *Sepia officinalis* cuttlebone (Figure 1, Table 1). Nonetheless, conversion kinetics is significantly different of that of the cephalopod cuttlebone. While in the cuttlebone replacement rates shoot up first and slow down at the end of the process, for *Porites* sp. aragonite and *Hyriopsis cumingii* prismatic aragonite we find at first (up to 4 days) a slow but steadily increasing dissolution of biological aragonite and precipitation of apatite (Figure 1a). Even though in *Porites* sp. and *Hyriopsis cumingii* prisms newly formed AP contents are high, the speed of conversion is not at all comparable to that of *Sepia officinalis* cuttlebone. In 14 days of alteration 25 and 38 wt % AP content is detected in *H. cumingii* prismatic shell layer and the *Porites* sp. skeleton, respectively. With progressive reaction time, between 4 and 9 days of reaction, conversion to AP in the coral skeleton stops, in contrast to prismatic aragonite in *H. cumingii*, where it still increases steadily. For the time period between 9 and 14 days of reaction, AP formation rates change again: it increases significantly for *Porites* sp., while it stops for *H. cumingii* prisms (Figure 1a).

In comparison to the cuttlebone, significantly less organics are occluded in the bivalve shell and the coral skeleton. Hence, there is significantly less surface area where the reaction fluid can react with the mineral. The steady formation of AP in the coral and the bivalve shell within the first four days of reaction can be attributed to dissolution of bioaragonite along outer surfaces of the skeletons. It is well visible (Figure 5) that already after 4 days of alteration the surface of *Porites* sp. becomes covered with a thick layer of AP crystals. Exactly at this point we find that conversion of aragonite to AP in *Porites* sp. stops (Figure 1a). As the coral skeleton is dense and has very little occluded organics the conversion rate to AP is slow compared to that of the exceedingly more porous and highly organic rich cuttlebone. However, it is much faster than the conversion rate of geologic aragonite. We know from previous work [57] that the reaction fluid enters the skeleton of *Porites* sp. through centers of calcification and that new mineral formation starts and extends from here into the skeleton. This characteristic is reflected in the conversion curve as well, with the sudden increase in AP content at later stages of reaction, between 9 and 14 days (Figure 1a), when AP formation in *Porites* sp. shoots up significantly.

The prismatic aragonite shell portion of *Hyriopsis cumingii* is slightly porous. We find pores between the columns as well as within the membranes that encase the aragonite of the columns (Figure A17). The reaction fluid enters the shell through these pores, infiltrates the entire prismatic shell portion and space between the columns as well as the inner parts of the columns. The network of biopolymer fibrils (Figure A10d–f) within the columns becomes easily decomposed as their fabric is delicate. Accordingly, the surface area where reaction fluid can react with the mineral becomes highly increased; the aragonite crystallites constituting the prisms are minute to small in size, easily dissolved and reprecipitated as apatite (Figure A10f). However, as Figure 1a shows, the conversion process for *H. cumingii* prisms stops abruptly at nine days of reaction. At this stage we find the surface of the columns to be covered with AP shielding the aragonite within the columns from reaction with the fluid. Even though more porosity is generated as the conversion progresses, an increase in tortuosity of the pathway that communicates the fluid with the remaining aragonite and a reduction in aragonite area exposed to the fluid might explain the slow-down of the reaction.

4.3.3. *Arctica islandica* Cross-Lamellar and *Hyriopsis cumingii* Nacreous Aragonite

The least conversion of bioaragonite to AP takes place in *Arctica islandica* cross-lamellar and *Hyriopsis cumingii* nacreous aragonite. In the latter we observe after 14 days of reaction less than 1 wt % of AP. In the shell of *Arctica islandica*, up to 7 days of conversion between 2 and 3 wt % and at 14 days of reaction only 11 wt % of AP formation, respectively (Figure 1, Table 1). As it is the case for *Porites* sp. and *Hyriopsis cumingii* prismatic aragonite, we find a slight increase in AP formation within the first 4 days of reaction for *Arctica islandica* aragonite as well. However, for *Arctica islandica* a time period follows (between 4 and 7 days) where AP formation ceases (Figure 1a). This is followed by a time span where we see a sudden increase in AP precipitation, especially along the seaward pointing rim of the shell (Figure 1a, Table 1). The shell of *Arctica islandica* is dense (especially shell parts next to the soft tissue of the animal) and consists of irregularly shaped mineral units that are embedded into a network of biopolymer fibrils (Figure A7 and Casella et al., 2017 [32]). The shell is crossed by growth lines, where the organic matter content is slightly increased (Figure A8). For the first 7 days of conversion treatment—where almost no AP formation occurs—most biopolymers within the shell become decomposed and, hence, the reaction fluid can permeate the entire hard tissue, not just its outer and inner surfaces. Due to decomposition of the biopolymers, there is a slight increase in mineral surface area, the amount of dissolved aragonite and that of newly formed apatite. For the final stage of alteration, we see for *Arctica islandica* a slight difference in AP formation kinetics between outer and inner shell layers. Along the seaward pointing shell part, conversion of bioaragonite to AP carries on (Figure 1a), while the aragonite that constitutes inward shell portions does not seem to transform further into apatite. In contrast to inner shell layers, the outer shell portion of *Arctica islandica* is porous (Figure A9 and Casella et al., 2017 [32]). Hence in these regions, the reaction fluid can enter the shell through the growth lines as well as through voids, the primary pores. These pores are largely absent in more inward shell parts. The latter is compact, up to 14 days of conversion there is not much space present for fluid permeation, aragonite dissolution and new AP precipitation (this study and Casella et al., 2017 [32]).

As it is well observable in Figure 1 and Table 1, *Hyriopsis cumingii* nacre is the most resistant biological hard tissue to aragonite dissolution and AP formation. Two facts account for this: (i) each nacre tablet is encased into a biopolymer membrane sheath (Figure A5c,d) that is obviously (ii) not affected by our conversion experiments due to: the used temperature, the chemical composition of our reaction fluid, and the time span of the experiment. Thus, the reaction fluid cannot infiltrate the shell except for the outer and inner surfaces of the shell. Hence, the tablets and their arrangement to stacks remains intact (Figure 3d). The behavior at conversion of this biogenic hard tissue resembles that of geologic aragonite most closely.

5. Conclusions

Biological hard tissues are hierarchical composites with unique microstructures. They interlink on many scales two distinct components: hard and brittle minerals with compliant biopolymers. Induced by evolutionary driven adaptation, modern skeletal microstructures and textures are highly diversified and are utilized as an additional means for the improvement of biomaterial functionality [58–63].

We discuss in this manuscript the effect of the composite nature of biocarbonate hard tissues and biogenic microstructures on the rate and kinetics of bioaragonite to apatite conversion in an open experimental set up. We deduce the following conclusions for the different biologic and non-biologic hard materials (Tables 1 and 2, Figure A11).

Table 2. Pattern of conversion for the investigated biogenic microstructures and for non-biological aragonite at different conversion times.

	0–4 Days	4–7 Days	7–9 Days	9–14 Days
Acicular aragonite *Porites* sp.	very strong increase	no change	no change	very strong increase
Prismatic aragonite *H. cumingii*	strong increase	very strong increase	very strong increase	no change
Cross-lamellar aragonite *A. islandica* outer layer	slight increase	no change	very strong increase	strong increase
Cross-lamellar aragonite *A. islandica* inner layer	slight increase	no change	very strong increase	no change
Nacreous aragonite *H. cumingii*	no change	no change	no change	no change
Geologic aragonite single crystal	-	no change	-	slight increase

	0–1 Hour	1–3 Hours	3–5 Hours	5–8 Hours	8–24 Hours
Granular aragonite *Sepia officinalis*	very strong increase	very strong increase	slight increase	strong increase	very strong increase

1. We find highly distinct rates and kinetics of conversion to apatite for the selected aragonitic biological hard tissues. This is dependent on the ability of the reaction fluid to access aragonite crystallites, which directly relates to the content and the extent of decomposition of biopolymers within the hard tissue, the extent of the newly formed surface area and the specific biological hard tissue macro- and microstructures.

2. When treated for up to, respectively, 1 and 14 days, a profound conversion of bioaragonite to apatite within mineral units and/or structural elements takes place in the cuttlebone of the cephalopod *Sepia officinalis* and in the prismatic columns of the bivalve *Hyriopsis cumingii*.

a. Conversion of *Sepia officinalis* aragonite occurs almost instantly. It is highly accelerated at the beginning and slows down towards the end of the conversion process.

b. Conversion to AP in *Hyriopsis cumigii* prisms is steadily increasing for almost the entire time span of the experiment. However, the process stagnates for the final stages of the experiment.

3. Even though having a large surface area given by the specific macrostructure, the acicular microstructure of the warm water coral *Porites* sp. gives a compact hard tissue. At the start of the conversion process only outer skeletal surfaces are subject to dissolution and conversion to AP. With progressive conversion, apatite formation accelerates quickly when the fluid enters the skeleton through the centers of calcification.

4. The cross-lamellar microstructure in *Arctica islandica* is, at first, highly resistant to phase conversion. It speeds slightly up when the network of organic biopolymers becomes destroyed, and the reaction fluid can permeate the entire hard tissue. The presence of pores within outer shell layers facilitates fluid infiltration even further and accelerates even more the conversion process for this part of the shell.

5. Nacreous aragonite in *Hyriopsis cumingii* is most resistant to conversion. Even though nacre tablets are encased by organic membranes, the latter are not decomposed easily, thus the aragonite of the nacreous shell layer is not attacked and remains intact.

6. All aragonite biominerals transform into AP at a much faster rate than geological aragonite. *Hyriopsis cumingii* nacre is the biomaterial that shows a conversion kinetics that resembles that of geologic aragonite most closely. This is consistent with the resistance to degradation of organic membranes in the former.

Supplementary Materials: The following are available online at http://www.mdpi.com/2075-163X/8/8/315/s1, X-ray diffractograms of all samples: *Sepia officinalis*, *H. cumingii* prismatic aragonite, *H. cumingii* nacreous aragonite, *Porites* sp., *A. islandica* inner layer, *A. islandica* outer layer, and Geologic aragonite.

Author Contributions: L.F.-D., S.V.-V. and M.G. conceived and designed the experiments; M.G. performed the experiments; M.G. and E.G. and M.N.Z. analyzed the data; A.Z. and X.Y. contributed to biochemical preparation, S.V.-V. contributed to the discussion. W.W.S. and S.V.-V. read the final version of the manuscript; L.F.-D., E.G. and M.G. wrote the paper.

Funding: This research was partially funded by projects CGL2016-77138-C2-1-P (MECC-Spain) and MAT2017-88148-R (MECC-Spain) (S.V.V. and L.F.-D.). M.G. is supported by the Deutsche Forschungsgemeinschaft, DFG Grant Gr 959/20-1,2.

Acknowledgments: We thank Christine Putnis for the invitation to publish our work within this Special Issue. Moreover, we want to thank the editors-in-house Jameson Chen and Jingjing Yang for handling the manuscript.

Conflicts of Interest: The authors declare no conflict of interest.

Appendix A

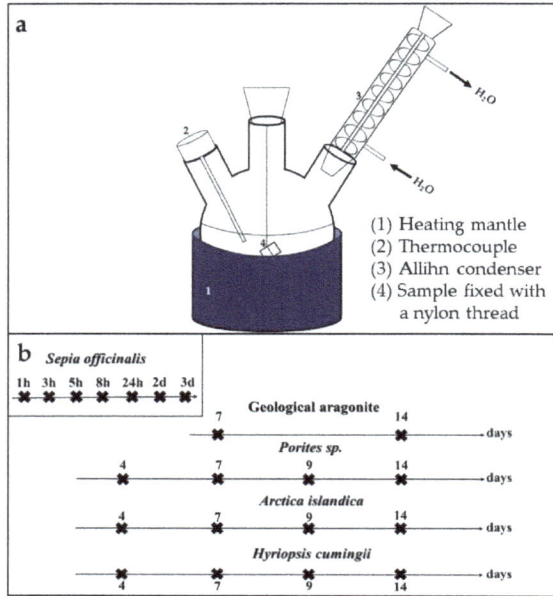

Figure A1. (**a**) Experimental setup; (**b**) overview of conducted conversion experiments. *Sepia officinalis* cuttlebone was transformed for 1 to 24 h and 2 and 3 days. *Porites* sp., *Arctica islandica* and *Hyriopsis cumingii* aragonite were transformed up to 14 days. Geological aragonite was transformed for 7 and 14 days. In (**a**): (1) Heating mantle, (2) Thermocouple, (3) Allihn condenser, (4) Sample fixed with a nylon thread.

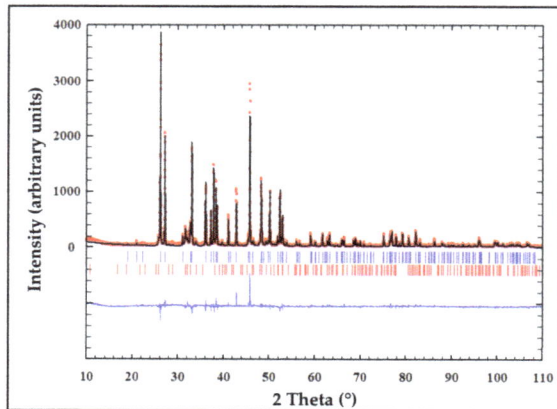

Figure A2. Exemplary Rietveld refinement plot for *Porites* sp. after conversion for one week. Red dots: data points, black line: calculated XRD profile, bottom blue line: difference of observed and calculated data, blue vertical bars: positions of aragonite diffraction peaks, red vertical bars: position of apatite diffraction peaks.

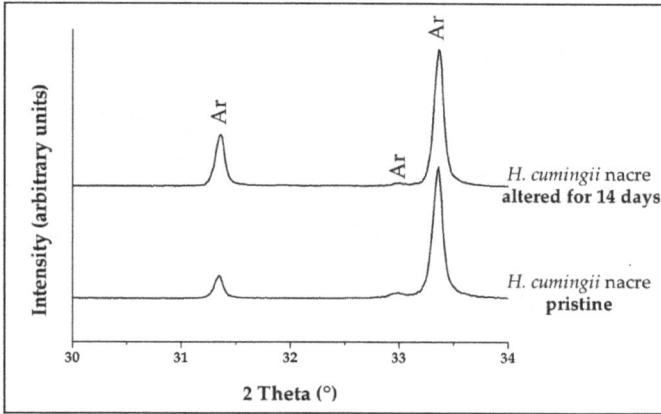

Figure A3. XRD diffractograms (30–34° 2θ) of pristine and altered *Hyriopsis cumingii* nacre. Conversion was carried out for 14 days.

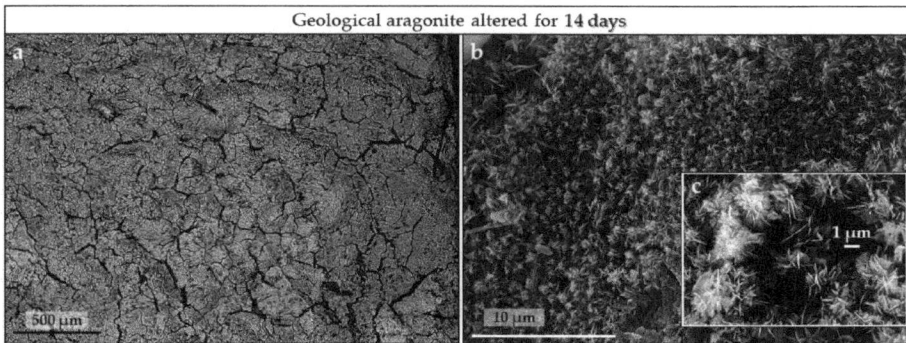

Figure A4. FE-SEM images of geological aragonite transformed for 14 days.

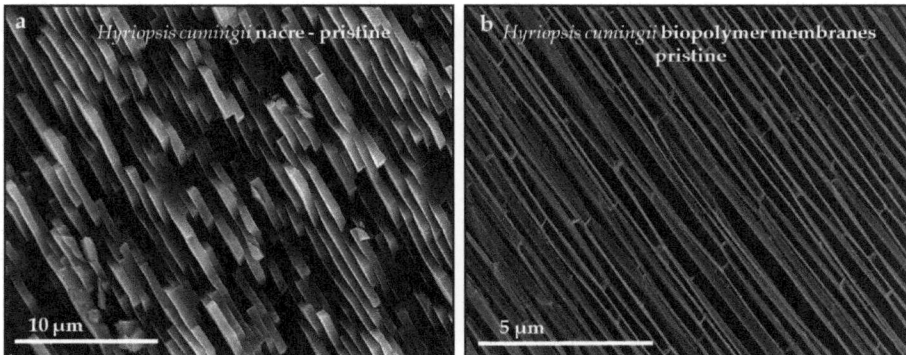

Figure A5. FE-SEM images of nacre tablets (**a**) and biopolymer sheaths (**b**) encasing the tablets.

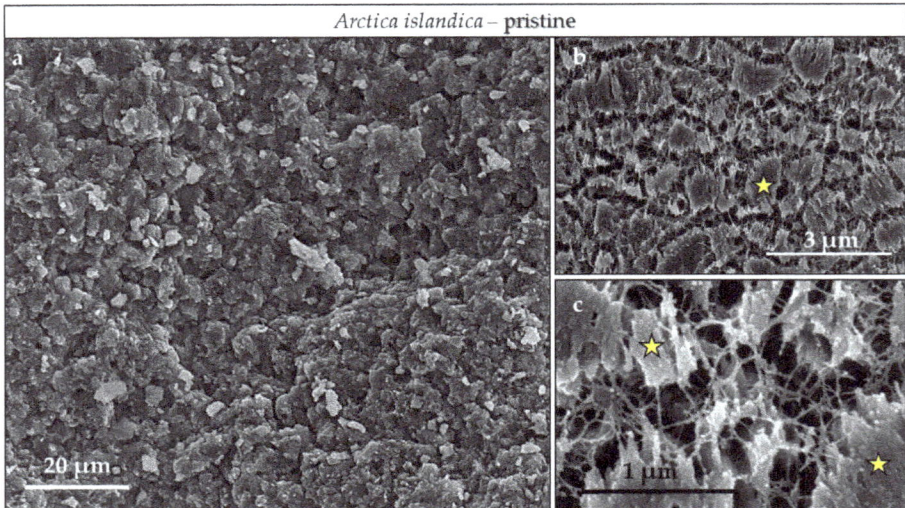

Figure A6. FE-SEM images of (**a**) aragonite crystallites and (**b**,**c**) mineral unit morphologies in pristine *Arctica islandica*, where mineral units are embedded into a network of biopolymer fibrils. Aragonitic mineral units are larger in size in the shell portion facing seawater relative to that part of the shell that is next to the soft tissue of the animal.

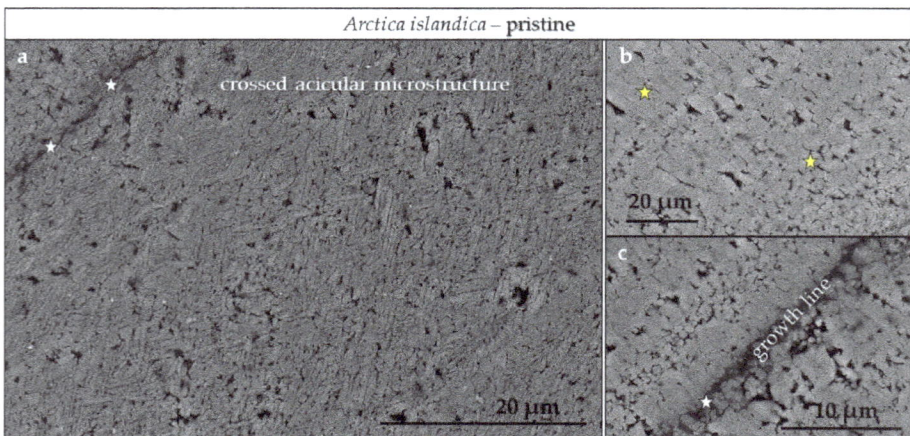

Figure A7. FE-SEM images of (**a**) the shell microstructure of pristine *Arctica islandica*; (**b**,**c**) variation in size and morphology of constituting mineral units and presence of growth lines (white stars in (**a**,**c**)) distinguished by an increased amount of biopolymer content.

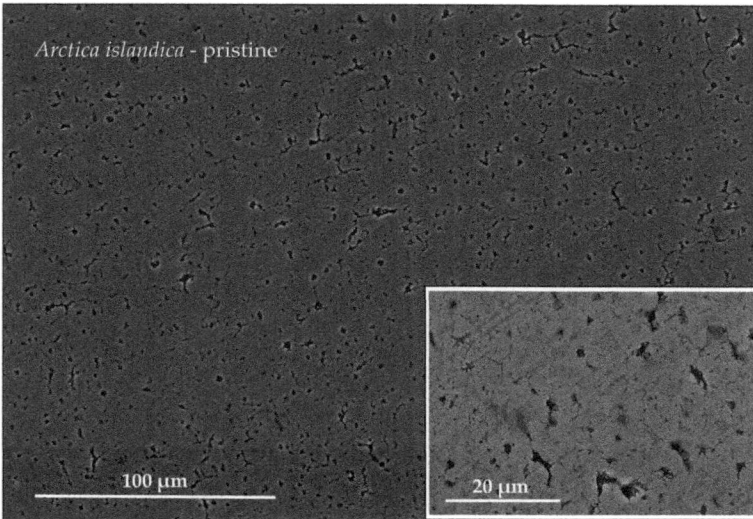

Figure A8. FE-SEM images of the portion of *Arctica islandica* shell that is next to seawater. This part of the shell is characterized by high porosity and larger sized mineral units compared to inner shell portions next to the soft tissue of the animal.

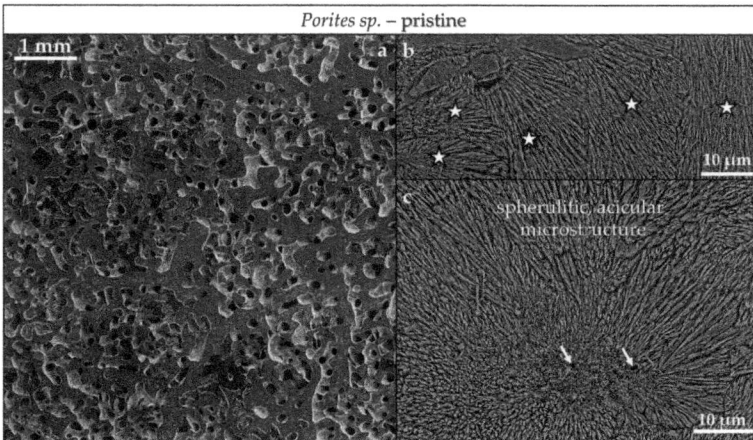

Figure A9. FE-SEM images depicting the macro- and microstructure of the skeleton of the modern coral *Porites* sp. (**a**) Due to its specific architecture comprising many thin vertical and transverse elements, the skeleton of *Porites* sp. has an exceedingly high surface area This does not imply that the skeleton is porous. The microstructure of *Porites* sp. (this study and Casella et al., 2018) comprises a multitude of differently sized spherulites, mineral units consisting of radially arranged acicles and fibrils (**b**,**c**). These nucleate at centers of calcification (white arrows in **c**) and grow radially outward increasing in length until they abut the adjoining spherulite.

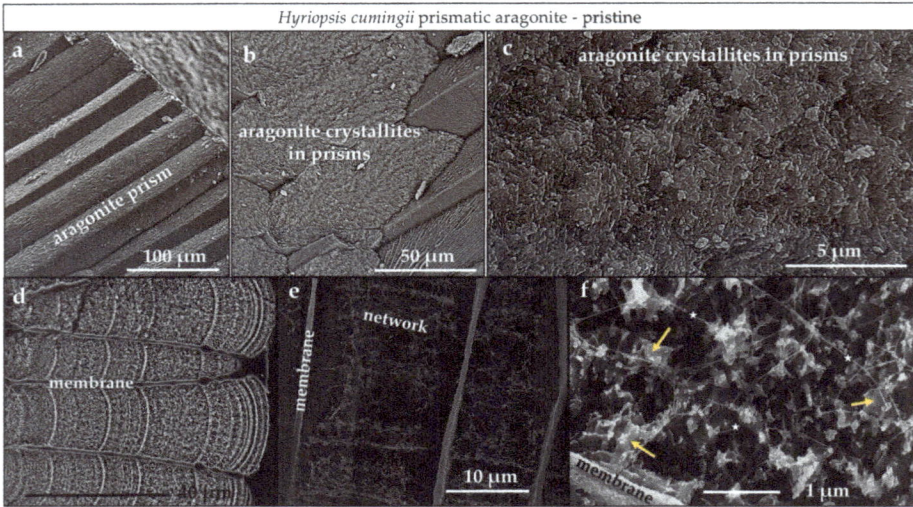

Figure A10. FE-SEM images depicting microscale features of pristine *Hyriopsis cumingii* prismatic aragonite. (**a**) Aragonitic prisms comprise the outer shell layer adjacent to seawater, with all prisms being encased by organic membranes (**d,e**). Each prism consists of aragonite crystallites (**b,c,f**), embedded in an irregular network of thin biopolymer fibrils (**e,f**).

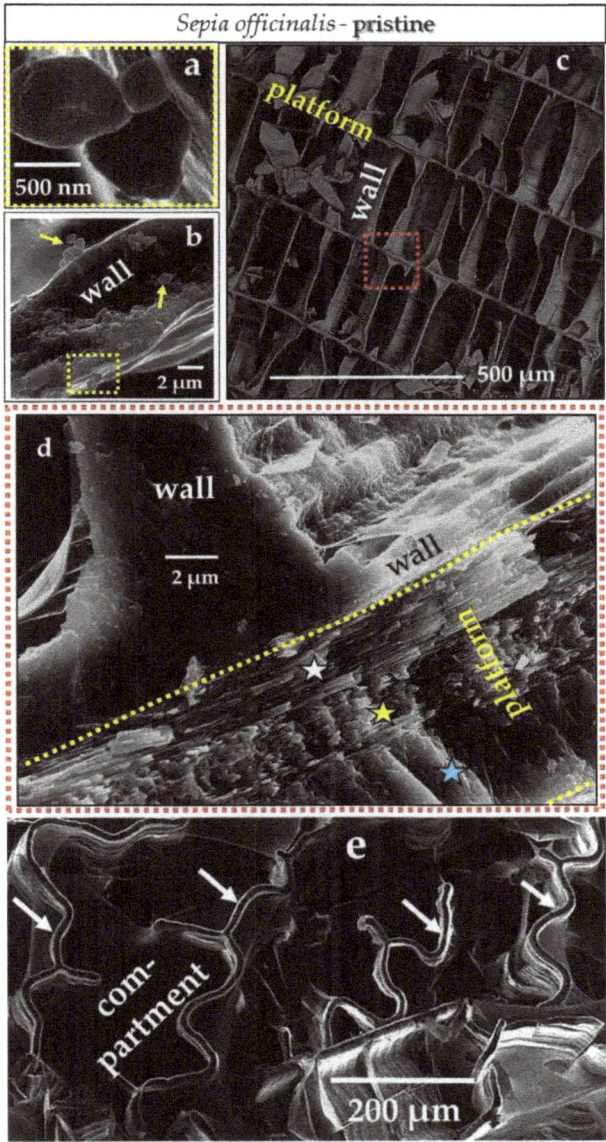

Figure A11. FE-SEM images visualizing the macro-, micro- and nanostructure of pristine *Sepia officinalis* cuttlebone. (**a**) The cuttlebone is a lightweight structure comprising horizontal platforms and vertical walls, this arrangement of structural elements renders stability but also induces the formation of compartments; (**b**) the vertical walls (see view from above) are stabilized from implosion by vertical stop ridges consisting of thick biopolymer membranes (see (**a**)). Both, the walls and ridges consist of nanoparticulate aragonite (**d**,**e**). The platforms comprise arrays of aragonite rods (**f**), with the stacks of rods showing a well-defined twisted arrangement (white and yellow stars in (**c**)).

Figure A12. FE-SEM images of arrays of aragonite rods (white stars in (**a**,**b**)) within pristine *Sepia officinalis* cuttlebone platforms. Each rod is a composite of a biopolymer matrix (or scaffold, yellow arrows in (**b**)) filled with aragonitic mineral.

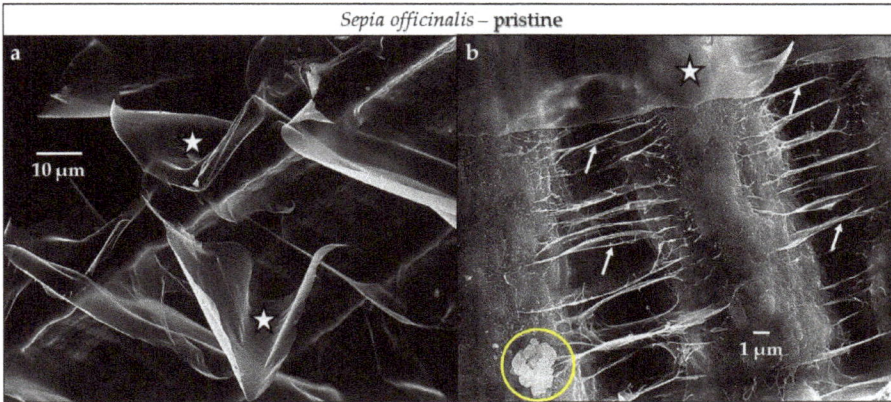

Figure A13. FE-SEM images of biopolymers occluded within the cuttlebone of pristine *Sepia officinalis*. Thick membranes line the walls and platforms (white stars in (**a,b**)). A network of biopolymer films is occluded within the major structural elements of the cuttlebone, e.g., in a wall (**b**). An accumulation of aragonite nanoparticles constituting the skeleton appear encircled in (**b**).

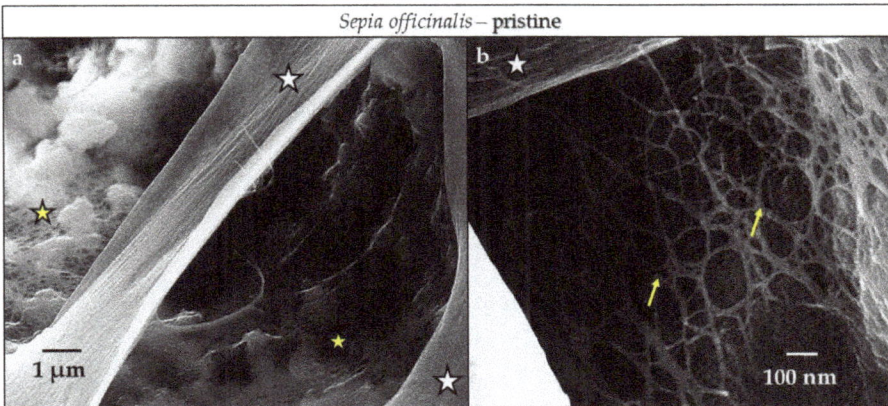

Figure A14. FE-SEM images of biopolymer membranes (white stars in (**a,b**)), biopolymer networks (yellow arrows in (**b**)) and biopolymer foams containing nanoparticulate aragonite (yellow stars in (**a**)) in the cuttlebone of pristine *Sepia officinalis*.

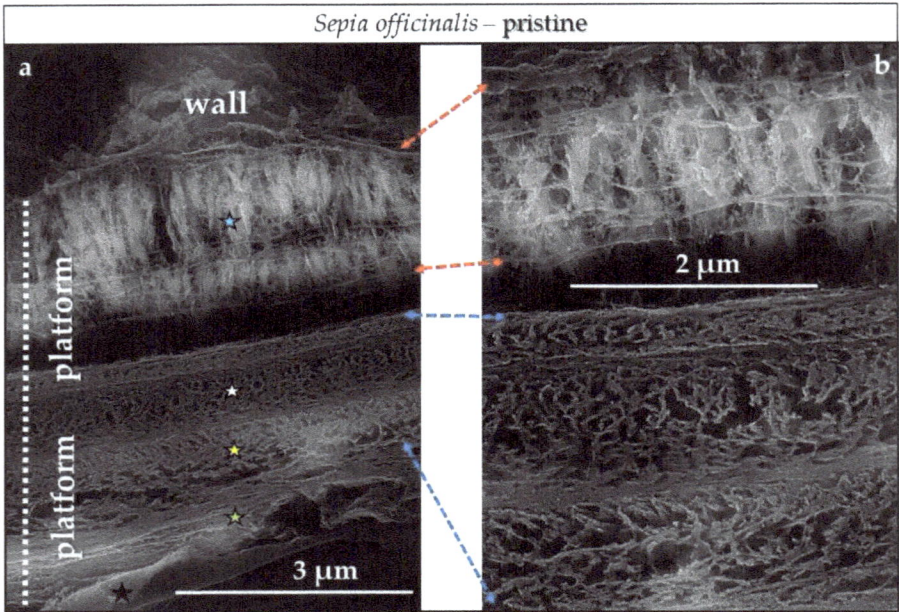

Figure A15. FE-SEM images of demineralized *Sepia officinalis* cuttlebone showing the organic component within the skeleton that encases aragonite crystallites and individual mineral units, e.g., aragonite rods, stacks of rods. The differently colored stars in (**a**) point to the different layers of a platform; see also Figure A11d. The black star in (**a**) indicates the presence of the thick biopolymer membrane that lines all outer surfaces of all structural elements within the cuttlebone; (**b**) Organic content within a platform shown with a slightly higher magnification, relative to that given in (**a**).

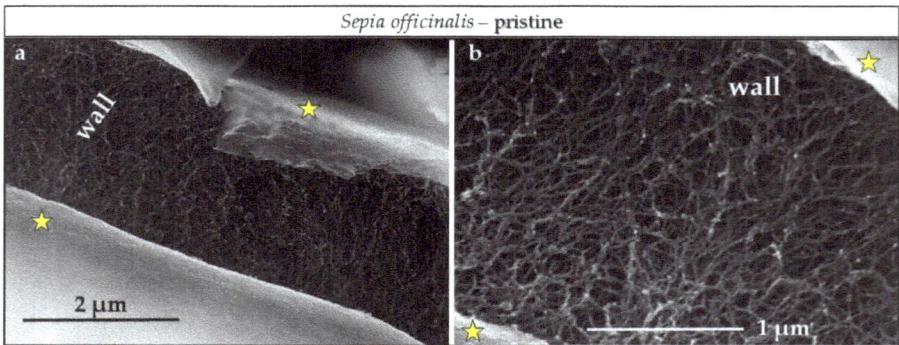

Figure A16. FE-SEM images of the biopolymer network that is occluded within the walls of the cuttlebone of *Sepia officinalis*. Yellow stars in (**a**,**b**) point to the protruding biopolymer membranes that line all outer surfaces of all structural elements within the cuttlebone.

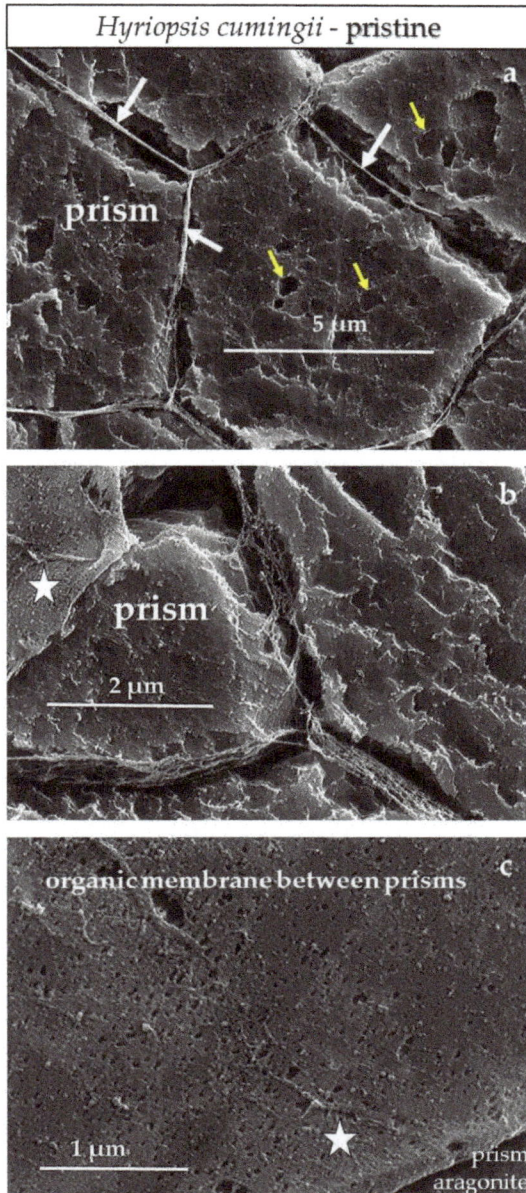

Figure A17. FE-SEM images of etched sample surfaces showing in top view; (**a**,**b**) the assemblage of prisms with occluded biopolymer membranes (white arrows in (**a**), white stars in (**b**,**c**) between them. Well visible is the porous nature of the membranes (**b**,**c**). Well visible is the internal structuring of prisms (e.g., (**a**)). Yellow arrows in (**a**) point to roundish mineral grains that are slightly misoriented to each other.

References

1. Habraken, W.; Habibovic, P. Calcium phosphates in biomedical applications: Materials for the future? *Materialstoday* **2016**, *19*, 69–87. [CrossRef]
2. Bohner, M. Resorbable biomaterials as bone graft substitutes. *Materialstoday* **2010**, *13*, 24–30. [CrossRef]
3. Bohner, M. Design of ceramic-based cements and putties for bone graft substitution. *Eur. Cells Mater.* **2010**, *20*, 1–12. [CrossRef]
4. Zhang, X.; Vecchio, K.S. Conversion of natural marine skeletons as scaffolds for bone tissue engineering. *Front. Mater. Sci.* **2013**, *7*, 103–117. [CrossRef]
5. Clarke, S.A.; Choi, S.Y. Osteogenic cell response to 3-D hydroxyapatite scaffolds developed via replication of natural marine sponges. *J. Mater. Sci. Mater. Med.* **2017**, *27*, 1–11. [CrossRef] [PubMed]
6. Rousseau, M. Nacre: A biomineral, a natural biomaterial, and a source of bio-inspiration. In *Highlights in Applied Mineralogy*; Heuss-Aßbichler, S., Amthauer, G., John, M., Eds.; De Gruyter: Berlin, Germany; Boston, MA, USA, 2018; pp. 285–300. ISBN 978-3-11-049734-2.
7. Roy, D.M.; Linnehan, S.K. Hydroxyapatite formed from Coral Skeletal Carbonate by Hydrothermal Exchange. *Nature* **1974**, *247*, 220–222. [CrossRef] [PubMed]
8. White, R.A.; Weber, J.N. Replamineform: A new process for preparing porous ceramic, metal, and polymer prosthetic materials. *Science* **1972**, *176*, 992–994. [CrossRef]
9. Silve, C.; Lopez, E. Nacre initiates biomineralization by human osteoblasts maintained in vitro. *Calcif. Tissue Int.* **1992**, *51*, 363–369. [CrossRef] [PubMed]
10. Berland, S.; Delattre, O. Nacre/bone interface changes in durable nacre endosseous implants in sheep. *Biomaterials* **2005**, *26*, 2767–2773. [CrossRef] [PubMed]
11. Westbroek, M.; Marin, F. A marriage of bone and nacre. *Nature* **1998**, *392*, 861–862. [CrossRef] [PubMed]
12. Fu, K.; Xu, Q. Characterization of a biodegradable coralline hydroxyapatite/calcium carbonate composite and its clinical implementation. *Biomed. Mater.* **2013**, *8*, 065007. [CrossRef] [PubMed]
13. Green, D.W.; Ben-Nissan, B. Natural and Synthetic Coral Biomineralization for Human Bone Revitalization. *Trends Biotechnol.* **2017**, *35*, 43–54. [CrossRef] [PubMed]
14. Pountos, I.; Giannoudis, P.V. Is there a role of coral bone substitutes in bone repair? *Injury* **2016**, *47*, 2606–2613. [CrossRef] [PubMed]
15. Zhang, G.; Brion, A. Nacre, a natural, multi-use, and timely biomaterial for bone graft substitution. *J. Biomed. Mater. Res. Part A* **2017**, *105*, 662–671. [CrossRef] [PubMed]
16. Reinares-Fisac, D.; Veintemillas-Verdaguer, S. Conversion of biogenic aragonite into hydroxyapatite scaffolds in boiling solutions. *CrystEngComm* **2017**, *19*, 110–116. [CrossRef]
17. Rodríguez-Carvajal, J.; Roisnel, T. Line broadening analysis using fullprof*: Determination of microstructural properties. *Mater. Sci. Forum* **2004**, *443–444*, 123–126. [CrossRef]
18. Jarosch, D.; Heger, G. Neutron diffraction refinement of the crystal structure of aragonite. *Tschermaks Mineral. Petrogr. Mitt.* **1986**, *35*, 127–131. [CrossRef]
19. Kay, M.I.; Young, R.A. Crystal structure of hydroxyapatite. *Nature* **1964**, *204*, 1050–1052. [CrossRef] [PubMed]
20. Dinnebier, R.E.; Billinge, S.J.L. *Powder Diffraction: Theory and Practice*; Royal Society of Chemistry: Cambridge, UK, 2008; pp. 300–303. ISBN 978-0-85404-231-9.
21. Thompson, P.; Cox, D.E. Rietveld refinement of Debye-Scherrer synchrotron X-ray data from Al_2O_3. *J. Appl. Crystallogr.* **1987**, *20*, 79–83. [CrossRef]
22. Kasioptas, A.; Geisler, T. Crystal growth of apatite by replacement of an aragonite precursor. *J. Cryst. Growth* **2010**, *312*, 2431–2440. [CrossRef]
23. Kannan, S.; Rocha, J.H.G. Fluorine-substituted hydroxyapatite scaffolds hydrothermally grown from aragonitic cuttlefish bones. *Acta Biomater.* **2007**, *3*, 243–249. [CrossRef] [PubMed]
24. Meejoo, S.; Maneeprakorn, W. Phase and thermal stability of nanocrystalline hydroxyapatite prepared via microwave heating. *Thermochim. Acta* **2006**, *447*, 115–120. [CrossRef]
25. Destainville, A.; Champion, E. Synthesis, characterization and thermal behavior of apatitic tricalcium phosphate. *Mater. Chem. Phys.* **2003**, *80*, 269–277. [CrossRef]
26. Raynaud, S.; Champion, E. Calcium phosphate apatites with variable ca/p atomic ratio i. Synthesis, characterisation and thermal stability of powders. *Biomaterials* **2002**, *23*, 1065–1072. [CrossRef]

27. Han, J.-K.; Song, H.-Y. Synthesis of high purity nano-sized hydroxyapatite powder by microwave-hydrothermal method. *Mater. Chem. Phys.* **2006**, *99*, 235–239. [CrossRef]

28. Ratner, B.; Hoffman, A. *Biomaterials Science. An Introduction to Materials in Medicine*, 3rd ed.; Academic Press: Cambridge, MA, USA, 2004; ISBN 978-0123746269.

29. González-Díaz, P.F.; Hidalgo, A. Infrared spectra of calcium apatites. *Spectrochim. Acta A* **1976**, *32*, 631–635. [CrossRef]

30. González-Díaz, P.F.; Santos, M. On the hydroxyl ions in apatites. *J. Solid State Chem.* **1977**, *22*, 193–199. [CrossRef]

31. Vandecandelaere, N.; Rey, C. Biomimetic apatite-based biomaterials: On the critical impact of synthesis and post-synthesis parameters. *J. Mater. Sci. Mater. Med.* **2012**, *23*, 2593–2606. [CrossRef] [PubMed]

32. Casella, L.A.; Griesshaber, E. Experimental diagenesis: Insights into aragonite to calcite conversion *Arctica islandica* shells by hydrothermal treatment. *Biogeosciences* **2017**, *14*, 1461–1492. [CrossRef]

33. Kasioptas, A.; Geisler, T. Polycrystalline apatite synthesized by hydrothermal replacement of calcium carbonate. *Geochim. Cosmochim. Acta* **2011**, *75*, 3486–3500. [CrossRef]

34. Schlosser, M.; Fröls, S. Combined hydrothermal conversion and vapor transport sintering of ag-modified calcium phosphate scaffolds. *J. Am. Ceram. Soc.* **2012**, *96*, 412–419. [CrossRef]

35. Putnis, A. Mineral replacement reactions: From macroscopic observations to microscopic mechanisms. *Mineral. Mag.* **2002**, *66*, 689–708. [CrossRef]

36. Putnis, A. Mineral replacement reactions. *Rev Mineral Geochem* **2009**, *70*, 87–124. [CrossRef]

37. Putnis, A.; Putnis, C.V. The mechanism of reequilibration of solids in the presence of a fluid phase. *J. Solid State Chem.* **2007**, *180*, 1783–1786. [CrossRef]

38. Ruiz-Agudo, E.; Putnis, C.V. Coupled dissolution and precipitation at mineral–fluid interfaces. *Chem. Geol.* **2014**, *383*, 132–146. [CrossRef]

39. Putnis, C.V.; Fernandez-Díaz, L. Ion partitioning and element mobilization during mineral replacement reactions in natural and experimental systems. In *Ion Partitioning in Ambient-Temperature Aqueous Systems*; Prieto, M., Stoll, H., Eds.; Mineralogical Society of Great Britain & Ireland: London, UK, 2011; ISBN 978-0-903-05626-7.

40. Pollok, K.; Putnis, C.V. Mineral replacement reactions in solid solution-aqueous solution systems: Volume changes, reactions paths and end-points using the example of model salt systems. *Am. J. Sci.* **2011**, *311*, 211–236. [CrossRef]

41. Putnis, A. Transient porosity resulting from fluid–mineral interaction and its consequences. *Rev. Mineral. Geochem.* **2015**, *80*, 1–23. [CrossRef]

42. Fernández-Díaz, L.; Pina, C.M. The carbonatation of gypsum: Pathways and pseudomorph formation. *Am. Mineral.* **2009**, *94*, 1223–1234. [CrossRef]

43. Xia, F.; Brugger, J. Mechanism and kinetics of pseudomorphic mineral replacement reactions: A case study of the replacement of pentlandite by violarite. *Geochim. Cosmochim. Acta* **2009**, *73*, 1945–1969. [CrossRef]

44. Xia, F.; Brugger, J. Three-dimensional ordered arrays of zeolite nanocrystals with uniform size and orientation by a pseudomorphic coupled dissolution-reprecipitation replacement route. *Cryst. Growth Des.* **2009**, *9*, 4902–4906. [CrossRef]

45. Eysel, W.; Roy, D.M. Topotactic reaction of aragonite to hydroxyapatite. *Z. Kristallogr. Cryst. Mater.* **1975**, *141*, 11–24. [CrossRef]

46. Roncal-Herrero, T.; Astilleros, J.M. Reaction pathways and textural aspects of the replacement of anhydrite by calcite at 25 °C. *Am. Mineral.* **2017**, *102*, 1270–1278. [CrossRef]

47. Cuesta Mayorga, I.; Astilleros, J.M. Epitactic Overgrowths of Calcite ($CaCO_3$) on Anhydrite ($CaSO_4$) Cleavage Surfaces. *Cryst. Growth Des.* **2018**, *18*, 1666–1675. [CrossRef]

48. Baig, A.A.; Fox, R.A. Relationships among Carbonated Apatite Solubility, Crystallite Size, and Microstrain Parameters. *Calcif. Tissue Int.* **1999**, *64*, 437–449. [CrossRef] [PubMed]

49. Ortoleva, P. Solute reaction mediated precipitate patterns in cross gradient free systems. *Z. Phys. B* **1982**, *49*, 149–156. [CrossRef]

50. Levi-Kalisman, Y.; Fallini, G. Structure of the nacreous organic matrix of a bivalve mollusk·shell examined in the hydrated state using cryo-TEM. *J. Struct. Biol.* **2001**, *135*, 8–17. [CrossRef] [PubMed]

51. Heinemann, F.; Launspach, M. Gastropod nacre: Structure, properties and growth-biological, chemical and physical basics. *Biophys. Chem.* **2011**, *153*, 126–1253. [CrossRef] [PubMed]

52. Checa, A.G.; Marcías-Sánchez, E. Organic membranes determine the pattern of the columnar prismatic layer of mollusk shells. *Proc. R. Soc. B* **2016**, *283*, 20160032. [CrossRef] [PubMed]

53. Gaspard, M.; Guichard, N. Shell matrices of recent rhynchonelliform brachiopods: Microstructures and glycosylation studies. *Trans. R. Soc. Edinb.* **2008**, *98*, 415–424. [CrossRef]

54. Avrami, M. Kinetics of Phase change. I General Theory. *J. Chem. Phys.* **1939**, *7*, 1103–1112. [CrossRef]

55. Avrami, M. Kinetics of Phase change. II Transformation-Time Relations for Random Distribution of Nuclei. *J. Chem. Phys.* **1940**, *8*, 212–224. [CrossRef]

56. Avrami, M. Granulation, Phase Change, and Microstructure Kinetics of Phase Change. III. *J. Chem. Phys.* **1941**, *9*, 177–184. [CrossRef]

57. Casella, L.A.; He, S. Assessment of hydrothermal alteration on micro- and nanostructures of biocarbonates: Quantitative statistical grain-area analysis of diagenetic overprint. *Biogeosciences* **2018**. [CrossRef]

58. Checa, A. Physical and Biological Determinants of the Fabrication of Molluscan Shell Microstructures. *Front. Mar. Sci.* (accepted).

59. Genin, G.M.; Kent, A. Functional grading of mineral and collagen in the attachment of tendon to bone. *Biophys. J.* **2009**, *97*, 976–985. [CrossRef] [PubMed]

60. Seidl, B.H.M.; Reisecker, C. Calcite distribution and orientation in the tergite exocuticle of the isopods *Porcellio scaber* and *Armadillidium vulgare* (Oniscidea, Crustacea)—A combined FE-SEM, polarized SCμ-RSI and EBSD study. *Z. Kristallogr.* **2012**, *227*, 777–792. [CrossRef]

61. Huber, J.; Fabritius, H.O. Function-related adaptations of ultrastructure, mineral phase distribution and mechanical properties in the incisive cuticle of mandibles of *Porcellio scaber* Latreille, 1804. *J. Struct. Biol.* **2014**, *188*, 1–15. [CrossRef] [PubMed]

62. Huber, J.; Griesshaber, E. Functionalization of biomineral reinforcement in crustacean cuticle: Calcite orientation in the partes incisivae of the mandibles of *Porcellio scaber* and the supralittoral species *Tylos europaeus* (Oniscidea, Isopoda). *J. Struct. Biol.* **2015**, *190*, 173–191. [CrossRef] [PubMed]

63. Griesshaber, E.; Yin, X. Patterns of mineral organization in carbonate biological hard materials. In *Highlights in Applied Mineralogy*; Heuss-Aßbichler, S., Amthauer, G., John, M., Eds.; De Gruyter: Berlin, Germany; Boston, MA, USA, 2018; p. 344, ISBN 978-3-11-049734-2.

© 2018 by the authors. Licensee MDPI, Basel, Switzerland. This article is an open access article distributed under the terms and conditions of the Creative Commons Attribution (CC BY) license (http://creativecommons.org/licenses/by/4.0/).

Article

Metasomatic Replacement of Albite in Nature and Experiments

Kirsten Drüppel [1],* and Richard Wirth [2]

[1]   Institute of Applied Geosciences, Karlsruhe Institute of Technology (KIT), Adenauerring 20b, 76131 Karlsruhe, Germany

[2]   Deutsches GeoForschungsZentrum, Telegrafenberg, 14473 Potsdam, Germany; wirth@gfz-potsdam.de

*   Correspondence: kirsten.drueppel@kit.edu; Tel.: +49-721-608-43326

Received: 1 April 2018; Accepted: 12 May 2018; Published: 17 May 2018

check for
updates

Abstract: Replacement of albite by sodium-rich, secondary phases is a common phenomenon, observed in different geological settings and commonly attributed to alkaline metasomatism. We investigated growth of nepheline and sodalite on albite in time series experiments between two and 14 days. A total of 42 hydrothermal experiments were performed in cold-seal hydrothermal vessels at a constant pressure of 4 kbar and 200–800 °C in the system SiO_2–Al_2O_3–NaCl–H_2O. To allow for fluid flow and material transport, a double-capsule technique was used; hereby, a perforated inner Pt capsule was filled with cleavage fragments of natural albite, whereas the shut outer Au capsule was filled with γ-Al_2O_3 and the NaCl–H_2O solution. Complete overgrowth of albite by sodalite and nepheline occurred after just two days of experiments. At high salinity ($\geq$17 wt % NaCl) sodalite is the stable reaction product over the whole temperature range whereas nepheline occurs at a lower relative bulk salinity than sodalite and is restricted to a high temperature of $\geq$700 °C. The transformation of albite starts along its grain margins, cracks or twin lamellae. Along the reaction front sodalite crystallizes as small euhedral and highly porous grains forming polycrystalline aggregates. Coarse sodalite dominates in the outermost domains of the reaction zones, suggesting recrystallization. Sodalite may contain fluid inclusions with trapped NaCl-rich brine, demonstrating that the interconnected microporosity provides excellent pathways for fluid-assisted material transport. Highly porous nepheline forms large, euhedral crystals with rectangular outline. Sodalite and nepheline in natural rock samples display only minor porosity but fluid and secondary mineral inclusions, pointing to coarsening of a previously present microporosity. The reaction interface between sodalite and albite in natural rock samples is marked by open channels in transmission electron microscopy. In many of the experiments, a zone of Si–H-rich, amorphous material is developed at the reaction front, which occurs at a temperature of up to of 750 °C as nanometer to 350 µm wide reaction zone around albite. This change in composition corresponds with the abrupt termination of the crystalline feldspar structure. The presence of sodalite as micro- to nanometer-sized, euhedral crystals within the amorphous zone demonstrates, that both the sodalite reaction rim and the amorphous material allow for fluid-assisted material transport between the crystalline albite (release of Si, Al) and the bulk fluid (H_2O, Na, Cl). This texture, moreover, suggests that the amorphous phase represents a metastable interstage reaction product, which is progressively replaced by sodalite and nepheline. Remarkably, product sodalite, nepheline, and the amorphous material largely inherit the trace element budget of the respective ancestor albite, indicating that at least part of the trace elements remained fixed during the reaction process. The observed reaction textures in both natural and experimental samples indicate an interfacial dissolution–reprecipitation mechanism. Results of our study bear important implications with respect to mineral replacement in the presence of a fluid phase, especially regarding the interpretation of trace element patterns of the product phases.

Keywords: albite; amorphous; analcime; dissolution–precipitation; hydrothermal experiments; metasomatism; nepheline; sodalite

1. Introduction

Replacement of a mineral by a more stable one is a widespread phenomenon in distinct geological environments and processes. This project focuses on the alkaline metasomatic transformation of albite in the presence of a NaCl-rich fluid phase. A major feature of this reaction type is material transport along grain and interphase boundaries in the hydrous fluids. When the reaction is completed the newly formed mineral had crystallized in the pre-existing volume of the ancestor mineral. In general, such mineral pseudomorphs preserve the external shape and accordingly the volume of the parent phase [1–6]. Another consequence of fluid–assisted mineral reactions, generally irrespective of the relative molar volumes of the parent and product phases, is the development of an interconnected microporosity in the product providing new pathways for circulating fluids [4,5,7,8]. The reaction produces a front between the educt and product phases. Despite the large number of studies concerning mineral replacement processes, the mechanisms involved are not fully understood. In part this is due to the complexity of natural mineral solid solutions and fluid compositions, when compared to common experimental setups. In addition, the physical and chemical conditions during the mineral replacement reaction in natural rocks are very difficult to estimate. Even if the peak P–T conditions of a rock can be calculated from geothermobarometry, it is often not clear at what stage and over what P–T interval pseudomorphs formed. In order to construct a replacement model, it is hence useful to combine microscopic observations and geochemical analyses of natural rock samples, for which the boundary conditions of the respective replacement reaction are known, with well-defined experimental investigations. This allows investigating experimentally the conditions under which a solid phase forms or changes as a function of the composition of the surrounding fluid and to further apply the results to more complex, natural systems.

The present experimental study explores growth of sodalite ($Na_8Al_6Si_6O_{24}Cl_2$), nepheline ($NaAlSiO_4$), and minor analcime ($NaAlSi_2O_6 \cdot H_2O$) on albite in the system SiO_2–Al_2O_3–$NaCl$–H_2O, mainly governed by dissolution, material transport from an external source and re-precipitation. In natural systems, these minerals typically occur in silica undersaturated igneous rocks such as nepheline syenites or alkali-metasomatized rocks. The chemical reactions describing these various transformations were proposed by [9] and then accepted by many scientists, e.g., [10–13]. According to observations of the alkali-metasomatic replacement of feldspars in experiments and natural metasomatic rock samples [9–14], experiments were performed at crustal pressures of 4 kbar, late-igneous to hydrothermal temperatures of 200–800 °C, and a fluid phase of variable salinity. We used a double-capsule technique, where an albite crystal, serving as source for Si, Na, and Al, is physically separated from γ-Al_2O_3 and material transport occurs via a H_2O–$NaCl$ fluid. After identification of the run products with X-ray diffraction (XRD) and high-resolution scanning electron microscopy (SEM), the composition of the coexisting phases was determined by electron microprobe analysis (EMPA). Internal microstructures like porosity or open pathways for element transport were studied with SEM and transmission electron microscopy (TEM), using the focused ion beam (FIB) method. Of special interest was the geometry and nature of the reaction front to characterize a possible influence of pre-existing heterogeneities like grain boundaries and microcracks on its origin and evolution. These methods allowed identifying the preferred pathways of the metasomatic solutions. The timing, nature and distribution of micropores and microcracks were used to decide whether these structures preferentially form along specific crystallographic orientations in albite or if they are mainly controlled by the grain boundaries of the phases involved. In addition, we investigated the behavior of trace elements during the replacement of plagioclase with in situ laser ablation ICPMS (LA-ICPMS) analysis. The knowledge about redistribution patterns of elements—in our case the spatial distribution of Sr and the light

REE—may help to interpret trace and rare earth element data with respect to natural mineral-forming processes. General conclusions of this project can be applied to other synthetic and natural systems, where similar replacement textures are observed.

2. Materials and Methods

2.1. Starting Materials

Cleavage fragments of natural albite from Swartbooisdrif, Namibia, Lomnitz, Poland, and Lovozero, Russia were used as educt phases for the hydrothermal experiments. A detailed description and analyses of the starting material are given in Section 3.1.

2.2. Experimental Setup and Procedure

A total of 42 hydrothermal experiments were performed in the system SiO_2–Al_2O_3–$NaCl$–H_2O (P: 4 kbar, T: 200–800 °C, 2–14 days; Table 1) in a conventional hydrothermal apparatus with horizontally arranged, Tuttle-type cold-seal vessels. For all experiments, a double-capsule technique was used to allow for fluid mobility. Hereby, a shut but perforated inner platinum capsule of 2.5 mm in diameter was filled with weighed amounts of cleavage fragments of natural albite of 0.5–1.2 mm in diameter (4–5 pieces) from different localities (i.e., Swartbooisdrif, Namibia, Lomnitz, Poland, Lovozero, Russia), whereas the shut outer capsule was filled with γ-Al_2O_3 and an aqueous brine of varying salinity (Figure 1). These inner capsules were placed in a welded outer gold capsule of 4.6 mm in diameter, filled with γ-Al_2O_3 and a mixture of analytical grade NaCl and distilled H_2O with varying salinity (9–75 wt % NaCl) as starting materials. A minor amount of NaOH (0.2 mol/L) was added to the fluid in order to promote solubility and transport of silica. The mixture was loaded into an Au capsule, the perforated Pt capsule with albite was placed on top, then the Au capsule was sealed by arc welding. The experimental design allows the saline fluid to migrate between the two capsules, hence promoting fluid-assisted mass transfer between the inner and the outer capsule. The capsule was placed in an oven at 110 °C for 12 h to check for tightness and was then put in externally heated cold-seal vessels.

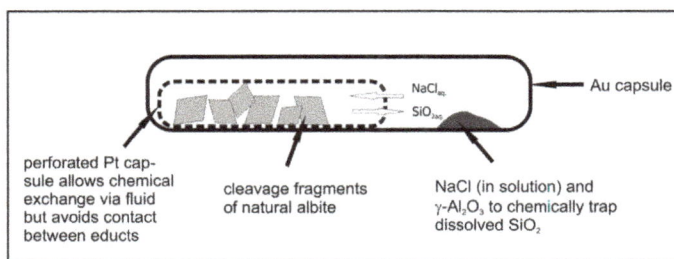

Figure 1. Experimental setup for the replacement of albite by sodalite and/or nepheline. The starting albite crystals are physically separated from γ-Al_2O_3 by a shut but perforated Pt capsule (2.5 mm in diameter), providing pathways for the bulk fluid, an H_2O–NaCl solution of variable salinity. Cleavage fragments of albite were in the order of 0.5–1.2 mm in diameter, the powdered γ-Al_2O_3 was placed in an outer Au capsule (4.6 mm in diameter) of approximately 2.5 to 3 cm in length (figure not to scale).

According to observations in natural rock samples, the vessels were pressurized by water to 4 kbar and heated under controlled pressure to 300–800 °C. The temperature during the experiments was controlled with NiCr-Ni thermocouples, with the temperature being accurate within ±5 °C. Fluid pressures were measured with Heise gauges with an estimated uncertainty of ±50 bar. The experiments were ended by switching off the power and cooling the autoclaves with compressed air to a temperature of <100 °C within 5 min, while maintaining the pressure.

Table 1. Starting materials, run conditions and products of hydrothermal experiments for the metasomatic transformation of albite to sodalite and nepheline.

Experiment	Starting Material	P (kbar)	T (°C)	d (days)	Salinity (wt %)	Solid Run Products
29	albite (Lomnitz), NaCl, -Al$_2$O$_3$	4	200	14	26	albite, halite, boehmite, sodalite
31	albite (Lomnitz), NaCl, -Al$_2$O$_3$	4	200	7	26	albite, halite, boehmite, sodalite
33	albite (Lomnitz), NaCl, -Al$_2$O$_3$	4	200	2	75	albite, halite, boehmite, sodalite, amorphous phase
13	albite (Lomnitz), NaCl, -Al$_2$O$_3$	4	300	6	39	albite, halite, corundum, boehmite, sodalite
20	albite (Lomnitz), NaCl, -Al$_2$O$_3$	4	300	7	39	albite, halite, corundum, boehmite, sodalite
LT-20	albite (Lomnitz), NaCl, -Al$_2$O$_3$	4	330	6	10	albite, halite, boehmite, analcime
LT-16	albite (Swartb.), NaCl, -Al$_2$O$_3$	4	350	6	17	albite, halite, boehmite, sodalite, analcime
LT-21	albite (Lomnitz), NaCl, -Al$_2$O$_3$	4	400	6	9	albite, halite, corundum, boehmite, analcime
07	albite (Swartb.), NaCl, -Al$_2$O$_3$	4	400	6	29	albite, halite, corundum, boehmite, sodalite
14	albite (Lomnitz), NaCl, -Al$_2$O$_3$	4	400	6	38	albite, halite, corundum, boehmite, sodalite, amorphous phase
LT-17	albite (Swartb.), NaCl, -Al$_2$O$_3$	4	450	6	17	albite, halite, corundum, boehmite
LT-02	albite (Swartb.), NaCl, -Al$_2$O$_3$	4	450	6	35	albite, halite, corundum, boehmite, sodalite, analcime
LT-05	albite (Swartb.), NaCl, -Al$_2$O$_3$	4	450	6	50	albite, halite, corundum, boehmite, sodalite
LT-22	albite (Lomnitz), NaCl, -Al$_2$O$_3$	4	500	6	9	albite, halite, corundum, boehmite
09	albite (Lomnitz), NaCl, -Al$_2$O$_3$	4	500	7	32	albite, halite, corundum, boehmite, sodalite, analcime, amorphous phase
LT-12	albite (Swartb.), NaCl, -Al$_2$O$_3$	4	550	6	15	albite, halite, corundum
LT-18	albite (Swartb.), NaCl, -Al$_2$O$_3$	4	550	6	17	albite, halite, corundum, boehmite
01	albite (Lovoz.), NaCl, -Al$_2$O$_3$	4	550	6	27	albite, halite, corundum
LT-03	albite (Swartb.), NaCl, -Al$_2$O$_3$	4	550	6	35	albite, halite, corundum, sodalite
15	albite (Lomnitz), NaCl, -Al$_2$O$_3$	4	550	6	39	albite, halite, corundum, sodalite
LT-04	albite (Swartb.), NaCl, -Al$_2$O$_3$	4	550	6	51	albite, halite, corundum, sodalite
10	albite (Lomnitz), NaCl, -Al$_2$O$_3$	4	600	7	25	albite, halite, corundum, sodalite
HT-10	albite (Lovoz.), NaCl, -Al$_2$O$_3$	4	600	6	30	albite, halite, corundum, sodalite
18	albite (Lomnitz), NaCl, -Al$_2$O$_3$	4	600	7	39	albite, halite, corundum, sodalite, amorphous phase
05	albite (Swartb.), NaCl, -Al$_2$O$_3$	4	600	6	41	albite, halite, corundum, sodalite
HT-12	albite (Lomnitz), NaCl, -Al$_2$O$_3$	4	600	6	42	albite, halite, corundum, sodalite
16	albite (Lomnitz), NaCl, -Al$_2$O$_3$	4	650	6	39	albite, halite, corundum, sodalite, amorphous phase
HT-08	albite (Lovoz.), NaCl, -Al$_2$O$_3$	4	650	6	44	albite, halite, corundum, sodalite
HT-04	albite (Lovoz.), NaCl, -Al$_2$O$_3$	4	650	6	55	albite, halite, corundum, sodalite
30	albite (Lomnitz), NaCl, -Al$_2$O$_3$	4	700	14	17	albite, halite, corundum, nepheline, amorphous phase
HT-13	albite (Lovoz.), NaCl, -Al$_2$O$_3$	4	700	6	17	albite, halite, corundum, nepheline
HT-11	albite (Lovoz.), NaCl, -Al$_2$O$_3$	4	700	6	27	albite, halite, corundum, sodalite, nepheline
34	albite (Lomnitz), NaCl, -Al$_2$O$_3$	4	700	2	28	albite, halite, corundum, sodalite, nepheline, analcime, amorphous phase
32	albite (Lomnitz), NaCl, -Al$_2$O$_3$	4	700	7	29	albite, halite, corundum
11	albite (Lomnitz), NaCl, -Al$_2$O$_3$	4	700	7	31	albite, halite, corundum, sodalite, amorphous phase
HT-14	albite (Lomnitz), NaCl, -Al$_2$O$_3$	4	700	6	34	albite, halite, corundum, sodalite
04	albite (Swartb.), NaCl, -Al$_2$O$_3$	4	700	6	42	albite, halite, corundum, sodalite, nepheline
HT-15	albite (Lovoz.), NaCl, -Al$_2$O$_3$	4	700	6	43	albite, halite, corundum, sodalite
HT-16	albite (Lovoz.), NaCl, -Al$_2$O$_3$	4	700	6	44	albite, halite, corundum, sodalite
12	albite (Lomnitz), NaCl, -Al$_2$O$_3$	4	750	7	40	albite, halite, corundum, sodalite, analcime, amorphous phase
HT-09	albite (Lovoz.), NaCl, -Al$_2$O$_3$	4	750	6	49	albite, halite, corundum, sodalite
HT-02	albite (Lovoz.), NaCl, -Al$_2$O$_3$	4	800	6	42	albite, halite, corundum, sodalite, nepheline
HT-06	albite (Lovoz.), NaCl, -Al$_2$O$_3$	4	800	6	62	albite, halite, corundum, sodalite

2.3. Analysis of the Solid Run Products

Characterization of the initial and experimental product phases was performed with several analytical and imaging methods. XRD patterns for powders of synthesis products were obtained by a PHILIPS PW 1050 diffractometer (Philips Analytical, Almelo, The Netherlands) with Bragg-Brentano geometry and a scintillation counter, using Ni-filtered CuKα radiation (40 kV, 30 mA) housed at the department of applied Geosciences, Technical University of Berlin, Germany. Phase analysis was carried out with PHILIPS X'Pert Plus 1.0 (1999).

Reaction textures were investigated with a high-resolution SEM (Type Hitachi S-4000, Tokyo, Japan), housed at the Center of Electron Microscopy (ZELMI), Technical University of Berlin, Germany. For qualitative investigations with the SEM, the solid run products were dispersed on a carbon-coated sample carrier and sputtered with gold. For detailed investigation of the reaction front and the microstructures, polished thin sections were prepared and investigated under back-scattered electron (BSE) mode. Semi-quantitative analyses of the compositions of the solid run products were performed with a conventional SEM (Type Hitachi S-2700, Tokyo, Japan), equipped with an Energy Dispersive Spectrometer (EDS) on polished, carbon-coated thin sections.

EMPA (Cameca, Gennevilliers, France) of the compositions of the educt and product phases was performed on a Cameca Camebax instrument at the ZELMI, TU Berlin, Germany. Natural and synthetic standards were used for instrument calibration. Mineral analyses were performed with an accelerating voltage of 15 kV, a beam current of 20 nA and an electron beam of 5 μm in diameter.

In situ LA-ICPMS trace element analyses of the initial albite and product phases of the hydrothermal experiments were performed to investigate the behavior of trace elements during metasomatism. The thin sections were analyzed with a 266 nm Nd:YAG laser (New Wave Research Inc., Merchantek Products, Fremont, CA, USA) connected to a quadrupole ICPMS (Agilent 7500i; Plasma power: 1320 W; Carrier gas flow: 1.11 L·min^{-1} (Ar); Plasma gas flow: 14.9 L·min^{-1} (Ar); auxiliary gas flow: 0.9 L·min^{-1} (Ar)) at the Institute of Mineralogy, Würzburg. Laser parameters used were a frequency of 10 Hz and an energy setting of 50% (0.9 mJ). Ablation patterns were 600 μm long lines, ablated with a scan speed of 10 μm/s. The diameter of the sample pit created by the laser is 50 μm. Data acquisition was done in time-resolved analysis mode with measurements of the instrument background (20 s) and of sodalite, albite, nepheline, and amorphous material (60 s). The certified reference material NIST 612 with the values of [15] was used as an external standard, whereas Si in albite and its replacement products, derived from the electron microprobe analyses, was used as internal standard. Raw counts for each element were solely corrected by subtracting the background counts and processed using the software GLITTER (Version 3.0; On-line Interactive Data Reduction for the LA-ICPMS, Macquarie Research Ltd., Sydney, Australia, 2000).

The geometry and nature of the reaction front were investigated by high-resolution TEM (FEI company, Hillsboro, OR, USA) at the GeoForschungsZentrum Potsdam (GFZ). Electron transparent foils for TEM were cut from selected areas across the reaction fronts in the naturally and experimentally reacted albite in polished thin sections by focused ion beam (FIB) milling in an ultra-high vacuum using a FEI FIB200 instrument (FEI Company, Hillsboro, OR, USA, see [16,17] for details). TEM investigations were performed on a FEI Tecnai G2 F20 X-Twin system incorporating a field-emission electron gun (operated at 200 kV), a Fishione high-angle annular dark-field detector (HAADF), a Gatan Tridiem imaging filter for acquisition of energy-filtered images and electron energy-loss spectroscopy (EELS) element mapping, and an EDAX Genesis X-ray analyzer with an ultrathin window. Characterization of the reaction front in natural and reacted samples included bright-field and dark-field imaging, and EDX analyses (in scanning-TEM mode).

3. Results

3.1. Composition of the Starting Materials

Cleavage fragments of natural albite of 0.5–1 mm in size comprise (1) metasomatic albite from Swartbooisdrif, Namibia; (2) fissure-grown albite from Lomnitz, Poland; and (3) fissure-grown albite from Lovozero, Russia. Namibian albite is the main mineral of a sodium-rich fenite, containing minor ankerite, sodalite, magnetite, and biotite. Albite is subhedral and microscopically clear, but may contain rare inclusions of biotite. Individual albite grains are up to 0.5 cm in size. Many albite grains do not display polysynthetic twin lamellae but, if present, twinning is on the albite law. Fissure-grown albites from both Poland and Russia form euhedral and glass-clear crystals. Poland albite is up to 1.5 cm in size and associated by calcite and quartz whereas medium-grained albite (1–5 mm) from Russia has grown as a grass-like crystal layer on alkali feldspar. Both albite types are unweathered and frequently twinned on the albite law.

The chemical composition of the different albites was determined by EMP analysis, averaging analyses from profiles of 10 to 15 points. Albite from all different localities is almost pure, homogeneous albite (Ab$_{99-100}$), with CaO, MgO, FeO, K$_2$O, and BaO contents at or below the detection limit (Table 2). Only minor Swartbooisdrif albite may show elevated Ca contents of up to 1 wt % (Ab$_{95}$). The three different albites differ regarding their trace element composition, analyzed by La-ICPMS (Table A1). Concentrations in Ca and Sr are highest in the Swartbooisdrif and lowest in the Lovozero albite, whereas Ba, Sc, Mn, Co, Ti, and the LREE contents decrease from Swartbooisdrif over Lovozero towards Lomnitz albite, the latter of which having LREE contents below the detection limit. Concentrations of Zn, Ga, Cu, and Y are highest in the Lovozero albite. Contents of the HREE (Gd to Lu) are below the detection limit in all samples.

Table 2. Representative EMP analyses of educt albite from Lomnitz, Lovozero, and Swartbooisdrif.

Sample	29	30	31	33	34	HT-08	HT-09	HT-10	LT-05	LT-07
Point	23	28	30	18	7	13	23	47	63	79
Mineral	albite Lomnitz	albite Lomnitz	albite Lomnitz	albite Lomnitz	albite Lomnitz	albite Lovozero	albite Lovozero	albite Lovozero	albite Swartb.	albite Swartb.
Analysis						wt %				
SiO$_2$	68.5	68.4	68.7	68.5	68.3	68.6	68.8	68.7	67.1	68.6
Al$_2$O$_3$	20.2	20.6	20.2	20.0	19.7	19.8	20.5	20.3	21.9	19.5
Fe$_2$O$_3$	0.05	0.00	0.00	0.00	0.03	0.01	0.04	0.01	0.04	0.02
CaO	0.02	0.05	0.06	0.02	0.02	0.02	0.01	0.00	0.99	0.02
Na$_2$O	11.4	11.5	11.7	11.6	11.3	12.2	12.4	11.7	10.8	12.2
K$_2$O	0.01	0.03	0.04	0.03	0.04	0.04	0.17	0.06	0.55	0.03
BaO	0.00	0.01	0.00	0.01	0.00	0.01	0.00	0.00	0.01	0.00
Sum	100.1	100.6	100.7	100.1	99.4	100.6	101.8	100.8	101.4	100.3

3.2. Experimental Run Products and Replacement Textures

A total of 42 hydrothermal experiments were performed in the system SiO$_2$–Al$_2$O$_3$–NaCl–H$_2$O (P: 4 kbar, T: 200–800 °C, 2–14 days) to provide a better understanding of the reaction processes related to the alkali-metasomatic replacement of albite. Run conditions, composition of the solid starting materials and salinities of the fluid phase are listed in Table 1. The replacement of albite never went to completion, so albite remained present in all experiments.

At high salinity sodalite was found to be the stable phase over the whole temperature range (Figure 2). As temperature rises, more NaCl is needed to stabilize sodalite. Nepheline occurred at a temperature >700 °C and lower average bulk salinity of the fluid. Remarkably, there is a discontinuity in the occurrence of nepheline in experiments performed at 700–800 °C. At 700 °C, the feldspathoid is observed at salinities of 17–28 wt % NaCl, then disappears, and appears again at a salinity of 42 wt % NaCl. It is not observed in experiments performed at 750 °C but in experiment HT-02 at 800 °C, 42 wt % NaCl.

Even during the shortest experiment of two days, complete overgrowths of sodalite and nepheline occurred on albite. Analcime is present as a minor phase in many of the experiments (at variable salinities of 9–40 wt % NaCl), where it forms part of the replacement assemblage in the inner capsule. In the outer capsule, originally filled with γ-Al$_2$O$_3$ and an aqueous brine of varying salinity, boehmite is preserved at low temperature, whereas corundum is found at T $\geq$ 550 °C. Nepheline exclusively grew around albite in the inner capsule and was not detected outside the inner capsule. Euhedral, isolated sodalite, on the other hand, is also frequently present in the outer capsule at high bulk salinity, testifying to transport of silica via the aqueous brine and its reaction with γ-Al$_2$O$_3$. Representative EMP analyses of the educt albite and the product phases of sodalite, nepheline, and analcime are listed in Tables 2 and 3, respectively.

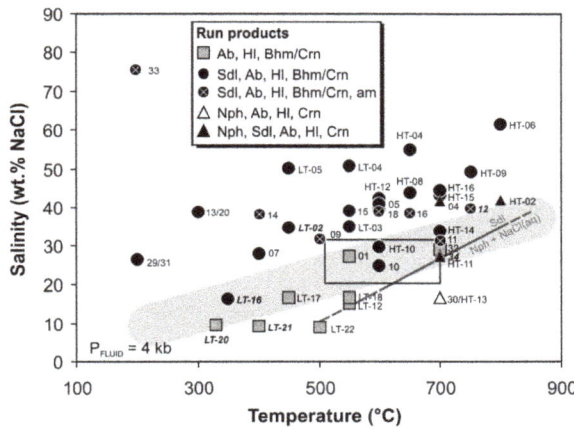

Figure 2. Salinity vs. temperature diagram for run products of the hydrothermal experiments. Rectangle marks the temperature and salinity conditions calculated for natural, sodalite-bearing rock samples of Namibia [18,19]. Samples marked bold-italic additionally contain analcime as reaction product. Grey solid and stippled line marks the experimentally derived course of the reaction 6 nepheline + 2NaCl(aq) = sodalite after [14]. Grey shaded area marks approximate course of the metasomatic replacement of albite by sodalite, following the reaction 6 albite + 2NaCl(aq) = sodalite + 12SiO$_2$(aq). (*Ab* albite, *Bhm* boehmite, *Crn* corundum, *Hl* halite, *Nph* nepheline, *Sdl* sodalite).

Table 3. Representative EMP analyses of product sodalite, nepheline, and analcime.

Sample	10	34	HT-04	HT-08	HT-10	LT-04	LT-05	LT-06	LT-07	LT-07	30	34	34	HT-11	HT-13	12	34	34
Point	16	14	8	16	48	59	66	68	78	80	27	3	8	35	31	8	10	11
Mineral					sodalite								nepheline				analcime	
Analysis								wt %										
SiO$_2$	38.3	36.6	37.5	37.2	37.1	37.1	37.7	38.0	37.7	37.1	47.8	49.9	47.4	46.5	40.7	56.8	55.3	55.0
Al$_2$O$_3$	30.5	31.8	32.7	32.6	32.0	32.3	32.4	31.7	32.5	32.5	34.6	32.3	33.1	34.0	35.9	23.1	25.9	26.0
Fe$_2$O$_3$	0.02	0.00	0.05	0.00	0.02	0.11	0.00	0.03	0.00	0.00	0.04	0.03	0.00	0.05	0.00	0.07	0.00	0.04
CaO	0.03	0.02	0.00	0.04	0.00	0.07	0.04	0.22	0.04	0.01	0.07	0.04	0.04	0.00	0.04	0.07	0.01	0.00
Na$_2$O	24.9	24.7	25.8	25.2	26.0	25.0	24.8	23.7	25.8	26.0	18.9	18.2	19.2	19.6	22.7	11.9	10.4	9.3
K$_2$O	0.00	0.00	0.07	0.14	0.18	0.05	0.06	0.06	0.07	0.08	0.05	0.01	0.00	1.29	1.63	0.02	0.02	0.05
Cl	9.15	8.66	6.30	6.41	6.27	6.30	6.36	6.26	6.11	6.17	0.00	0.02	0.00	0.02	0.15	0.41	0.37	0.34
Sum	102.8	101.8	102.4	101.6	101.6	101.0	101.3	100.0	102.2	101.9	101.4	100.5	99.8	101.4	101.1	92.3	91.9	90.7

The transformation of albite to sodalite starts along the grain margins and, if present, along cracks (Figure 3a) and propagates until a complete pseudomorph of sodalite after albite has formed. The product sodalite forms euhedral crystals, in short experiments displaying a pseudohexagonal shape due to fast growth (Figure 3b). Along the embayed reaction fronts sodalite crystallizes as very fine grained, highly porous grains forming polycrystalline aggregates with randomly oriented crystallites. Larger grains of massive sodalite dominate in the outermost domains of the reaction zones (Figure 3c), suggesting coarsening of early sodalite aggregates. An increasing run duration also involves coarsening of the grains, resulting in formation of massive sodalite along the reaction front. If present, albite twin lamellae are preferentially dissolved during the reaction process. Hereby, the reaction zones propagate from twin planes towards the center of individual lamellae. Polycrystalline sodalite aggregates partially fill the former lamellae but display no detectable crystallographic relationship with albite (Figure 3d). Along the outermost margins of the reaction rim, euhedral sodalite grew into the fluid and newly nucleated sodalite grains occur on the surfaces of previously crystallized sodalite even after long run times of 14 days. This texture indicates fluid transport through the sodalite rim and immediate reaction at the sodalite–fluid interface. Newly formed sodalite is often extremely porous (Figure 3e) with the interconnected microporosity providing excellent pathways for fluid-assisted material transport from and towards the reaction front. In five of the 42 experiments, sodalite contains numerous fluid inclusions, containing the trapped NaCl-rich aqueous fluid.

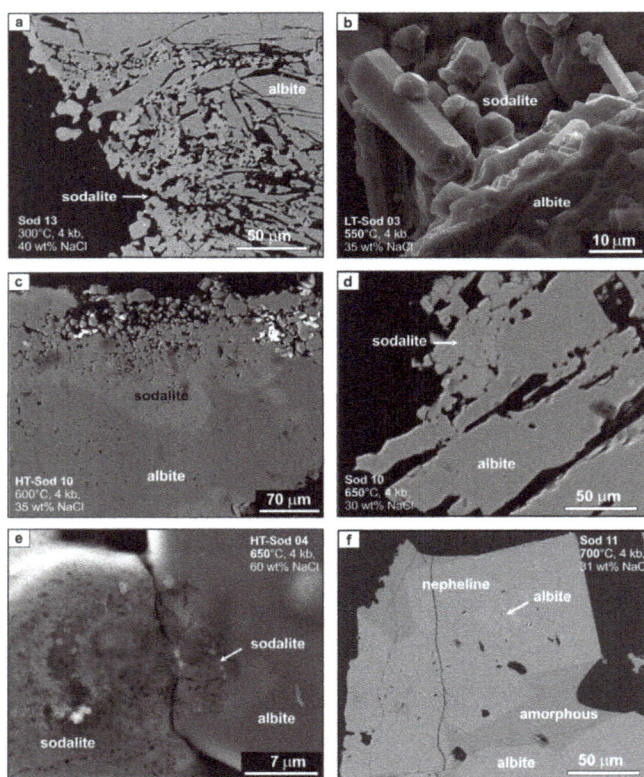

Figure 3. Reaction textures during the metasomatic replacement of albite (SEM images). (**a**) Sodalite replacing albite along the grain margins and cracks; (**b**) Hexagonal sodalite prisms, formed due to fast growth of sodalite; (**c**) Embayed reaction front with sodalite replacing albite; (**d**) Sodalite replacing albite along twin lamellae; (**e**) Sodalite is characterized by a high microporosity; (**f**) Nepheline crystals with rectangular outline replace albite at a high temperature of >650 °C.

At lower salinities of the fluid and a temperature of >700 °C, nepheline forms at the expense of albite. Like sodalite, nepheline displays a high porosity but, in contrast to sodalite, forms large, crystals with rectangular outlines of up to 150 µm in length that do not display any pseudomorphic relationship with the ancestor albite (Figure 3f). Small irregular patches of the original albite are preserved as inclusions in the newly formed nepheline.

3.3. Amorphous Material Formation

A conspicuous feature in many of the experiments is a zone of amorphous material, which developed at the reaction front between albite and newly formed sodalite and/or nepheline (Figure 4). The highly porous amorphous material is present during every stage of the experiments and at almost every experimental condition (even up to a temperature of 750 °C) as several nanometers to 350 µm wide reaction zone around albite relicts, separating it from sodalite and nepheline. A reaction rim of polycrystalline sodalite aggregates surrounds the amorphous material (Figure 4a,b). In extreme cases, the amorphous material may form an almost complete pseudomorph after albite (Figure 4b). Numerous cracks in the amorphous material point to a gel-like nature and high amounts of volatiles before cooling and drying of the samples (Figure 4a–c). Investigation with TEM reveals that the reaction interface between both sodalite and albite and the amorphous material is always sharp. Occasionally, an open channel is developed between albite and the amorphous zone, suggesting a structural discontinuity and lack of adhesion between the two phases; this channel may be filled by precipitates of the migrating fluid (i.e., mainly NaCl; Figure 4e,f). When compared to the ancestor albite, the amorphous layer is uniformly depleted in Na and, to a minor degree, in Al and enriched in Si and H. Chlorine contents are close to or below the detection limit. The drastic change in composition at the boundary with the amorphous layer corresponds with the abrupt termination of the crystalline feldspar structure. The presence of sodalite as micro- to nanometer-sized, euhedral crystals and diffuse Schlieren in the amorphous matrix is indicative of in situ nucleation and growth (Figure 4c,d). Their formation, however, requires transport of Al of the crystalline albite and migration of the NaCl-rich aqueous fluid into the Si-rich amorphous layer and suggests that material transport via the fluid phase is the most important process of this reaction type.

3.4. Behavior of Trace Elements during Replacement

To investigate the behavior of trace elements during metasomatism, albite from three different localities (Swartbooisdrif, Namibia, Lomnitz, Poland, Lovozero, Russia) each of which displaying its own characteristic trace element pattern, were used as educt phases. Remarkably, both the product sodalite and nepheline as well as the amorphous material inherit the trace element budget of the respective albite, displaying similar trace element patterns (Figure 5, Tables A1 and A2). Following this, the trace element contents of the solid product phase are mainly controlled by the bulk trace element budget of the reacting phase and not by fluid–mineral fractionation coefficients. It can be concluded that at least part of the trace elements remained fixed in the amorphous phase during the reaction process.

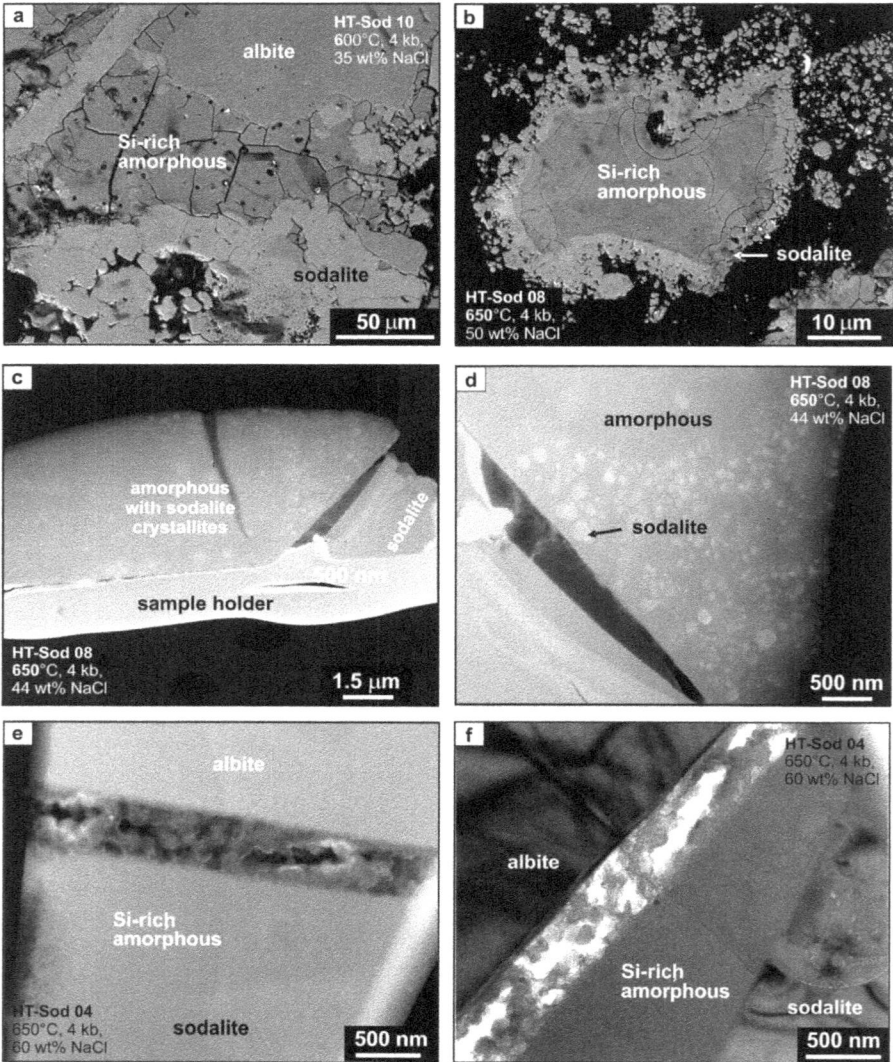

Figure 4. Amorphous material formation (**a**,**b**: SEM images; **c**–**f**: TEM images). (**a**) Amorphous material formation at the reaction front between albite and newly formed sodalite; (**b**) Amorphous material pseudomorphing the shape of former albite surrounded by newly formed sodalite; (**c**) TEM foil of the interphase boundary between sodalite and the amorphous phase in (**b**); A sharp contact is developed between sodalite and the amorphous material. Sodalite crystallization occurs irregularly, both within sharply defined areas and irregular Schlieren; (**d**) Detail of (**e**); (**e**) TEM foil of the interphase boundary between the educt albite and the newly formed sodalite separated by an open channel filled with NaCl crystals and a broad zone of Si-rich amorphous material; (**f**) Bright field image of (**e**).

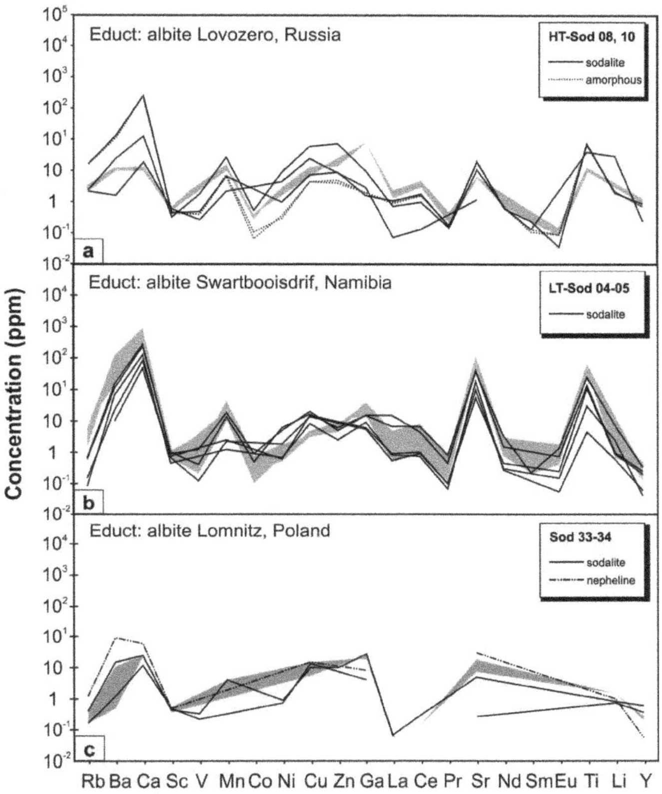

Figure 5. Laser-ICP-MS data for selected trace elements of different samples of educt albite (grey area) and the corresponding product sodalite, nepheline, and amorphous material of the hydrothermal experiments. (**a**) Experiments with fissure-grown albite from Lovozero, Russia; (**b**) Experiments with metasomatic albite from Swartbooisdrif, Namibia; (**c**) Experiments with fissure-grown albite from Lomnitz, Poland.

4. Discussion

4.1. Reaction Textures in the Experiments

Our hydrothermal experiments at 4 kbar, 200–800 °C in the system SiO_2–Al_2O_3–$NaCl$–H_2O show that albite is replaced by sodalite as the stable reaction product over the whole temperature range at high salinity, following Equation (1).

$$6NaAlSi_3O_8 + 2NaCl(aq) = Na_8Al_6Si_6O_{24}Cl_2 + 12SiO_2(aq) \tag{1}$$
$$\text{albite} \qquad\qquad\qquad \text{sodalite}$$

The formation of nepheline replacing albite (Equation (2)) occurs at high temperature but lower average bulk salinity of the fluid, also leading to an increase of Si in the solution.

$$NaAlSi_3O_8 = NaAlSiO_4 + 2SiO_2(aq) \tag{2}$$
$$\text{albite} \qquad \text{nepheline}$$

In part of the experiments (T: 700–800 °C, salinity: 27–42 wt % NaCl), both nepheline and sodalite were formed. In natural rock samples, the feldspathoids commonly display a replacement relationship (see Section 4.3), indicating progress of Equation (3).

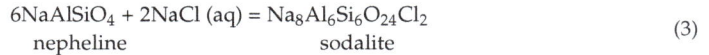

$$6NaAlSiO_4 + 2NaCl\ (aq) = Na_8Al_6Si_6O_{24}Cl_2$$
$$\text{nepheline} \qquad\qquad\qquad \text{sodalite} \tag{3}$$

The fact that nepheline occurs discontinuously in the temperature-salinity range of the experiments may indicate, that part of the nepheline represents an interstage product, which is preserved or not, depending on the completion of the reaction. Textural relationships observed in our experiments, however, indicate that nepheline and sodalite rather coexist in this temperature-salinity interval, forming small, euhedral grains, crystallizing next to each other. This interpretation is in agreement with the experimental results of [14], who also observed a transition zone at intermediate salinities, where sodalite and nepheline coexist.

Analcime is a minor phase mainly present in experiments performed at low temperature <500 °C but also occurs as texturally late, accessory phase at temperatures of 700–750 °C, possibly formed during cooling. It commonly replaces albite, as in Equation (4).

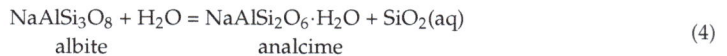

$$NaAlSi_3O_8 + H_2O = NaAlSi_2O_6 \cdot H_2O + SiO_2(aq)$$
$$\text{albite} \qquad\qquad \text{analcime} \tag{4}$$

Equations (1), (2) and (4) demonstrate, that the dissolution of albite leads to a strong enrichment of Si in the solution. While the dissolved silica will be removed from the system in natural examples of metasomatism, in the experiments it remains in the system and progressively reacts with Al and Na to give nepheline, sodalite, and minor analcime.

A complete overgrowth of sodalite or nepheline on albite already occurred after two days, given the physicochemical conditions suitable for their formation. With increasing duration of the experiment (up to 14 days), the educt albite is increasingly consumed. Sodalite crystallizes as very fine grained, highly porous grains forming polycrystalline aggregates with randomly oriented crystallites displaying no detectable crystallographic relationship with albite, whereas nepheline forms large, euhedral crystals with rectangular outline. Both sodalite and nepheline are highly porous and may contain fluid inclusions with trapped NaCl-rich brine, demonstrating that the interconnected microporosity provides excellent pathways for fluid-assisted material transport.

A conspicuous feature of many of the experiments is a zone of highly porous, amorphous material, which occurs up to a temperature of 750 °C. If present, the amorphous material developed at the reaction front, ranging from several nanometers to a 350-μm reaction zone around albite. When compared to the latter, the amorphous layer is uniformly depleted in Na and Al and enriched in Si and H. This change in composition corresponds with the abrupt termination of the crystalline feldspar structure. The presence of sodalite as euhedral, micro- to nanometer-sized, euhedral crystals within the amorphous zone is indicative of its in situ nucleation and growth. Remarkably, both product sodalite and the amorphous material largely inherit the trace element budget of the respective ancestor albite, displaying almost indistinguishable trace element patterns, hence indicating that at least part of the trace elements remained fixed during the reaction process.

4.2. The Role of the Amorphous Phase

Amorphous layers, mainly developed between the fluid–solid interface and the unaltered mineral, are a common phenomenon of plagioclase alteration experiments under atmospheric conditions, in particular weathering experiments at acid to neutral pH [20–23]. Like the amorphous material developed in our alteration experiments, these near-surface amorphous regions are enriched in Si and H and show depletion in other metals, in particular the interstitial cations and Al [20–24]. The origin of the amorphous layer has been a subject of research and debate for years. It has been discussed in terms of preferential leaching of cations due to differences in bond strength (i.e., nonstoichiometric dissolution) and interdiffusion with H⁺ [25]. In recent years, authors have shown that the composition and nature of the amorphous material analyzed in their samples cannot be explained by a volume diffusion process but could be modeled assuming an interfacial dissolution–precipitation process (i.e., [24,26,27]). The findings in our experiments support the latter hypothesis. The sharp chemical and structural interface between albite and the amorphous layer, observed in our experiments, and the lack of albite structure relicts in the amorphous material indicates, that the latter represents an individual phase rather than a surface altered zone controlled by interdiffusion, like a leached layer. Following this, elements released from the albite structure (i.e., Si, Al, Na) may precipitate directly at the reaction front, forming a metastable reaction product, e.g., a hydrated silica gel. The fact that sodalite and nepheline (both NaCl-rich) subsequently crystallize in situ from the Si-rich and Na- and Al-poor amorphous material demonstrates that neither the outermost sodalite reaction rims nor the amorphous zone act as protective layers but allow for transport of aqueous chemical reactants and products between the crystalline reaction front (release of Si) and the bulk fluid (H_2O, Na, Cl, Al). Our findings are in agreement with an interface-coupled dissolution-precipitation model (see [6], for a review). The observation of sodalite and nepheline nucleating and growing within the amorphous layer strongly suggests that the amorphous layer represents a metastable intermediate phase rather than a final reaction product. Presumably, the amorphous material would have reacted with excess alumina and NaCl in the capsule to give sodalite and nepheline and be consumed if the experiment duration had been longer.

4.3. Reaction Textures in Natural Metasomatic Rock Samples

The replacement textures and chemical signatures of the run products of the experiments are remarkably similar to those of natural metasomatic rock samples. In natural fenite samples from NW Namibia, sodalite formed at the expense of anorthosite plagioclase during the subsolidus stage [18]. Alkali metasomatism occurred as a result of the interaction of anorthosites with sodium-rich, carbonatite-derived fluids. Mineral equilibria combined with fluid inclusions data prove that sodalite formed at c. 4 kbar and c. 700 ± 70 °C [18]. In syenites, also bordering the carbonatites, albite was progressively replaced by nepheline, followed by the formation of texturally late sodalite and cancrinite. The sodalitization of nepheline in the nepheline syenite occurred under a similar temperature (c. 775–700 °C) like the formation of sodalite overgrowth on albite [19]. The SiO_2-undersaturated bulk compositions of both the metasomatized anorthosites and syenites provide evidence that fenitization included removal of silica. While in nature, silica is removed from the system, in the experiments silica cannot be removed and reacts with alumina and Na to give nepheline, sodalite, and minor analcime.

The texture of syenites neighboring carbonatite at Swartbooisdrif, NW Namibia, is inhomogeneous with fine-grained K-feldspar-rich zones alternating with irregular zones, mainly composed of nepheline crystals of up to 5 cm in length, preserving numerous albite inclusions. At the direct contact to the carbonatite dykes pegmatoidal, euhedral nepheline crystals of up to 15 cm across can be observed (Figure 6a), which are virtually free of albite inclusions. The increase of the nepheline crystal size at the contact to the carbonatite dyke and the replacement of nepheline by sodalite may be attributed to the late influx of Na-rich carbonatitic fluids, triggering nepheline growth. In these zones, nepheline exhibits a pale blue tint, from its partial replacement by sodalite along grain margins and irregular cracks (Figure 6b).

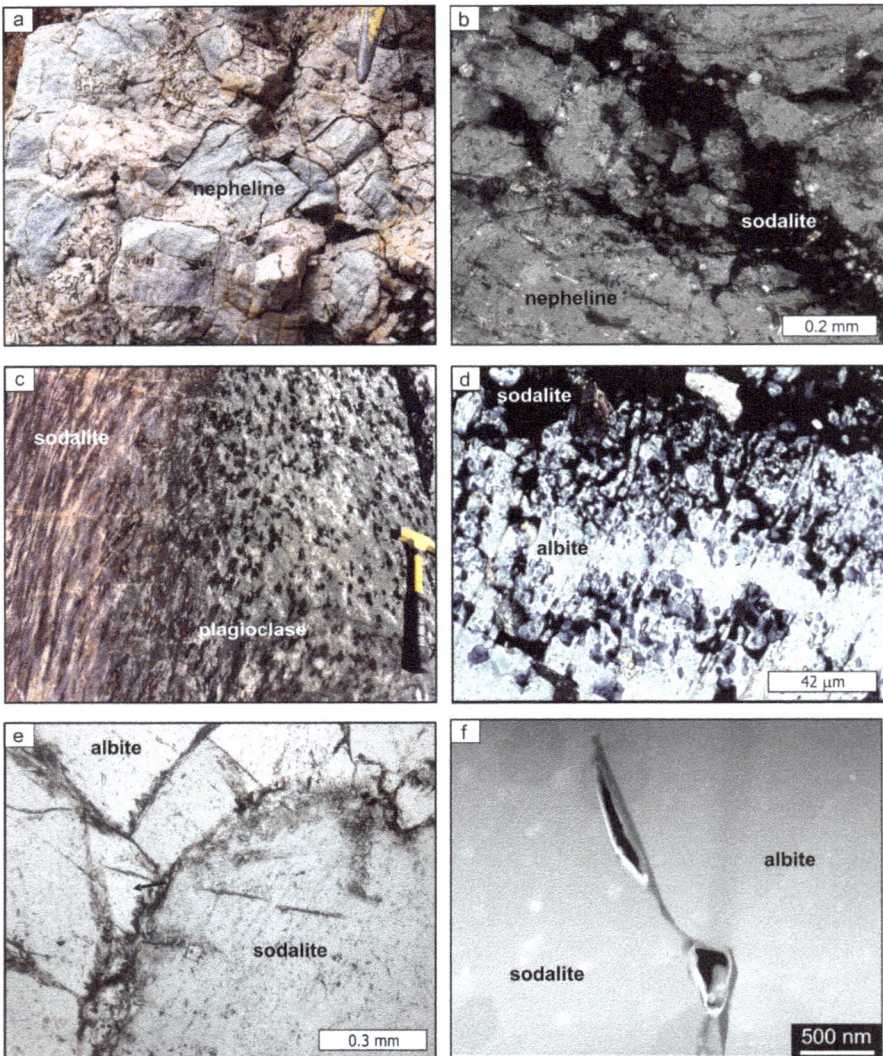

Figure 6. Natural examples of alkaline metasomatism in NW Namibia (**b**,**d**,**e**: microphotographs; **f**: TEM image); (**a**) Syenite direct contact to the carbonatite dykes contains euhedral nepheline of up to 15 cm across; (**b**) The pale blue tint of nepheline in (**a**) results from its partial replacement by sodalite along grain margins and irregular cracks; (**c**) Alkali-metasomatism of anorthosites, Namibia, is characterized by the replacement of plagioclase by sodalite; (**d**) Sodalite replaces albite along ist margins and the polysynthetic twin lamellae (sodalite isotropic); (**e**) Metasomatic sodalite of the natural rock samples from Namibia contains numerous fluid inclusions; (**f**) TEM-foil of the phase boundaries between albite and sodalite: an open channel is developed.

Sodalitization in the fenites starts along the grain margins of plagioclase and, if present, along cracks or albite twin lamellae (Figure 6c,d). The reaction fronts then progressively enter the plagioclase cores as embayed reaction zones. In most cases, replacement of plagioclase by sodalite went to completion. The sodalite displays only minor porosity but fluid and secondary mineral inclusions, pointing to coarsening of a previously present microporosity (Figure 6e). The reaction interface between sodalite and albite, analyzed with TEM, is sharp and marked by an open channel (Figure 6f). Amorphous material was not observed but might have previously been present and then reacted to solid phases during long-term re-equilibration of the natural rock samples, in accordance with the interpretation of the amorphous material of our experiments as a metastable interstage product.

The major and trace element compositions of natural sodalite and plagioclase have been investigated with the aid of an electron microprobe and LA-ICPMS. Both sodalite and plagioclase are the almost pure sodium endmembers, generally displaying Fe_2O_3, CaO, K_2O, S and F contents <0.1 wt %. The REE contents of sodalite, determined by LA-ICPMS, are generally low (ΣREE: 2.5–3.3 ppm), whereas remarkably high values of As (422–657 ppm), Ti (38–57 ppm), Sn (14–51 ppm) and Ba (57–92 ppm) have been recorded. The Rb and Sr contents of sodalite are in the range of 9–13 ppm and 10–14 ppm, respectively. Regarding their REE patterns, all analyzed sodalites are characterized by an enrichment of the light REE (La_{cn}/Yb_{cn}: 4.6–12.1) and flattening in the heavy REE. All REE-patterns display a strong positive Eu anomaly (Eu_{cn}/Eu_{cn}^*: 1.5–3.3) although the CaO contents of sodalite are close to or below the detection limit. These REE patterns are similar to those obtained for plagioclase of the anorthosites, indicating that sodalite inherited its trace element budget from plagioclase.

4.4. Implications for the Reaction Process

In general, the observed reaction textures in the model system indicate an interfacial dissolution-precipitation mechanism, with the shape of the parent albite being preserved. In all experiments, cracks as well as grain and interphase boundaries act as preferred pathways for the aqueous solutions. The Si-rich and Na–Al-poor composition of the amorphous layer, developed as an intermediate metastable phase between the unaltered albite and the precipitating sodalite or nepheline, indicates that Al and Na went into the bulk solution during the early stage of the experiment. Since both sodalite and nepheline are subsequently nucleating and growing within the amorphous layer, aluminum has to be transported from the bulk fluid phase to the amorphous layer demonstrating that part of the γ-Al_2O_3 is dissolved and mobilized. The high porosity of the metastable amorphous material as well as the newly formed sodalite and nepheline allow for continuous transport of aqueous chemical reactants and products between the crystalline reaction front (release of Si, Al, Na) and the bulk fluid (H_2O, Na, Cl) (Figure 7). In case of sodalite, even fluid inclusions are frequent, providing evidence for excellent pathways for material transport via the fluid phase. In addition, the presence of open channels at the boundary between albite and the amorphous layer, containing precipitates of the circulating fluid, strongly suggest that the fluid permeated both the sodalite rims and the thick altered layer. The observed variable major element composition of the amorphous layer of different experiments is presumably mainly influenced by differences in the pH values of the bulk fluid, which underwent dramatic changes during the onset of sodalite crystallization (the composition of the educt phases, i.e., albite and γ-Al_2O_3, is almost similar with respect to the major element composition in all experiments). The formation of an amorphous layer as interstage product during the reaction process can moreover explain the observed behavior of the trace elements (i.e., inheritance of trace element patterns) during metasomatism. In situ LA-ICPMS analyses of the trace element contents of the educt albite and the product phases sodalite, nepheline and the amorphous phase indicate, that the alkaline metasomatism does not result in a complete equilibration and exchange between the solid phases involved in the reaction. Only minor compositional heterogeneities exist on the grain and sub-grain scale. The trace element budget of the product sodalite, nepheline and the amorphous phase is apparently inherited from the respective precursor albite. Due to its high viscosity when compared to the aqueous fluid, the amorphous material, containing the bulk trace element budget of the ancestor albite, remained fixed in position and underwent only minor exchange of trace elements with the circulating fluid. Following this, the composition of the metasomatic product phases provides direct information about the precursor mineral and allows us to deduce the corresponding metasomatic reaction. Another possible explanation for the apparent lack of fractionation of any trace element during replacement may be the similarity in crystal structure of the reactants and products.

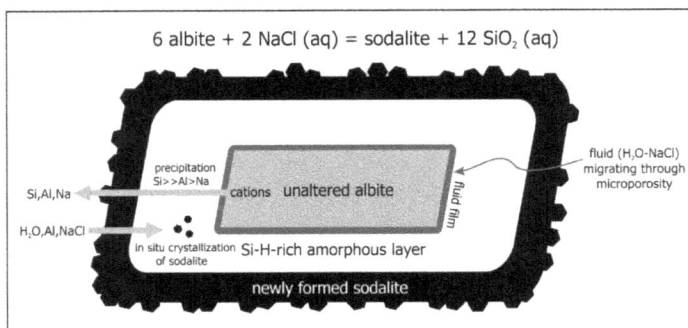

Figure 7. Model of transport mechanism for the hydrothermally driven reaction (Equation (1)): 6 albite + 2NaCl (aq) = sodalite + 12SiO₂ (aq).

The coupled interfacial dissolution–reprecipitation mechanism we observed, with amorphous material as a metastable interstage product, may be a more universal phenomenon. Amorphous altered layers are already described from a number of environments, i.e., among others, chemical weathering, radioactive waste storage, biomineralization, electrochemistry, and medical sciences, mostly related to low-T processes. Taken into account the high-T stability of the amorphous phase (up to 750 °C) present in the early stages of the reaction process in our experiments this process may be a more universal process. Accordingly, results of our experiments bear important implications with respect to mineral replacement in the presence of a fluid in many geological environments, especially regarding the interpretation of trace element patterns and isotopic ratios of the product phases.

Author Contributions: K.D. conceived, designed, and performed the experiments and analyzed the XRD, EMP, and SEM data; R.W. prepared the TEM foils and performed and interpreted the TEM analyses; K.D. wrote the paper.

Acknowledgments: We wish to thank Christa Zecha (TU Berlin) for careful sample preparation. The assistance of Francois Galbert (TU Berlin) during EMP analysis is greatly appreciated. We are grateful to Helene Brätz and Reiner Klemd (University of Erlangen) for their help and advice with the LA-ICP-MS measurements. We wish to thank Ulrich Gernert (TU Berlin) for his assistance during SEM work, and for his ideas to improve our analytical approach. The authors would like to thank the anonymous reviewers for their valuable comments and suggestions to improve the quality of the paper. Our work was sponsored by the Deutsche Forschungsgemeinschaft (grant DR 744/3-1; DFG-Research Unit FOR 741 "Nanoscale Processes and Geomaterials Properties"), which is gratefully acknowledged. We acknowledge support by Deutsche Forschungsgemeinschaft and Open Access Publishing Fund of Karlsruhe Institute of Technology.

Conflicts of Interest: The authors declare no conflict of interest.

Appendix A

Table A1. Representative LA-ICPMS analyses of albite.

Sample	LT-05	LT-05	LT-05	LT-05	LT-04	LT-04	LT-04	LT-04	LT-04	HT-10	HT-10	33	34
	Swarth.									Lovozero		Lomnitz	
	ppm												
Li	7.79	5.28	<0.38	0.849	<0.37	<0.39	2.55	0.670	1.33	<0.38	<0.39	1.47	<0.37
Be	1.14	<0.58	0.981	<0.76	<0.93	<0.82	<0.74	<0.89	0.619	25.6	30.9	2.39	1.20
Ca	7169	2911	7174	7323	8763	8673	929	9019	8564	12.5	9.85	28.9	24.7
Sc	0.911	1.14	0.844	1.05	0.998	1.10	0.671	0.752	0.660	0.549	0.770	0.585	0.396
Ti	395	358	393	481	661	278	30.7	451	332	10.1	13.6	<4.7	<2.7
V	0.235	1.97	0.224	0.375	0.961	<0.15	0.124	0.204	<0.11	<0.12	<0.15	<0.23	<0.14
Cr	<2.1	<1.7	<2.4	<2.1	<2.2	<2.5	2.45	<2.3	<1.9	<2.9	<2.5	6.83	<2.1
Mn	19.2	13.2	25.7	42.6	45.8	11.2	2.21	22.5	21.7	10.5	17.3	3.84	<0.54
Co	0.309	0.102	0.862	0.543	0.761	0.188	<0.12	0.274	0.203	0.349	0.391	<0.22	<0.12
Ni	0.480	<0.35	1.05	1.85	1.09	<0.49	1.88	0.970	0.669	3.04	<0.62	<0.89	<0.43
Cu	3.13	5.15	3.65	3.49	3.34	3.73	13.1	3.41	3.60	13.8	6.92	14.8	5.50
Zn	4.89	<2.3	5.17	5.52	8.16	4.56	<2.8	6.13	4.71	13.7	<3.5	<4.9	<2.9
Ga	35.7	17.0	34.8	33.3	34.4	34.0	5.70	40.0	32.9	84.7	83.2	20.2	25.8
Rb	6.21	3.77	1.35	2.75	2.58	1.61	0.154	2.65	4.29	2.14	3.37	0.367	<0.06
Sr	1055	597	1054	1099	967	1007	101	1168	1168	5.97	6.79	7.61	18.5
Y	0.137	0.334	0.194	0.485	0.324	0.275	0.042	0.343	0.185	0.937	1.36	0.204	<0.03
Zr	<0.07	<0.06	<0.08	0.157	<0.08	<0.08	0.321	<0.10	<0.07	12.1	15.0	0.458	<0.07
Nb	<0.04	<0.03	<0.05	0.047	<0.04	<0.05	<0.03	<0.05	<0.03	1.11	1.31	0.625	<0.03
Sn	<0.37	0.466	<0.39	0.450	<0.39	0.432	<0.36	<0.43	0.617	<0.38	<0.41	<0.62	<0.37
Cs	<0.03	<0.03	<0.03	<0.03	<0.03	<0.05	<0.03	<0.03	<0.03	<0.04	<0.03	<0.07	<0.04
Ba	1043	311	711	581	776	641	21.8	1284	679	9.66	10.5	11.1	0.472
La	3.11	0.498	3.65	3.53	4.00	3.48	0.525	4.87	4.36	1.37	2.46	0.053	<0.03
Ce	5.28	0.980	6.03	6.45	6.88	6.99	0.926	7.95	7.43	3.14	5.14	0.144	<0.03
Pr	0.506	0.125	0.637	0.551	0.761	0.802	0.100	0.780	0.731	0.231	0.472	<0.05	<0.03
Nd	2.13	0.731	2.59	2.28	2.61	2.82	0.300	3.15	2.82	0.775	1.89	<0.34	<0.15
Sm	0.285	0.204	0.248	<0.23	0.489	0.310	<0.23	<0.27	0.207	<0.23	0.575	<0.41	<0.21
Eu	1.51	0.395	1.62	1.43	1.49	1.67	0.146	1.82	1.48	0.082	0.143	<0.11	<0.05
Gd	<0.21	<0.18	0.251	<0.19	0.319	0.261	<0.18	<0.31	<0.19	<0.23	<0.25	<0.05	<0.22
Tb	<0.03	0.036	<0.04	0.037	<0.04	<0.04	<0.03	<0.04	<0.03	0.043	<0.03	<0.05	<0.03
Dy	<0.15	<0.09	<0.15	0.215	<0.13	<0.19	<0.14	<0.17	<0.13	<0.17	0.383	<0.33	<0.13
Ho	<0.03	<0.03	<0.03	<0.03	<0.04	<0.04	<0.04	<0.04	<0.03	<0.04	0.040	<0.06	<0.03
Er	<0.09	<0.09	<0.11	<0.12	<0.11	<0.11	<0.11	<0.13	<0.09	<0.12	<0.11	<0.18	<0.11
Tm	<0.03	<0.03	<0.04	<0.03	<0.04	<0.04	<0.04	<0.05	<0.03	<0.04	<0.03	0.053	<0.03
Yb	<0.17	<0.14	<0.16	<0.19	<0.18	<0.21	<0.18	<0.21	<0.17	<0.19	<0.19	<0.28	<0.16
Lu	<0.04	<0.03	<0.04	<0.03	<0.03	<0.05	<0.03	<0.04	<0.03	<0.05	0.049	<0.07	<0.03
Hf	<0.11	<0.11	<0.11	<0.13	<0.14	<0.14	<0.13	<0.16	<0.11	<0.14	<0.14	<0.24	<0.13
Ta	<0.02	<0.03	<0.03	<0.03	<0.03	<0.04	0.075	<0.04	<0.03	0.091	0.137	0.145	<0.03
Pb	3.61	1.95	4.00	4.68	3.48	2.84	3.96	5.94	5.86	1.02	0.266	1.21	<0.09
Th	<0.06	<0.04	<0.05	<0.05	<0.06	<0.06	<0.05	<0.06	<0.06	0.976	1.65	<0.09	<0.04
U	<0.04	<0.04	<0.05	<0.05	<0.05	<0.06	<0.05	<0.06	<0.04	0.255	0.238	<0.09	<0.05

Table A2. Representative LA-ICPMS analyses of sodalite, nepheline, and the amorphous phase.

Sample	LT-05	LT-05	LT-05	LT-04	LT-04	LT-04	HT-08	HT-08	HT-08	HT-10	HT-10	33	34	34
	Swartbooisdrif							Lovozero					Lomnitz	
	Sdl	Sdl	Sdl	Sdl	Sdl	Sdl	Sdl	am	am	Sdl	Sdl	Sdl	Sdl	Nph
						ppm								
Li	<0.32	0.867	0.989	2.55	0.594	1.20	31.4	2.25	2.31	<0.48	2.10	0.834	0.720	0.972
Be	<0.69	<0.58	<1.2	<0.74	<0.55	<0.66	2.34	-0.21	-0.23	<0.88	<0.38	3.99	<0.82	<1.1
Ca	87.6	513	2535	929	1122	310	121	2324	2364	19.4	2466	26.1	11.6	67.6
Sc	0.385	0.756	0.960	0.671	0.462	0.391	0.327	0.501	0.442	0.592	0.449	0.454	0.521	0.477
Ti	7.82	111	130	30.7	4.44	5.81	40.8	66.0	71.5	<4.2	74.1	<5.4	<4.5	<3.9
V	<0.12	<0.15	0.392	0.124	<0.11	<0.11	1.78	0.389	0.415	0.274	0.483	0.344	0.232	<0.16
Cr	<2.1	<2.4	<2.9	2.45	<1.6	<1.7	<4.1	<0.68	<0.65	<2.8	1.13	<3.4	14.5	<2.9
Mn	0.464	2.57	7.05	2.21	1.23	<0.36	28.0	6.91	7.18	2.14	6.74	4.26	<0.87	<0.84
Co	<0.11	<0.13	0.514	<0.12	<0.07	0.162	0.549	0.064	0.106	<0.06	<0.06	<0.23	<0.12	<0.18
Ni	<0.46	0.607	6.34	1.88	0.649	<0.46	9.21	0.365	0.307	4.42	1.04	0.943	0.763	<0.63
Cu	16.0	15.1	16.4	13.1	8.43	12.2	59.4	4.57	4.60	25.4	7.27	10.4	13.6	15.2
Zn	<2.9	<2.9	5.93	<2.8	2.40	<2.6	77.1	5.20	4.01	<3.9	8.81	9.58	<3.2	<3.8
Ga	5.15	5.85	15.0	5.70	9.42	3.60	9.80	1.96	1.93	3.05	1.67	26.6	4.26	7.89
Rb	<0.06	<0.07	0.690	0.154	0.080	<0.06	2.44	16.9	16.8	2.12	16.3	0.396	0.164	1.20
Sr	6.20	57.4	411	101	174	18.9	10.6	19.2	19.4	1.23	20.9	5.11	0.260	29.7
Y	<0.04	0.208	0.249	0.042	0.062	0.040	0.248	0.718	0.736	<0.04	0.871	0.610	0.378	0.056
Zr	<0.07	0.187	<0.12	0.321	<0.06	<0.07	3.14	52.0	51.1	0.341	60.6	1.77	1.16	0.158
Nb	0.098	3.03	0.523	3.88	<0.03	<0.04	7.74	0.219	0.244	0.191	0.190	1.72	11.4	0.122
Sn	1.05	0.457	<0.54	<0.36	0.340	<0.34	2.63	0.270	0.196	0.487	0.453	0.682	<0.39	<0.49
Cs	<0.05	<0.03	<0.06	<0.03	<0.03	<0.03	<0.08	0.164	0.193	<0.04	0.175	0.056	<0.04	<0.06
Ba	8.15	10.4	145	21.8	64.3	2.95	24.1	116	120	1.58	130	15.3	1.21	94.2
La	0.123	0.906	15.1	0.525	0.806	0.088	0.749	0.972	0.907	0.072	1.07	0.066	<0.04	<0.04
Ce	0.060	1.04	4.31	0.926	0.712	0.070	1.05	1.66	1.64	0.129	1.80	0.294	<0.04	<0.04
Pr	<0.18	0.103	0.449	0.100	0.070	<0.02	0.143	0.179	0.189	<0.03	0.159	<0.05	<0.03	<0.03
Nd	<0.18	0.466	1.5	0.300	0.273	<0.14	0.663	0.636	0.656	<0.24	0.636	<0.28	<0.19	<0.26
Sm	<0.23	<0.25	<0.35	<0.23	<0.21	<0.22	0.139	0.138	0.109	<0.27	0.256	<0.32	<0.22	<0.35
Eu	<0.06	0.251	0.214	0.146	0.057	<0.05	<0.14	0.090	0.095	<0.08	0.035	<0.08	<0.06	<0.09
Gd	<0.19	0.736	1.39	<0.18	<0.16	<0.21	<0.39	0.127	0.102	<0.28	0.099	<0.29	<0.23	<0.26
Tb	<0.03	<0.21	<0.24	<0.03	<0.02	<0.03	<0.06	0.016	0.022	<0.04	0.014	<0.05	<0.03	<0.04
Dy	<0.15	<0.03	<0.04	<0.14	<0.11	<0.14	<0.22	0.121	0.080	<0.18	0.158	<0.21	<0.15	<0.16
Ho	<0.03	<0.16	<0.05	<0.04	<0.03	<0.03	<0.08	0.021	0.034	<0.03	0.030	<0.04	<0.03	<0.05
Er	<0.11	<0.04	<0.05	<0.11	<0.08	<0.08	<0.19	0.092	0.065	<0.13	0.082	<0.17	<0.12	<0.14
Tm	<0.03	<0.09	<0.17	<0.04	<0.03	<0.03	<0.07	-0.01	0.017	<0.04	0.016	<0.05	<0.04	<0.04
Yb	<0.16	<0.18	<0.22	<0.18	<0.13	<0.17	<0.29	0.116	<0.06	<0.24	0.106	<0.24	<0.16	<0.22
Lu	<0.05	<0.03	<0.05	<0.03	<0.03	<0.03	<0.08	<0.01	0.013	<0.04	<0.02	<0.05	<0.03	<0.05
Hf	<0.12	<0.12	<0.19	<0.13	<0.09	<0.12	<0.27	1.22	1.36	<0.18	1.65	<0.21	<0.15	<0.17
Ta	<0.03	1.23	2.21	0.075	<0.02	<0.03	0.098	0.013	0.027	<0.04	<0.02	0.198	0.193	<0.03
Pb	5.75	3.52	56.2	3.96	5.25	1.10	17.8	3.78	3.82	2.94	4.81	1.55	0.846	1.76
Th	<0.05	<0.06	0.158	<0.05	<0.04	<0.05	<0.11	0.230	0.230	<0.07	0.325	<0.07	0.085	<0.07
U	<0.05	<0.04	0.155	<0.05	<0.04	<0.04	<0.11	0.155	0.145	<0.06	0.157	<0.07	<0.05	<0.07

(am = amorphous phase; Nph = nepheline; Sdl = sodalite).

References

1. Merino, E.; Wang, Y.; Wang, Y.; Nahon, D. Implications of pseudomorphic replacement for reaction–transport modelling in rocks. *Mineral. Mag.* **1994**, *58*, 599–601. [CrossRef]
2. Merino, E.; Dewers, T. Implications of replacement for reaction–transport modelling. *J. Hydrol.* **1998**, *209*, 137–146. [CrossRef]
3. Cesare, B. Multi-stage pseudomorphic replacement of garnet during polymetamorphism: 1. Microstructures and their interpretation. *J. Metamorph. Geol.* **1999**, *17*, 723–734. [CrossRef]
4. Putnis, A. Mineral replacement reactions: From macroscopic observations to microscopic mechanisms. *Mineral. Mag.* **2002**, *66*, 689–708. [CrossRef]
5. Putnis, C.V.; Mezger, K. A mechanism of mineral replacement: Isotope tracing in the model system KCl–KBr–H$_2$O. *Geochim. Cosmochim. Acta* **2004**, *68*, 2839–2848. [CrossRef]
6. Putnis, A. Mineral replacement reactions. *Rev. Mineral. Geochem.* **2009**, *70*, 87–124. [CrossRef]
7. Walker, F.D.L.; Lee, M.R.; Parsons, I. Micropores and micropermeable texture in alkali feldspars: Geochemical and geophysical implications. *Mineral. Mag.* **1995**, *59*, 507–536. [CrossRef]
8. Putnis, A.; Mauthe, G. The effect of pore size on cementation in porous rocks. *Geofluids* **2001**, *1*, 37–41. [CrossRef]
9. Finch, A.A. Conversion of nepheline to sodalite during subsolidus processes in alkaline rocks. *Mineral. Mag.* **1991**, *55*, 459–463. [CrossRef]
10. Sharp, Z.D.; Helffrich, G.R.; Bohlen, S.R.; Essene, E.J. The stability of sodalite in the system NaA1SiO$_4$-NaCl. *Geochim. Cosmochim. Acta* **1989**, *53*, 1943–1954. [CrossRef]
11. Balassone, G.; Bellatreccia, F.; Mormone, A.; Biagioni, C.; Pasero, M.; Petti, C.; Mondillo, N.; Fameli, G. Sodalite-group minerals from the Somma-Vesuvius volcanic complex, Italy: A case study of K-feldspar-rich xenoliths. *Mineral. Mag.* **2012**, *76*, 191–212. [CrossRef]
12. Upadhyay, D. Alteration of plagioclase to nepheline in the Khariar alkaline complex, SE India: Constraints on metasomatic replacement reaction mechanisms. *Lithos* **2012**, *155*, 19–29. [CrossRef]
13. Dumańska-Słowik, M.; Heflik, W.; Pieczka, A.; Sikorska, M.; Dąbrowa, Ł. The transformation of nepheline and albite into sodalite in pegmatitic mariupolite of the Oktiabrski Massif (SE Ukraine). *Spectrochim. Acta Part A Mol. Biomol. Spectrosc.* **2015**, *150*, 837–845. [CrossRef] [PubMed]
14. Kotelnikov, A.; Zhornyak, L. Stability of sodalite under hydrothermal conditions. *Geochem. Int.* **1995**, *32*, 87–90.
15. Pearce, N.J.G.; Perkins, W.T.; Westgate, J.A.; Gorton, M.P.; Jackson, S.E.; Neal, C.R.; Chenery, S.P. A compilation of new and published major and trace element data for NIST SRM 610 and NIST SRM 612 glass reference materials. *Geostand. Newslett. J. Geostand. Geoanal.* **2007**, *21*, 115–144. [CrossRef]
16. Wirth, R. Focused Ion Beam (FIB): A novel technology for advanced application of micro– and nanoanalysis in geosciences and applied mineralogy. *Eur. J. Mineral.* **2004**, *16*, 863–877. [CrossRef]
17. Wirth, R. Focused Ion Beam (FIB) combined with SEM and TEM: Advanced analytical tools for studies of chemical composition, microstructure and crystal structure in geomaterials on a nanometre scale. *Chem. Geol.* **2009**, *261*, 217–229. [CrossRef]
18. Drüppel, K.; Hoefs, J.; Okrusch, M. Fenitizing processes induced by ferrocarbonatite magmatism at Swartbooisdrif, NW Namibia. *J. Petrol.* **2005**, *46*, 377–406. [CrossRef]
19. Drüppel, K. Petrogenesis of the Mesoproterozoic Anorthosite, Syenite and Carbonatite Suites of NW Namibia and Their Contribution to the Metasomatic Formation of the Swartbooisdrif Sodalite Deposits. Ph.D. Thesis, University of Würzburg, Würzburg, Germany, 2003.
20. Nesbitt, H.W.; Muir, I.J. SIMS depth profiles of weathered plagioclase and processes affecting dissolved Al and Si in some acidic soil solutions. *Nature* **1988**, *334*, 336–338. [CrossRef]
21. Casey, W.H.; Westrich, H.R.; Arnold, G.W.; Banfield, J.E. The surface chemistry of dissolving labradorite feldspar. *Geochim. Cosmochim. Acta* **1989**, *53*, 821–832. [CrossRef]
22. Hellmann, R.; Dran, J.-C.; Della Mea, G. The albite–water system Part III. Characterization of leached and hydrogen–enriched layers formed at 300 °C using MeV ion beam techniques. *Geochim. Cosmochim. Acta* **1997**, *61*, 1575–1594. [CrossRef]
23. Schweda, P.; Sjöberg, L.; Södervall, U. Near–surface composition of acid–leached labradorite investigated by SIMS. *Geochim. Cosmochim. Acta* **1997**, *61*, 1985–1994. [CrossRef]

24. Hellmann, R.; Penisson, J.M.; Hervig, R.L.; Thomassin, J.H.; Abrioux, M.F. An EFTEM/HRTEM high–resolution study of the near surface of labradorite feldspar altered at acid pH: Evidence for interfacial dissolution–reprecipitation. *Phys. Chem. Miner.* **2003**, *30*, 192–197. [CrossRef]

25. Weissbart, E.J.; Rimstidt, J.D. Wollastonite incongruent dissolution and leached layer formation. *Geochim. Cosmochim. Acta* **2000**, *64*, 4007–4016. [CrossRef]

26. Labotka, T.C.; Cole, D.R.; Fayek, M.; Riciputi, L.R.; Stadermann, F.J. Coupled cation and oxygen–isotope exchange between alkali feldspar and aqueous chloride solution. *Am. Mineral.* **2004**, *89*, 1822–1825. [CrossRef]

27. Hellmann, R.; Wirth, R.; Daval, D.; Barnes, J.-P.; Penisson, J.-M.; Tisserand, D.; Epicier, T.; Florin, B.; Hervig, R.L. Unifying natural and laboratory chemical weathering with interfacial dissolution–reprecipitation: A study based on nanometer–scale chemistry of fluid–silicate interfaces. *Chem. Geol.* **2012**, *294–295*, 203–216. [CrossRef]

© 2018 by the authors. Licensee MDPI, Basel, Switzerland. This article is an open access article distributed under the terms and conditions of the Creative Commons Attribution (CC BY) license (http://creativecommons.org/licenses/by/4.0/).

minerals

MDPI

Review

Mineral Transformations in Gold–(Silver) Tellurides in the Presence of Fluids: Nature and Experiment

Jing Zhao * and Allan Pring

Chemical and Physical Sciences, College of Science and Engineering, Flinders University, Bedford Park, Adelaide, SA 5042, Australia; allan.pring@flinders.edu.au
* Correspondence: jing.zhao@flinders.edu.au

Received: 16 January 2019; Accepted: 4 March 2019; Published: 9 March 2019

check for updates

Abstract: Gold–(silver) telluride minerals constitute a major part of the gold endowment at a number of important deposits across the globe. A brief overview of the chemistry and structure of the main gold and silver telluride minerals is presented, focusing on the relationships between calaverite, krennerite, and sylvanite, which have overlapping compositions. These three minerals are replaced by gold–silver alloys when subjected to the actions of hydrothermal fluids under mild hydrothermal conditions ($\leq$220 °C). An overview of the product textures, reaction mechanisms, and kinetics of the oxidative leaching of tellurium from gold–(silver) tellurides is presented. For calaverite and krennerite, the replacement reactions are relatively simple interface-coupled dissolution-reprecipitation reactions. In these reactions, the telluride minerals dissolve at the reaction interface and gold immediately precipitates and grows as gold filaments; the tellurium is oxidized to Te(IV) and is lost to the bulk solution. The replacement of sylvanite is more complex and involves two competing pathways leading to either a gold spongy alloy or a mixture of calaverite, hessite, and petzite. This work highlights the substantial progress that has been made in recent years towards understanding the mineralization processes of natural gold–(silver) telluride minerals and mustard gold under hydrothermal conditions. The results of these studies have potential implications for the industrial treatment of gold-bearing telluride minerals.

Keywords: gold–(silver) tellurides; natural porous gold; interface-coupled dissolution–reprecipitation; hydrothermal method; calaverite; krennerite; sylvanite

---

1. Introduction

Gold–(silver) tellurides are important accessory minerals, carrying a significant proportion of the gold endowment in some low to medium temperature hydrothermal vein deposits. Gold–(silver) telluride minerals have become one of the most important sources of gold in the world. The Golden Mile deposit in Kalgoorlie, Western Australia, has been an economically important gold–(silver) telluride deposit for over a century; it contained approximately 1450 tons gold, of which approximately 20% was in the form of tellurides [1]. Other notable modern and historic gold deposits carrying significant amounts of the gold as tellurides include Cripple Creek, Colorado (~875 tons gold) [2]; Emperor, Fiji (~360 tons of gold, 10–50% occurring as tellurides) [3,4]; and Săcărîmb, Romania [5]. Another important example is the recently discovered Sandaowanzi gold deposit on the northeastern edge of the Great Xing'an Range, Heilongjiang Province, North East China, with a total reserve of $\geq$25 tons of gold and an average grade of 15 g/t [6–9]. We believe that this is the first case of a major gold deposit in which the gold telluride minerals are the dominant ore, with more than 95% of recovered gold occurring as tellurides.

Eight gold–(silver) tellurides have been described and are currently recognized as valid minerals: calaverite, krennerite, sylvanite, petzite, muthmannite, empressite, hessite, and stuetzite. A summary of the characteristics and physical properties of the main gold (and/or silver) telluride minerals is presented in Table 1 and the compositions of these minerals are shown in Figure 1. The gold-rich telluride species—calaverite, krennerite and sylvanite—are the most common and economically important minerals of the group, with a chemical composition of $Au_{1-x}Ag_xTe_2$. Cabri [10] gave the following compositional fields for these minerals: Calaverite 0 to 2.8 wt % Ag ($0 \leq x \leq 0.11$); krennerite 3.4 to 6.2 wt % Ag ($0.14 \leq x \leq 0.25$); and sylvanite 6.7 to 13.2 wt % Ag ($0.27 \leq x \leq 0.50$). A more recent work by Bindi et al. [11] showed that calaverite and sylvanite can have overlapping compositional fields, and share a similar layered structural topology (as shown in Figure 2). The Ag content of calaverite, sylvanite, and krennerite has been linked to its substitution for Au and stabilization of the complex modulated structures adopted by these minerals [11,12]. The incommensurately modulated structure of calaverite was determined by Bindi et al. [11] and its modulations are related to the distribution of Au^{3+} and Au^+ and the substitution of Ag^+ for Au^+. In krennerite, Ag and Au are ordered to avoid Ag–Te–Ag linkages [12]. Sylvanite occurs in two forms, one is a commensurately modulated superstructure based on the calaverite sub-cell and the other is an incommensurately modulated form [13]. On a historical note, calaverite was the first mineral, or compound, to be recognized to have an incommensurately modulated structure. It was identified by morphological crystallographers in 1901 as their attempts to index crystal faces required a model which had intergrowing lattices [14]. The other five telluride minerals listed in Table 1 are much less important in gold production and four of them contain more silver than gold.

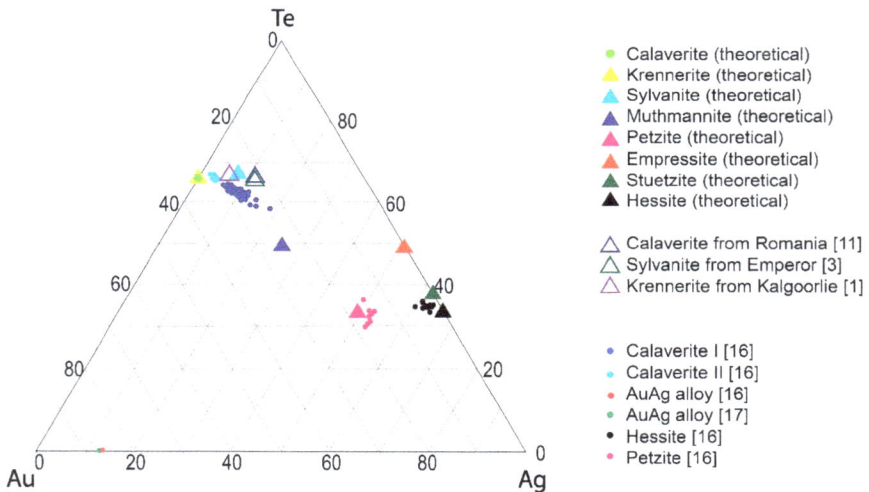

Figure 1. Ternary diagrams of Au–(Ag)–Te system (atom%), showing compositions of gold–(silver) tellurides from mineral database [15] and references [1,3,8,11,16,17]. Compositions of synthetic gold–(silver) tellurides [16,17] are shown as small colored dots.

Gold–(silver) telluride minerals in gold deposits are considered refractory ores from a mineral processing perspective, as they are not efficiently leachable in cyanide solutions. Therefore, additional processing steps are required to improve gold recovery when tellurides are present in the ore (e.g., [18,19]). Fine grinding and pretreatments (normally roasting gold tellurides at temperatures ≥ 800 °C) are generally utilized to improve gold recovery. These methods are energy-intensive and raise environmental issues due to the release of Te species into the atmosphere. An alternative strategy for gold recovery from telluride ores is needed for deposits rich in these refractory gold ores.

(A) Sylvanite

(B) Calaverite

● Au   ● Ag   ●● Te

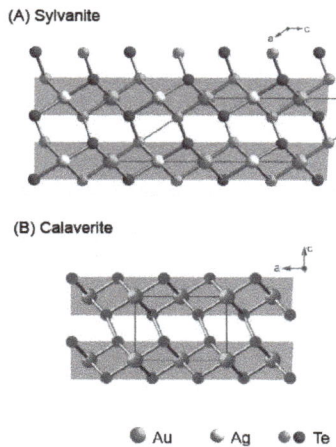

Figure 2. Projections of the crystal structures of sylvanite (**A**) and calaverite (**B**). Crystal structure data for the minerals are from references [11,20].

Table 1. Characteristics and physical properties of the main gold–(silver) tellurides.

Mineral	Chemical Formula	Color	Density (g/cm³)	Hardness	Composition wt % Au	Ag	Te
Calaverite	$AuTe_2$	Silver white to brassy yellow	9.04	2.5–3	43.6	0	56.4
Krennerite	$(Au_{1-x},Ag_x)Te_2$	Silver white to blackish yellow	8.53	2.5	43.6	0	56.4
Sylvanite	$AuAgTe_4$	Steely gray to silver gray	7.9–8.3 (8.1)	1.5–2	34.4	6.3	59.4
Muthmannite	$(Ag,Au)Te_2$	Blackish yellow, grayish white	-	2.5	34.3	19.2	46.5
Petzite	Ag_3AuTe_2	Bright steel gray to iron black	8.7–9.14	2.5	25.4	41.7	32.9
Empressite	$AgTe$	Bronze, light bronze	7.5–7.6	3.5	0	46.3	53.7
Stuetzite	$Ag_{5-x}Te_3$, (x = 0.24–0.36)	Gray, dark bronze	8	3.5	0	57.0	43.0
Hessite	Ag_2Te	Lead gray, steel gray	7.2–7.9	1.5–2	0	62.8	37.2

Note: Data is from [15].

2. Gold–(Silver) Tellurides in Nature and Their Alteration

The economic importance of gold–(silver) telluride minerals in gold deposits has meant that they have received significant attention from geologists and mineralogists. More than 100 occurrences have been reported worldwide. The International Geoscience Programme project IGCP-486 was undertaken from 2003 to 2008 and focused on the interplay between mineralogy and ore genesis of telluride minerals [4]. The project directly contributed to a summary of the distribution of gold–(silver) telluride-bearing deposits and a better understanding of the formation of these deposits. Gold–(silver) telluride deposits normally contain a dozen or more different telluride and selenide minerals and present complex ore textures. An example is seen in the ores of the recently discovered Sandaowanzi gold deposit, where sylvanite is the most abundant gold-bearing mineral and together with petzite and krennerite accounts for >60% of the total tellurides by volume [7,8]. The mixtures of gold–(silver) telluride minerals were explored in both vein ores and disseminated ores [7]. As shown in Figure 3, gold and krennerite coexist with petzite and stuetzite, or form symplectic intergrowths with sylvanite. The size of individual telluride grains at this deposit can be up to 3 cm in diameter. In these textures (Figure 3A), stuetzite is irregularly shaped and randomly distributed as patches within petzite symplectites. Native gold and krennerite occur in close association as a mineral pair and are often included within petzite–stuetzite symplectites. Native gold also occurs as isolated grains along intragranular cracks in the telluride grains. The various combinations of gold tellurides (Figure 3B) have been attributed to retrograde reactions [21], and Liu et al. [8] suggested the formation of telluride assemblages at Sandaowanzi is related to the breakdown of early telluride phases (e.g., γ-phase and χ-phase of Cabri [10]). In this deposit, isolated gold grains occur in a "bamboo shoot-like" morphology

(Figure 3C), with the filaments being 3 to 5 μm in diameter and 10 to 15 μm in length. Gold also occurs in irregular patches within cavities in gold tellurides (Figure 3D).

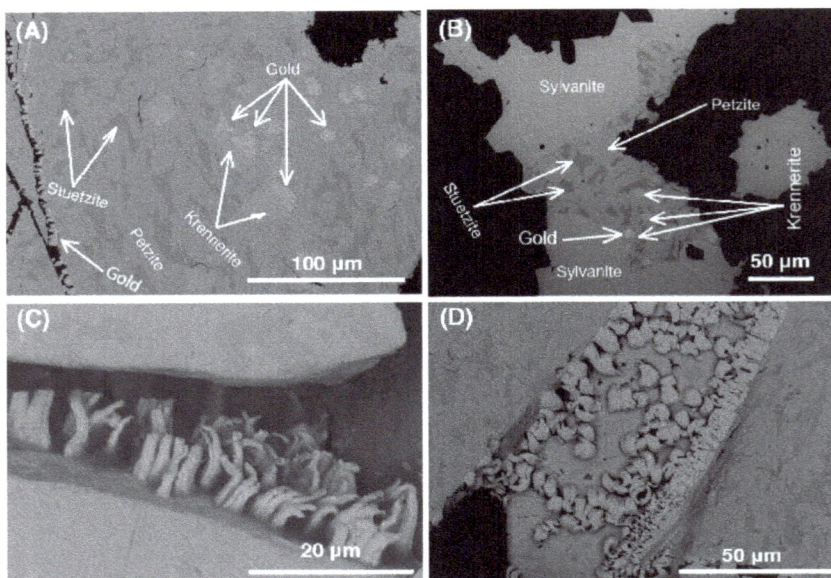

Figure 3. Mineralogy and microstructures of Au–(Ag) telluride ores at Sandaowanzi deposit [7,8]. (**A**) Native gold along intragranular cracks in the telluride grains and native gold–krennerite pair included in petzite–stuetzite symplectite. (**B**) Native gold and krennerite patches contained a petzite–stuetzite symplectite in association with sylvanite. (**C**) Bamboo shoot-like native gold grains along intragranular cracks in telluride grains. (**D**) Irregular shaped native gold grains within a cavity in gold tellurides.

The alteration of gold–(silver) tellurides to fine wires, or spongy gold, is well known [22] and the gold product is sometimes called "mustard gold" because of its distinctive appearance in reflected light (Figure 4) [23]. The formation of mustard gold at the Dongping Mines (Hebei Province, China) has been linked to the decomposition of calaverite by selective leaching of tellurium while leaving the gold alloy in the cavity formed by the alteration reaction [24,25]. This type of pseudomorphic alteration was also documented by Palache et al. [26]. The occurrence of microporous gold has also been observed under cold climatic conditions, such as at the Aginskoe low-sulfidation epithermal deposit in Central Kamchatka, Russia. In this deposit, calaverite is the main Au telluride mineral and it has been partially replaced by porous gold [27]. By comparing the textures of microporous gold from this natural occurrence with those obtained experimentally via the dealloying of gold–(silver) tellurides [16,17,28,29], Okrugin et al. [30] confirmed that natural microporous gold can form via the replacement of telluride minerals and assessed the role that hydrothermal fluids may play in the formation of microporous gold.

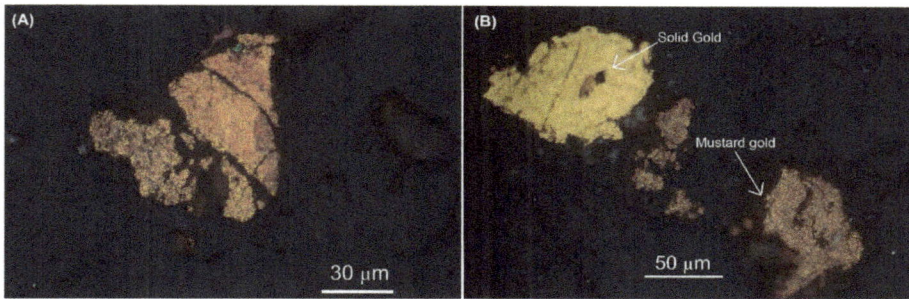

Figure 4. Native gold from the Gaching ore occurrence (Maletoyvayam ore field), Kamchatka, Russia (polished sections in reflected light) (imaged by N. Tolstykh). (**A**) Color of mustard gold is brown yellow to brown under reflected light. (**B**) Porous gold (mustard gold) is observed along with a homogeneous gold grain (solid gold).

3. Mineral Replacement Reactions of Gold–(Silver) Tellurides in the Presence of Fluids

There are limited reliable thermodynamic data for gold–(silver) tellurides due to the compositional overlap and structural complexity of the main mineral phases and therefore the difficulty in calculating meaningful phase diagrams that represent observed assemblages in Au–Ag–Te systems. Many studies on tellurium-bearing systems have focused only on the binary subsystems of Au–Te and Ag–Te. Since the 1960s, several experimental [10,31,32] and theoretical studies (e.g., [33]) have been conducted on the Au–Ag–Te system. Cabri [10] conducted a systematic investigation in the Au–Ag–Te ternary system, to determine the equilibrium phase relations in the mineralogically important area of the ternary system and phase changes in the assemblages over a range of temperatures. However, it should be noted that Cabri's study was performed using traditional dry sealed tube methods rather than under hydrothermal conditions. Zhang et al. [34] evaluated the stability of calaverite and hessite and discussed it in the context of the stability of other minerals in the Au–Ag–Te system. The calculated stability of hessite and calaverite were used to explain the physicochemical conditions of formation of the Gies and Golden Sunlight gold–(silver) telluride deposits in Montana, USA. Wang et al. [35] contributed new thermodynamic data for the Au–Te system, while McPhail [36] and Grundler et al. [37–39] studied the complexation and transport of tellurium in hydrothermal fluids.

The mineral replacement reactions of gold–(silver) tellurides in the presence of fluids have been explored in recent years. In a study of the kinetics and mechanism of mineral replacement reactions, Zhao et al. [28] investigated the replacement of calaverite by porous gold over a wide range of hydrothermal conditions. The transformation proceeds in a pseudomorphic manner via a coupled dissolution-reprecipitation (CDR) reaction mechanism. While the gold precipitates locally and preserves the shape of the original calaverite grain, the tellurium is selectively removed and lost to the bulk solution. Zhao et al. [16] further investigated the transformation of sylvanite to Au–Ag alloy by exploring the roles of temperature and fluid composition. The reaction follows a complex path, where CDR reactions interact with solid-state diffusion processes, and results in complex textures. This complexity is due to the fact that sylvanite has a higher Ag content, which results in the formation of a metastable Ag-rich, Te-depleted calaverite I phase. To achieve equilibrium, the metastable phase breaks down to stable calaverite II plus phase-χ. Phase-χ subsequently breaks down to hessite and petzite. To further investigate the effects of Ag in the parent crystal for the reaction path of Au–Ag tellurides during replacement, Xu et al. [17] designed a set of hydrothermal experiments using krennerite under similar conditions to those used by Zhao et al. [16,24]. The results show that krennerite transformed to Au–Ag alloy in a pseudomorphic manner very similar to calaverite and distinct from sylvanite. The reaction paths of these three reactions are summarized in Figure 5. In the next section we will review in detail these three comprehensive studies of the transformation of calaverite, krennerite, and sylvanite to gold–silver alloys by CDR reactions, focusing on the product textures, reaction mechanism, and the kinetics of the oxidative leaching of Te.

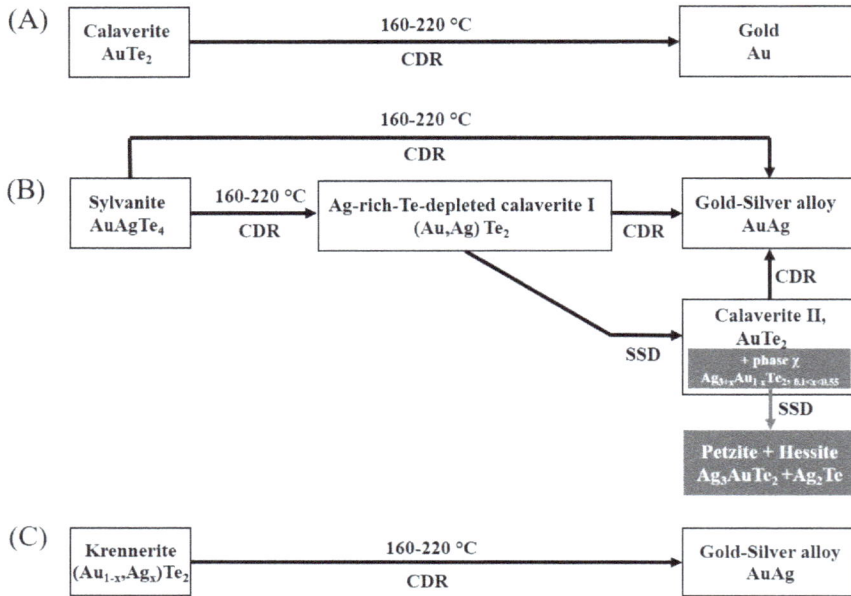

Figure 5. Overview of the proposed reaction paths of the hydrothermal reaction for calaverite (**A**), sylvanite (**B**) and krennerite (**C**). CDR stands for coupled dissolution reprecipitation and SSD stands for solid state diffusion.

3.1. Product Textures

When calaverite grains are heated in a series of 0.2 M buffer solutions (ranging from $pH_{25\,°C}$ 2 to 12) at 220 °C, Te is selectively removed from the calaverite, leaving a rim of porous gold (Figure 6A) [28]. The gold filaments produced grow perpendicular to the surface of calaverite. The gold filaments have diameters ranging from 200 to 500 nm, with lengths up to ~25 μm (Figure 6B). Texturally, they are randomly-oriented gold crystals, forming generally dendritic aggregates (Figure 6C). This texture is most likely due to repeated twinning on {111}, which is common in reticulated and dendritic gold aggregates [26]. The morphology of the gold sponge does not vary significantly with solution pH and temperature, but the extent of the reaction depends on the solubility of Te^{4+} in solution, and this is pH-dependent (Figure 7) [28,39]. The textural features of the replacement of calaverite by gold are consistent with a pseudomorphic replacement reaction proceeding via an interface-coupled dissolution reprecipitation (ICDR) process [40–42].

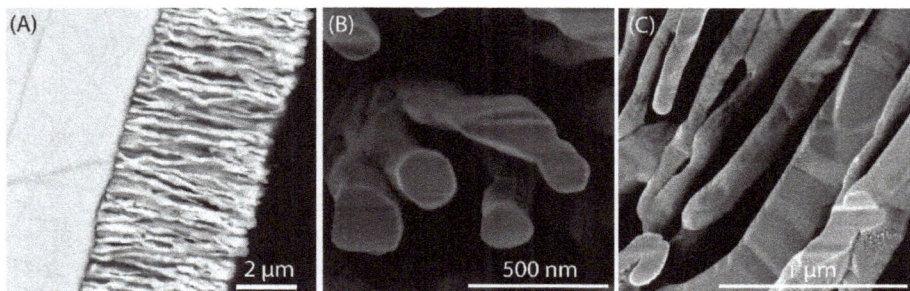

Figure 6. (**A**) Back-scattered electron image of cross section of partially-reacted calaverite showing the phase boundary between the porous gold product and the parent calaverite (solid grain). High-magnification images of gold showing three-dimensional structure of gold filaments, which were cut perpendicular (**B**) and parallel (**C**) to the long axis of the gold filaments.

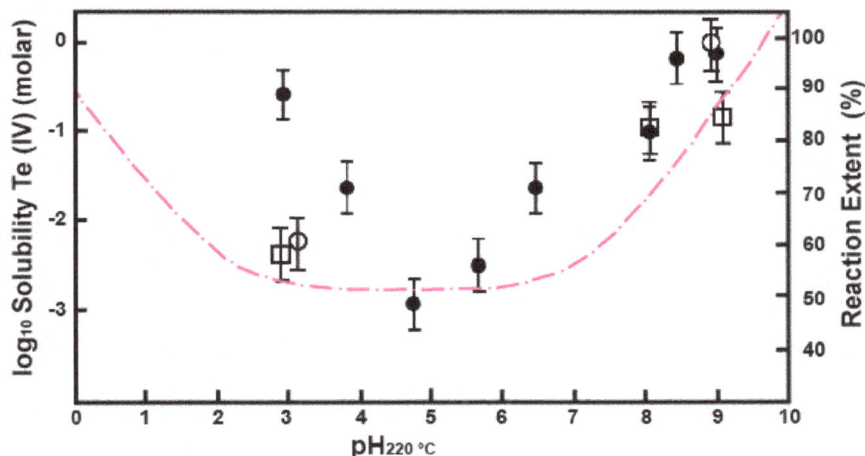

Figure 7. The curve of estimated solubility of Te(IV) in water at 220 °C is shown as pink dashed line (data from reference [28,38,39]). Solid circles stand for the reaction extent of the replacement of calaverite [28]. Hollow circles and squares stand for the replacement of krennerite [17] and sylvanite [16], respectively. Errors of the reaction extent (3 − σ; ± 6%) are plotted at each point. Reaction extent observed experimentally corresponded well to the solubility of tellurium.

The replacement of krennerite is similar to calaverite [17], proceeding via the ICDR reaction mechanism. An Au–Ag alloy of wormlike filaments was produced due to higher silver contents in krennerite (Figure 8). Natural krennerite normally contains 3.4 to 6.2 wt % Ag ($0.14 \leq x \leq 0.25$), compared to calaverite which contains 0 to 2.8 wt % Ag ($0 \leq x \leq 0.11$) [10]. The krennerite used in a study by Xu et al. [17] had a composition $Au_{0.82}Ag_{0.18}Te_{2.00}$, and an Au:Ag ratio of 4.6. The average composition of the product is $Au_{0.85}Ag_{0.15}$, and Au:Ag is ~5.7, which is slightly higher than that of the parent krennerite. The increase of the Au:Ag ratio is due to the dissolution of Ag in the reaction fluid and in textural terms for the Au–Ag alloy filaments have diameters ranging from 200 to 1000 nm. As the reaction proceeds, Au–Ag alloy wires also develop locally, having diameters up to 5 μm and lengths ranging from 25 μm to 200 μm and longer.

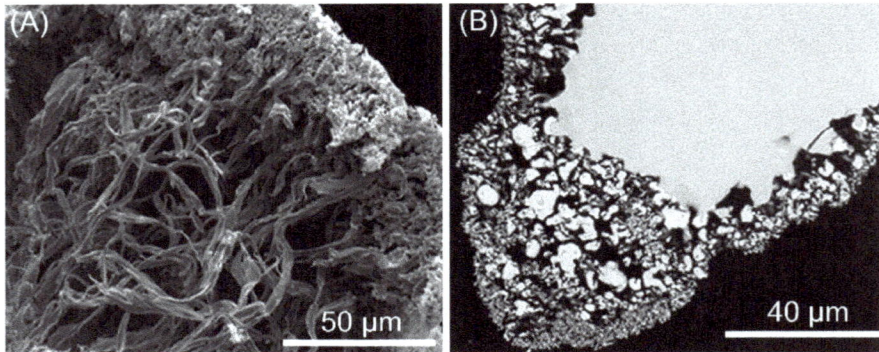

Figure 8. (**A**) Secondary electron image showing the highly porous Au–Ag alloy in the shape of filaments. (**B**) Backscattered electron image of cross section of partially-reacted krennerite grains showing larger Au–Ag alloy particles coexisting with fine-grained Au–Ag alloy in the resultant gold rim (imaged by W. Xu).

Compared to calaverite and krennerite, sylvanite generally contains significantly higher Ag contents (6.7 to 13.2 wt % Ag, illustrated by Cabri [10]). In the study by Zhao et al. [16] the sylvanite had a composition of $Au_{0.63}Ag_{0.36}Te_{2.00}$ which corresponds to 9.2 wt % Ag. In contrast to the replacement of calaverite and krennerite, sylvanite was replaced by an assemblage of products and the resulting textures are complex. In addition to Au–Ag alloy ($Au_{0.87}Ag_{0.13}$), a range of other phases formed as intermediate products, including petzite (($Au_{0.92}Ag_{3.15}$)Te_2), hessite ($Ag_{1.89}Au_{0.07}Te$), and two compositions of calaverite. The calaverite I phase has an Ag-rich, Te-depleted composition, ($Au_{0.78}Ag_{0.22}$)$Te_{1.74}$, which is similar to natural krennerite, but its XRD pattern is close to natural calaverite. Calaverite II has a normal calaverite composition of ($Au_{0.93}Ag_{0.07}$)Te_2. The calaverite I phase is porous while calaverite II lacks obvious signs of porosity in SEM images. The texture of a partially-reacted sylvanite grain is shown in Figure 9. The Au–Ag alloy rim is composed of wormlike Au–Ag alloy particles (Figure 9A), with diameters ranging from 200 to 1000 nm. Wire gold has also developed locally (up to 5 μm in diameter, 25 μm in length; Figure 9A). The rim of the grain is highly porous, with the Au–Ag alloy growing loosely on the surface and along cracks within the sylvanite (Figure 9B). Relatively large gaps were observed between the alloy rim and the particle. Underneath the Au–Ag alloy rim (Figure 9C,D), sylvanite is replaced by assemblages of calaverite I and a mixture of petzite and hessite. Petzite and hessite occur intimately mixed either as small patches or inclusions within calaverite I, or adjacent to grains of calaverite II. Au–Ag alloy and calaverite II are observed together within petzite-hessite lamellae, which is similar to the textures of natural tellurides at the Sandaowanzi deposit.

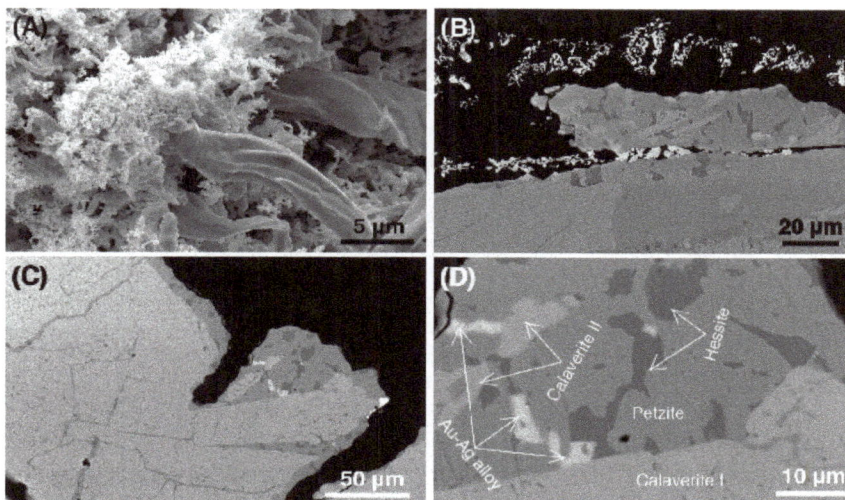

Figure 9. (**A**) Secondary electron images showing the micro Au–Ag alloy wires growing on the surface of a partially-reacted sylvanite grain. (**B,C**) Backscattered electron images of cross section of partially-reacted sylvanite grains showing a range of products after the replacement reaction. (**D**) Zoomed in image of Figure 9C, showing the textures of the calaverite II, petzite, hessite, and Au–Ag alloy; petzite and hessite occur intimately mixed either as small patches or inclusions within calaverite I.

3.2. Reaction Mechanism

Under oxidizing conditions, gold–(silver) tellurides are ultimately replaced by gold or Au–Ag alloy, while the Te is eventually lost to bulk solution and some is precipitated in the form of $TeO_2(s)$ particles on the outer surface of gold/Au–Ag alloy. The selective removal of Te from gold–(silver) tellurides is often referred to as leaching, a process conventionally considered as a solid-state diffusion-driven mechanism. In this case, it proceeds in a pseudomorphic manner via an interface-coupled dissolution–reprecipitation (ICDR) mechanism (summarized in Figure 10). The distinctive textural outcome of a CDR reaction is that the product phase of gold or Au–Ag alloy preserves the external dimension of the parent mineral. The scale of pseudomorphism in the replacement of gold–(silver) tellurides by gold or Au–Ag alloy varies from nanometer scale (e.g., the replacement of calaverite) to a few micrometers (e.g., the replacement of sylvanite). The textural features indicate that the dissolution of gold–(silver) telluride is the rate-controlling step, which is closely coupled with the precipitation rate of the products in both space and time scales [40]. The coupling between parent and product minerals is controlled by the solution chemistry at the reaction front. The porosity is strong textural evidence for a CDR reaction. The reaction is sustained by continuous mass transport through open pathways for the influx of fluid and solutes (e.g., the oxidant) to the reaction interface and the removal of dissolved Te and Ag from the reaction interface (e.g., [43,44]). The abundant porosity of the product phases is associated with negative volume changes; although systems with positive volume changes still exhibit porosity, it is often very fine grained [45]. The overall volume change is determined by the changes in molar volume as well as the solubility of the parent and product phases within a given solution [46]. The former parameter plays a role in the extent of the volume change, but the latter determines the sign of the volume change [41,47]. The solubility of each phase is a function of the grain size, fluid composition, temperature, and pressure, among other variables, and hence will likely evolve as the replacement reaction proceeds [47]. Pollok et al. [46] defined the change in volume by considering not only molar volumes but the relative solubilities of the parent and product:

$$\Delta V = 100 \times \left(\frac{n_p V_{m,p} - n_d V_{m,d}}{n_d V_{m,d}} \right), \tag{1}$$

where n_p and n_d are the number of moles of the product precipitated and the parent dissolved, and $V_{m,p}$ and $V_{m,d}$ are the molar volumes of the precipitating and dissolving phases. Considering that metallic gold or Au–Ag alloy is the final product of the replacement reaction, the molar volume changes (ΔV) in the three replacement reactions range from −79% to −85% (details are listed in Table 2). The processes include (i) the dissolution of gold–(silver) tellurides, (ii) the oxidation of the Te to a soluble Te(IV) complex and the transportation of Te from the reaction front to the bulk hydros solution, and (iii) the precipitation of gold/Au–Ag alloy. When natural gold–(silver) tellurides are treated with solution, the dissolution of the mineral occurs at the reaction front, resulting in the formation of aqueous Au, Ag, and Te complexes. To decide the nature of the predominant aqueous species of Au, Ag, and Te, and discuss the relative solubilities of each element as a function of pH, Zhao et al. [16] calculated simple diagrams of $\log fO_2(g)$ vs. pH for the system containing the same amounts of Au, Ag, and Te added to the solution by the dissolution of sylvanite at 200 °C for a solution containing 0.01 M chloride. The results illustrate that the dominant Te aqueous species is $H_2TeO_3(aq)$ under acidic to slightly basic ($pH_{200\,°C}$ 2–7) conditions and $HTeO_3^-$ under more basic conditions. Ag is mainly present as AgCl(aq) under acidic conditions, but the dominant Ag aqueous species is $Ag(OH)_2^-$ under basic conditions. Taking $O_2(aq)$ as the oxidant and assuming Au immobility, the overall reaction of gold tellurides to Au–Ag alloy can be described as below:

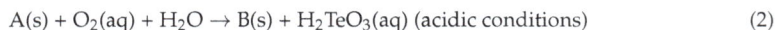

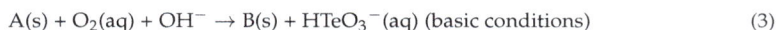

$$A(s) + O_2(aq) + H_2O \rightarrow B(s) + H_2TeO_3(aq) \text{ (acidic conditions)} \tag{2}$$

$$A(s) + O_2(aq) + OH^- \rightarrow B(s) + HTeO_3^-(aq) \text{ (basic conditions)} \tag{3}$$

where A is the parent phase of gold–(silver) tellurides, and B is the solid product phase or phases. In the replacements of calaverite and krennerite, B represents the single product of gold or Au–Ag alloy. In transformation of sylvanite, B represents both calaverite I and the Au–Ag alloy. Once the concentrations of Te and Ag in solution reach a critical state, the reaction switches and sylvanite dissolution is coupled to the precipitation of calaverite I. This indicates that this reaction is controlled by the amount of Te and Ag in solution. Calaverite I is an unstable phase, which further breaks down to calaverite II and phase χ ($Ag_{3+x}Au_{1-x}Te_2$, 0.1 < x < 0.55) via exsolution which may be fluid-catalyzed [48]. Both products of calaverite and phase χ subsequently transform to Au–Ag alloy by Reactions 2 or 3. Phase χ breaks down to a fine intergrowth of petzite and hessite during the quenching of the autoclaves from the reaction temperature (160 to 220 °C) to room temperature [16]. Cabri [10] reported that phase χ breaks down to a mixture of petzite and hessite at 105 °C. It is unclear whether the breakdown of calaverite I to calaverite II plus phase χ is really a solid-state diffusion-controlled reaction, or a fluid-catalyzed breakdown reaction in which the highly porous calaverite I undergoes a recrystallization and unmixing reaction driven by a reduction of internal surface area. The lack of porosity in the calaverite II points to solid-state exsolution. Such reactions have recently been studied in the breakdown of the bornite–digenite solid solution [48,49]. Zhao et al. [50] synthesized bornite–digenite solid solution (*bdss*) by replacing chalcopyrite under hydrothermal conditions. The results demonstrated that its composition principally depended on the temperature of the reaction rather than solution composition. Upon quenching, the unquenchable nanoscale porosity within the *bdss* system coalesces into fluid inclusions, specifically along grain boundaries, which catalyze the breakdown of the unstable *bdss* to exsolve digenite [48] or chalcopyrite [49], depending on the solution condition.

The transformation of sylvanite proceeds by a complex pathway combining dissolution–reprecipitation, fluid-catalyzed unmixing, and solid-state processes, which all compete during different stages of the reaction. The interplay of different reaction mechanisms results in complex textures, which could easily be misinterpreted in terms of complex multi-episodic geological evolution.

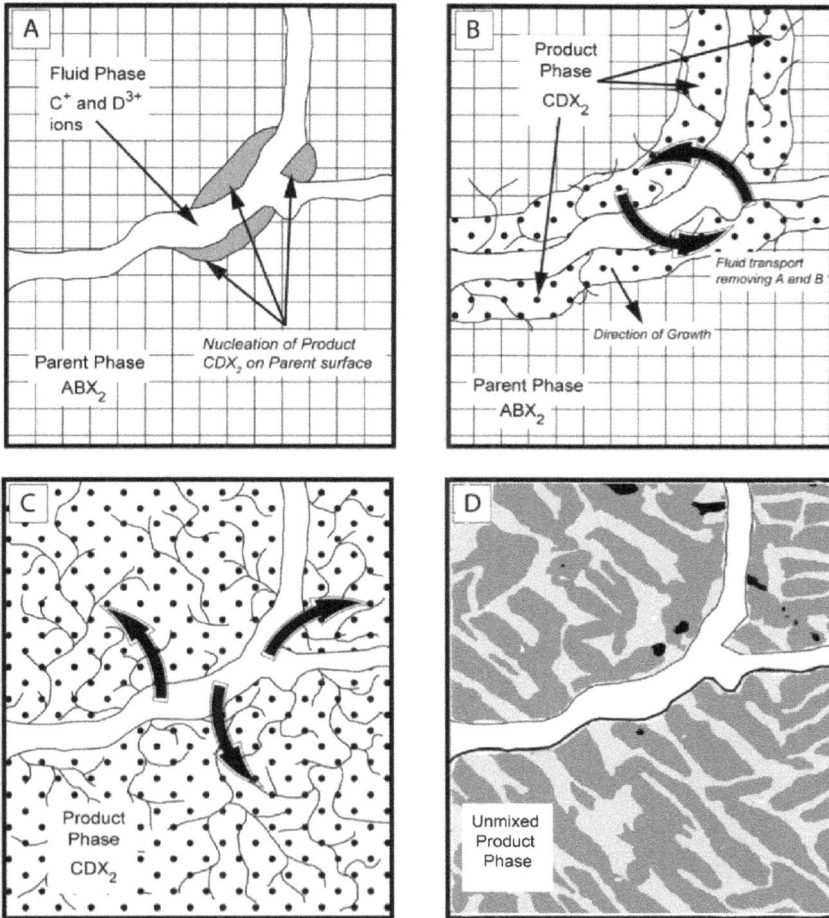

Figure 10. A diagrammatic representation of interface-coupled dissolution–reprecipitation. (**A**) Fluid containing C^+ and D^{3+} ions goes through cracks within the parent phase of ABX_2. ABX_2 phase dissolves at the reaction surface and CDX_2 product forms on the surface. (**B**) Fluid transports through pores and cracks to reaction interface adding C^+ and D^{3+}. (**C**) Curved arrows show direction of fluid flow. (**D**) Porosity anneals out over time and fluid inclusions form from the solution trapped within the crystal. Unstable product phase of CDX_2 exsolves into two new phases upon quenching, catalyzed by the fluid within the porosity and along the grain boundaries.

Table 2. Summary of experimental replacement reactions of gold–(silver) tellurides under hydrothermal conditions.

Parent Mineral	Average Composition	Au:Ag:Te	Mechanism	Overall Reaction	Products	ΔVm *
Calaverite	$Au_{0.94}Ag_{0.05}Te_{2.00}$	47:2.5:100	CDR	Calaverite + $2\,O_2$(aq) + $2\,H_2O$ → Au + $2\,H_2TeO_3$(aq)	Porous gold	−79.32%
Sylvanite	$Au_{0.63}Ag_{0.36}Te_{2.00}$	31.5:17.75:100	CDR + exsolution	Sylvanite + $2.07\,O_2$(aq) + $1.87\,H_2O$ + $0.27\,Cl^-$ → $0.72\,Au_{0.87}Ag_{0.13}$ + $0.27\,AgCl$(aq) + $2\,H_2TeO_3$(aq)	Porous Au–Ag alloy, $Au_{0.87}Ag_{0.13}$ / Petzite, $(Au_{0.92}Ag_{3.15})Te_2$ / Hessite, $Ag_{1.89}Au_{0.07}Te$ / Calaverite I, $(Au_{0.78}Ag_{0.22})Te_{1.74}$ / Calaverite II, $(Au_{0.93}Ag_{0.07})Te_2$	−84.92%
Krennerite	$Au_{0.82}Ag_{0.18}Te_{2.00}$	41:9:100	CDR	Krennerite + $2.01\,O_2$(aq) + $1.98\,H_2O$ + $0.04\,H^+$(aq) + $0.04\,Cl^-$ → $0.96\,Au_{0.85}Ag_{0.15}$ + $0.04\,AgCl$(aq) + $2\,H_2TeO_3$(aq)	Porous Au–Ag alloy, $Au_{0.85}Ag_{0.15}$	−80.19%

Note: * Volume change for each reaction was calculated using the Equation $\Delta V = 100 \cdot \left(\frac{n_p V_{m,p} - n_d V_{m,d}}{n_d V_{m,d}} \right)$, where n_p and n_d are the number of moles of product precipitated and parent dissolved, and $V_{m,p}$ and $V_{m,d}$ are the molar volumes of the precipitating and dissolving phases [46]. Molar volume of each phase equals the molar mass (M) divided by the mass density (ρ). The density of the starting minerals is listed in Table 1, and the average density of sylvanite is $8.1\ g/cm^3$. The density of gold–silver alloy was calculated using the data of gold density ($19.32\ g/cm^3$) and silver density ($10.49\ g/cm^3$). The calculated density of $Au_{0.87}Ag_{0.13}$ alloy is $17.41\ g/cm^3$ and $17.15\ g/cm^3$ for $Au_{0.85}Ag_{0.15}$.

4. Applications and Implications

The morphology of the nanoscale spongy gold wire produced by the replacement of calaverite under hydrothermal conditions is remarkably similar to the mustard gold samples [24,25] and the microporous gold samples [30] found in nature (Figure 11). Although the fluids used in the experiments are probably more aggressive than those found in nature, the similarity in the textures of porous gold may reflect similar processes of formation. The natural microporous gold found at the Aginskoe deposit, Central Kamchatka epithermal district [30], has remarkably similar textures to these synthetic spongy gold filaments in terms of both morphology and size. As shown in Figure 11, each grain of microporous gold from Aginskoe consists of aggregates of fine fibers that are about 30–300 nm in diameter and ≥ 5 μm in length. Andreeva et al. [27] indicated calaverite is the main Au telluride at Aginskoe and it is a likely precursor in this case. It often displays partial alteration to porous gold. The mustard gold found at the Dongping Mines (Hebei Province, China) is microporous gold aggregate [24] with slightly coarser textures. Mustard gold typically has the same Au:Ag ratio as the calaverite in the deposit, and the formation corresponds to the selective leaching of tellurium from calaverite. The recrystallization of gold may occur during or after the decomposition of calaverite, resulting in porous gold filaments. The textures of an ICDR product are normally related to the chemistry of the fluid, as demonstrated in the replacement of leucite ($KAlSi_2O_6$) by analcime ($NaAlSi_2O_6 \cdot H_2O$) [51]. The coarsening of the structure may also occur upon the completion of the replacement reaction, which is driven by surface energy reduction to create self-similar microstructures with ever-increasing filament size.

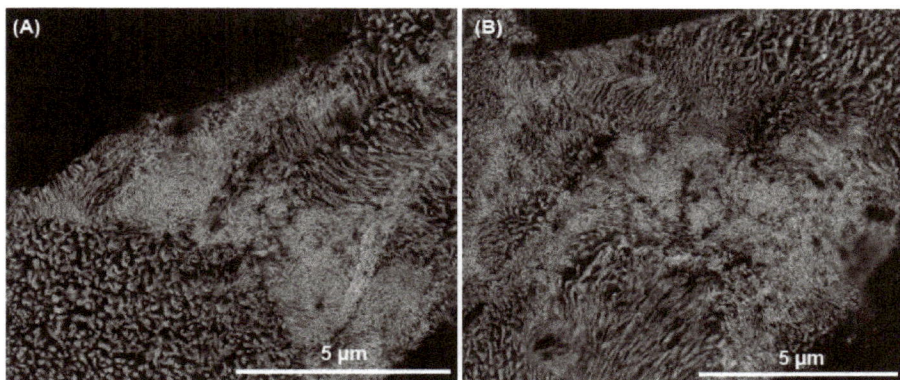

Figure 11. The natural microporous gold found at the Aginskoe deposit, Central Kamchatka epithermal district (imaged by Barbara Etschmann). (**A**,**B**) Microporous gold grain consists of aggregates of small fibers. The diameters of the gold fibers vary in a single grain.

The increasing knowledge base of the controls of interface-coupled dissolution–reprecipitation reactions over recent years is leading to an improved understanding of mineralization processes of natural systems. As shown in Figure 3, the typical textures of gold–(silver) tellurides from the Sandaowanzi gold deposit are krennerite–gold intergrowths imbedded within stuetzite–petzite symplectites. Here, the natural krennerite has a similar composition to synthetic calaverite I, and the composition of stuetzite is similar to that of synthetic hessite. According to the calculated bulk composition of the complex texture in Figure 3A, the precursor composition might have been a more Ag-rich but Te-depleted sylvanite than that used by Zhao et al. [16]. The complex textures observed at the Sandaowanzi deposit are remarkably similar to those synthesized (Figure 9) by Zhao et al. [16] by replacing sylvanite under hydrothermal conditions, implying broad similarities in the formation conditions. The experimental studies on the replacement of sylvanite by gold [17] indicate that the formation of the two/three-phase symplectites at Sandaowanzi gold deposit, are related to the replacement of the gold–(silver) telluride precursor via an interface-coupled dissolution–reprecipitation. The precursor of the mineralization was formed by the upwelling of early mineralization fluids from a deep sub-alkaline magmatic source and the mineral precipitation at the near-surface faulting during the cooling [7,8]. The precursor reacts with meteoric water infiltrating within fractures along the quartz boundaries at mild temperatures, leading to the formation of gold, the precipitation of different gold–(silver) telluride mineral associations, and the different types of solid solutions (e.g., phase χ). The results of sylvanite replacement reaction directly explain the generation sequence of gold–(silver) tellurides in nature. Native gold, Au–Ag tellurides, and Ag tellurides at Sandaowanzi deposit are all the products of precursor replacement reactions, which were formed at the same time-scale but distributed in different layers of quartz matrix formed at different stages. The results also explain that some of the later mineral assemblages (e.g., hessite–petzite and stuetzite–petzite) represent the breakdown of metastable solid solutions during cooling rather than the initial reaction conditions of ore formation. In the experiments, the products are layered by the sequence of the reactions. From the surface of the sylvanite grain to the core, the layers of products show a general trend of increasing Ag telluride abundances together with a decrease in the abundance of Au-dominant tellurides. At the Sandaowanzi gold deposit, the high-grade vein ores are mainly distributed at the +130 m level, which is in the center of two low-grade disseminated mineralization zones along the margins of the orebody. According to the experimental results, the +130 m level could be the starting surface of reaction. The majority of the Te and Ag were dissolved into solution during alteration processes and transferred to the low-grade disseminated mineralization zone by fluids, eventually forming small particles of silver tellurides and other tellurides (e.g., HgTe and PbTe) within the matrix of very fine-grained quartz (μm scale).

The recent studies on gold–(silver) tellurides under hydrothermal conditions show that these minerals can be transformed to gold/Au–Ag alloy relatively rapidly (within hours) under all conditions (even in Milli-Q water) at moderately elevated temperatures (~200 °C). This process could be added as a preliminary treatment in ore processing before the traditional cyanide process. For gold-bearing tellurides, the overall reactions provide an efficient and less toxic alternative to pretreatment by roasting.

5. Outlook

Porous gold is a form of gold with significant technological potential due to its low density, high strength and large specific surface area. The dramatic increase in attention to this material over the last two decades is due to its many potential applications in areas such as catalysis, energy storage, and sensor technology. A number of methods for synthesizing this material have been developed—for example, de-alloying, templating, electrochemical, and self-assembling. De-alloying of gold metal alloys is currently the most widely-used method. This approach fabricates the porous gold structure by selectively dissolving the less noble components from a gold alloy. To make a porous gold sponge by de-alloying, Au–Ag alloy is firstly synthesized and then the Au–Ag alloy is treated with a high-concentration nitrate solution [52]. This is a two-step process, excluding any purification of the gold source. However, a similar structure of porous gold can be produced by replacing natural gold–(silver) tellurides using a hydrothermal method over a wide range of solutions under mild conditions, the reaction being completed within days, depending on the solution composition. This single-step method appears to have the advantage of allowing fine-tuning of the nature of the porous gold, as the dissolution of a gold telluride occurs over a much wider range of solution conditions than does that of a simple Au–Ag alloy. It is also possible to use natural gold telluride minerals as well as synthetic gold tellurides. A comprehensive experimental study of the controls on the texture of the porous gold obtained via such a route is required to optimize the reaction conditions, manipulate the morphology of the porous gold sponge, and test this porous gold as a functional material in terms of catalysis, energy storage, and sensor technology.

Minerals **2019**, *9*, 167

Author Contributions: Writing—Original Draft Preparation, J.Z.; Writing—Review & Editing, A.P.

Funding: This work has been made possible by the financial support of the Australian Research Council (grants DP140102765, DP1095069, and DP170101893).

Acknowledgments: We thank Joel Brugger and Barbara Etschmann for their input into the original work. We thank Junlai Liu from China University of Geosciences for his contributions to Figure 3 and Nadezhda Tolstykh from VS Sobolev Institute of Geology and Mineralogy of Siberian Branch of Russian Academy of Sciences (SB RAS) for his contributions to Figure 4. We also thank all the editors and three anonymous referees. Grateful acknowledgements also to Philippa Horton, who helped edit the revision.

Conflicts of Interest: The authors declare no conflict of interest.

References

1. Shackleton, J.M.; Spry, P.G.; Bateman, R. Telluride mineralogy of the Golden Mile deposit, Kalgoorlie, Western Australia. *Can. Miner.* **2003**, *41*, 1503–1524. [CrossRef]
2. Kelley, K.D.; Romberger, S.B.; Beaty, D.W.; Pontius, J.A.; Snee, L.W.; Stein, H.J.; Thompson, T.B. Geochemical and geochronological constraints on the genesis of Au-Te deposits at Cripple Creek, Colorado. *Econ. Geol.* **1998**, *93*, 981–1012. [CrossRef]
3. Ahmad, M.; Solomon, M.; Walshe, J.L. Mineralogical and geochemical studies of the Emperor gold telluride deposit, Fiji. *Econ. Geol.* **1987**, *82*, 234–270. [CrossRef]
4. Cook, N.J.; Ciobanu, C.L.; Spry, P.G.; Voudouris, P.; the participants of IGCP-486. Understanding gold-(silver)-telluride-(selenide) mineral deposits. *Episodes* **2009**, *32*, 249–263.
5. Cook, N.J.; Ciobanu, C.L.; Capraru, N.; Damian, G.; Cristea, P. Mineral assemblages from the vein salband at Sacarimb, Golden Quadrilateral, Romania: II. Tellurides. *Geochem. Miner. Petrol.* **2005**, *43*, 56–63.
6. Tran, M.D.; Liu, J.L.; Hu, J.J.; Zou, Y.X.; Zhang, H.Y. Discovery and geological significance of Sandaowanzi telluride type gold deposit in the northern Daxing'anling, Heilongjiang, China. *Geol. Bull. China* **2008**, *27*, 584–589.
7. Liu, J.L.; Bai, X.D.; Zhao, S.J.; Tran, M.D.; Zhang, Z.C.; Zhao, Z.D.; Zhao, H.B.; Lu, J. Geology of the Sandaowanzi telluride gold deposit of the northern Great Xing'an Range, NE China: Geochronology and tectonic controls. *J. Asian Earth Sci.* **2011**, *41*, 107–118. [CrossRef]
8. Liu, J.L.; Zhao, S.J.; Cook, N.J.; Bai, X.D.; Zhang, Z.C.; Zhao, Z.D.; Zhao, H.B.; Lu, J. Bonanza-grade accumulations of gold tellurides in the Early Cretaceous Sandaowanzi deposit, northeast China. *Ore Geol. Rev.* **2013**, *54*, 110–126. [CrossRef]
9. Zhai, D.; Liu, J. Gold-telluride-sulfide association in the Sandaowanzi epithermal Au-Ag-Te deposit, NE China: Implications for phase equilibrium and physicochemical conditions. *Miner. Petrol.* **2014**, *108*, 853–871. [CrossRef]
10. Cabri, L.J. Phase relations in the Au-Ag-Te system and their mineralogical significance. *Econ. Geol.* **1965**, *60*, 1569–1605. [CrossRef]
11. Bindi, L.; Arakcheeva, A.; Chapuis, G. The role of silver on the stabilization of the incommensurately modulated structure in calaverite, AuTe$_2$. *Am. Miner.* **2009**, *94*, 728–736. [CrossRef]
12. Dye, M.D.; Smyth, J.R. The crystal structure and genesis of krennerite, Au$_3$AgTe$_8$. *Can. Miner.* **2012**, *50*, 363–371. [CrossRef]
13. Van Tendeloo, G.; Amelinckx, S.; Gregoriades, P. Electron microscopic studies of modulated structures in (Au, Ag) Te$_2$: III. Krennerite. *J. Solid State Chem.* **1984**, *53*, 281–289. [CrossRef]
14. Smith, G.F.H. On the remarkable problem presented by crystalline development of calaverite. *Miner. Mag.* **1901**, *13*, 122–150.
15. Mineral Database. Available online: http://webmineral.com/data/ (accessed on 8 March 2019).
16. Zhao, J.; Xia, F.; Pring, A.; Brugger, J.; Grundler, P.V.; Chen, G. A novel pre-treatment of calaverite by hydrothermal mineral replacement reactions. *Miner. Eng.* **2010**, *23*, 451–453. [CrossRef]
17. Zhao, J.; Brugger, J.; Xia, F.; Ngothai, Y.; Chen, G.; Pring, A. Dissolution-reprecipitation vs. solid-state diffusion: Mechanism of mineral transformations in sylvanite, (AuAg)$_2$Te$_4$, under hydrothermal conditions. *Am. Miner.* **2013**, *98*, 19–32. [CrossRef]
18. Kongolo, K.; Mwema, M.D. The extractive metallurgy of gold. *Hyperfine Interact.* **1998**, *111*, 281–289. [CrossRef]

19. Grosse, A.C.; Dicinoski, G.W.; Shaw, M.J.; Haddad, P.R. Leaching and recovery of gold using ammoniacal thiosulfate leach liquors (a review). *Hydrometallurgy* **2003**, *69*, 1–21. [CrossRef]

20. Hodge, A.M.; Hayes, J.R.; Caro, J.A.; Biener, J.; Hamza, A.V. Characterization and Mechanical Behavior of Nanoporous Gold. *Adv. Eng. Mater.* **2006**, *8*, 853–857. [CrossRef]

21. Ciobanu, C.L.; Cook, N.J.; Damian, G.H.; Damian, F.L.; Buia, G. Telluride and sulphosalt associations at Scrâmb. In *Gold-Silver-Telluride Deposits of the Golden Quadrilateral, South Apuseni Mts*; IAGOD: Alba Iulia, Romania, 2004; Volume 12, pp. 145–186.

22. Wilson, A.F. Origin of quartz-free gold nuggets and supergene gold found in laterites and soils—A review and some new observations. *Aust. J. Earth Sci.* **1984**, *31*, 303–316.

23. Tolstykh, N.; Vymazalova, A.; Tuhy, M.; Shapovalova, M. Conditions of formation of Au-Se-Te mineralization in the Gaching ore occurrence (Maletoyvayam ore field), Kamchatka, Russia. *Miner. Mag.* **2018**, *82*, 649–674. [CrossRef]

24. Petersen, S.B.; Makovicky, E.; Li, J.L.; Rose-Hansen, J. Mustard gold from the Dongping Au–Te deposit, Hebei Province, People's Republic of China. *Neues Jahrb. Mineral. Mon.* **1999**, *8*, 337–357.

25. Li, J.L.; Makovicky, E. New studies on mustard gold from the Dongping Mines, Hebei Province, China: The tellurian, plumbian, manganoan, and mixed varieties. *Neues Jahrb. Mineral. Abh.* **2001**, *176*, 269–297.

26. Palache, C.; Berman, H.; Frondel, C. *Dana's System of Mineralogy I*; Wiley: New York, NY, USA, 1944.

27. Andreeva, E.D.; Matsueda, H.; Okrugin, V.M.; Takahashi, R.; Ono, S. Au–Ag–Te mineralization of the low-sulfidation epithermal Aginskoe deposit, Central Kamchatka, Russia. *Resour. Geol.* **2013**, *63*, 337–349. [CrossRef]

28. Zhao, J.; Brugger, J.; Grundler, P.V.; Xia, F.; Chen, G.; Pring, A. Mechanism and kinetics of a mineral transformation under hydrothermal conditions: Calaverite to metallic gold. *Am. Miner.* **2009**, *94*, 1541–1555. [CrossRef]

29. Xu, W.; Zhao, J.; Brugger, J.; Chen, G.; Pring, A. Mechanism of mineral transformations in krennerite, Au_3AgTe_8, under hydrothermal conditions. *Am. Miner.* **2013**, *98*, 2086–2095. [CrossRef]

30. Okrugin, V.M.; Andreeva, E.; Etschmann, B.; Pring, A.; Li, K.; Zhao, J.; Griffiths, G.; Lumpkin, G.R.; Triani, G.; Brugger, J. Microporous gold: Comparison of textures from Nature and experiments. *Am. Miner.* **2014**, *99*, 1171–1174. [CrossRef]

31. Markham, W.L. Synthetic and natural phases in the system Au–Ag–Te, Part 1 and 2. *Econ. Geol.* **1960**, *55*, 1148–1178. [CrossRef]

32. Legendre, B.; Souleau, C.; Hancheng, C. Le système ternaire or-argent-tellure. *Bulletin de al Société Chimique de France Partie* **1980**, *1*, 197–204. (In French)

33. Afifi, A.M.; Kelly, W.C.; Essene, E.J. Phase relations among tellurides, sulfides, and oxides: I. Thermochemical data and calculated equilibria; II. Applications to telluride-bearing ore deposits. *Econ. Geol.* **1988**, *83*, 377–394. [CrossRef]

34. Zhang, X.; Spry, P.G. Petrological, mineralogical, fluid inclusion, and stable isotope studies of the Gies gold-silver telluride deposit, Judith Mountains, Montana. *Econ. Geol.* **1994**, *89*, 602–627. [CrossRef]

35. Wang, J.H.; Lu, X.G.; Sundman, B.; Su, X.P. Thermodynamic reassessment of the Au-Te system. *J. Alloys Compd.* **2006**, *407*, 106–111. [CrossRef]

36. McPhail, D.C. Thermodynamic properties of aqueous tellurium species between 25 and 350 °C. *Geochim. Cosmochim. Acta* **1995**, *59*, 851–866.

37. Grundler, P.V.; Brugger, J.; Meisser, N.; Ansermet, S.; Borg, S.; Etschmann, B.; Testemale, D.; Bolin, T. Xocolatlite, $Ca_2Mn_2{}^{4+}Te_2O_{12} \cdot H_2O$, a new tellurate related to kuranakhite: Description and measurement of Te oxidation state by XANES spectroscopy. *Am. Miner.* **2008**, *93*, 1911–1920. [CrossRef]

38. Grundler, P.V.; Pring, A.; Brugger, J.; Spry, P.G.; Helm, L. Aqueous solubility and speciation of Te(IV) at elevated temperatures. *Geochim. Cosmochim. Acta* **2009**, *73*, A472.

39. Grundler, P.V.; Brugger, J.; Etschmann, B.E.; Helm, L.; Liu, W.; Spry, P.G.; Tian, Y.; Testemale, D.; Pring, A. Speciation of aqueous tellurium(IV) in hydrothermal solutions and vapors, and the role of oxidized tellurium species in Te transport and gold deposition. *Geochim. Cosmochim. Acta* **2013**, *120*, 298–325. [CrossRef]

40. Putnis, A. Mineral replacement reactions. *Rev. Mineral. Geochem.* **2009**, *70*, 87–124. [CrossRef]

41. Putnis, A.; Austrheim, H. Mechanisms of metasomatism and metamorphism on the local mineral scale: The role of dissolution-reprecipitation during mineral reequilibration. In *Metasomatism and the Chemical Transformation of Rock*; Harlov, D.E., Austrheim, H., Eds.; Springer: Berlin, Germany, 2013.

42. Altree-Williams, A.; Pring, A.; Ngothai, Y.; Brugger, J. Textural and compositional complexities resulting from coupled dissolution-reprecipitation reactions in geomaterials. *Earth Sci. Rev.* **2015**, *150*, 628–651. [CrossRef]

43. Putnis, A.; Putnis, C.V. The mechanism of reequilibration of solids in the presence of a fluid phase. *J. Solid State Chem.* **2007**, *180*, 1783–1786. [CrossRef]

44. Xia, F.; Brugger, J.; Chen, G.; Ngothai, Y.; O'Neill, B.; Putnis, A.; Pring, A. Mechanism and kinetics of pseudomorphic mineral replacement reactions: A case study of the replacement of pentlandite by violarite. *Geochim. Cosmochim. Acta* **2009**, *73*, 1945–1969. [CrossRef]

45. Zhao, J.; Brugger, J.; Chen, G.; Ngothai, Y.; Pring, A. Experimental study of the formation of chalcopyrite and bornite via the sulfidation of hematite: Mineral replacements with a large volume increase. *Am. Miner.* **2014**, *99*, 343–354. [CrossRef]

46. Pollok, K.; Putnis, C.V.; Putnis, A. Mineral replacement reactions in solid solution-aqueous solution systems: Volume changes, reactions paths and end-points using the example of model salt systems. *Am. J. Sci.* **2011**, *311*, 211–236. [CrossRef]

47. Ruiz-Agudo, E.; Putnis, C.V.; Putnis, A. Coupled dissolution and precipitation at mineral-fluid interfaces. *Chem. Geol.* **2014**, *383*, 132–146. [CrossRef]

48. Zhao, J.; Brugger, J.; Grguric, B.A.; Ngothai, Y.; Pring, A. Fluid enhanced coarsening of mineral microstructures in hydrothermally synthesized bornite-digenite solid solution. *ACS Earth Space Chem.* **2017**, *1*, 465–474. [CrossRef]

49. Li, K.; Brugger, J.; Pring, A. Exsolution of chalcopyrite from bornite-digenite solid solution: An example of a fluid-driven back-replacement reaction. *Miner. Depos.* **2018**, *53*, 903–908. [CrossRef]

50. Zhao, J.; Brugger, J.; Ngothai, Y.; Pring, A. The replacement of chalcopyrite by bornite under hydrothermal conditions. *Am. Miner.* **2014**, *99*, 2389–2397. [CrossRef]

51. Xia, F.; Brugger, J.; Ngothai, Y.; O'Neill, B.; Chen, G.; Pring, A. Three-dimensional ordered arrays of zeolite nanocrystals with uniform size and orientation by a pseudomorphic coupled dissolution-reprecipitation replacement route. *Cryst. Growth Des.* **2009**, *9*, 4902–4906. [CrossRef]

52. Pertlik, F. Kristallchemie natürlicher Telluride I: Verfeinerung der Kristallstruktur des Sylvanits; $AuAgTe_4$. *Tschermaks Mineralogische und Petrographische Mitteilungen* **1984**, *33*, 203–212. (In German) [CrossRef]

© 2019 by the authors. Licensee MDPI, Basel, Switzerland. This article is an open access article distributed under the terms and conditions of the Creative Commons Attribution (CC BY) license (http://creativecommons.org/licenses/by/4.0/).

MDPI

St. Alban-Anlage 66

4052 Basel

Switzerland

Tel. +41 61 683 77 34

Fax +41 61 302 89 18

www.mdpi.com

Minerals Editorial Office

E-mail: minerals@mdpi.com

www.mdpi.com/journal/minerals

www.ingramcontent.com/pod-product-compliance
Lightning Source LLC
Chambersburg PA
CBHW051843210326
41597CB00033B/5753